THE TELEVISION WILL BE REVOLUTIONIZED

D0036212

The Television Will Be Revolutionized

Second Edition

Amanda D. Lotz

NEW YORK UNIVERSITY PRESS
New York and London

NEW YORK UNIVERSITY PRESS
New York and London
www.nyupress.org

© 2014 by New York University
All rights reserved

References to Internet websites (URLs) were accurate at the time of writing.
Neither the author nor New York University Press is responsible for URLs that
may have expired or changed since the manuscript was prepared.

Library of Congress Cataloging-in-Publication Data
Lotz, Amanda D.
The television will be revolutionized / Amanda D. Lotz. — Second edition.
pages cm   Includes bibliographical references and index.
ISBN 978-1-4798-6573-4 (hardback) — ISBN 978-1-4798-6525-3 (pb)
1. Television broadcasting. 2. Television broadcasting—United States. 3. Television—
Technological innovations. 4. Television broadcasting—Technological innovations. I. Title.
PN1992.5.L68   2014
384.55'0973—dc23   2014015124

New York University Press books are printed on acid-free paper,
and their binding materials are chosen for strength and durability.
We strive to use environmentally responsible suppliers and materials
to the greatest extent possible in publishing our books.

Manufactured in the United States of America

Also available as an ebook

For Wes, Sayre, and Calla

CONTENTS

PREFACE

Every book has its own story about the process of its creation—call it a book's biography. Part of the biography of the first edition of *The Television Will Be Revolutionized* is that I returned the proofs to New York University Press three days after the birth of my first child, in July 2007. So watching that child head off to kindergarten in the fall of 2012 gave the age of the book an uncommon physical form. Oddly, it was easier to fathom that it was time for kindergarten than that the insights of the book were now five years old.

Much has happened in the television industry in these intervening years, though much has also remained unchanged. When the first edition went to press, I was playing around on the beta version of Hulu—though there was little content, YouTube was still a start-up, the broadcast networks had clunky interfaces for streaming the very little content they made available, and there was nothing but movies or advertorials available through video on demand. A cable executive interviewed for this edition noted that the speed with which broadband-delivered video became ubiquitous caught even most industry insiders by surprise. Those on the forefront had tried out the slow-streaming, blocky, early ventures in this area and didn't think much of the experience. Then it all seemed to change overnight. Netflix, HBO GO, smart-phones, tablets, TV Everywhere, oh my! The vast proliferation of content through various broadband providers is addressed in all of the book's revised chapters, as it has affected every aspect of the industry and provides the biggest paradigmatic adjustment from the first edition.

The revised second edition updates developments throughout the television industry over the last seven years, focuses the book more on broad frameworks for thinking about these changes than on cataloging a multiplicity of initiatives and experiments, and makes general improvements upon certain aspects that I found suboptimal when I revisited it. In some places, details about the conditions of the industry in 2006 remain because they provide a valuable benchmark of the transition that allows the book to be a history of the present, while other details have been eliminated. Though many consider the book to be about the post-network era, I saw the contribution of the first edition primarily as a calling into existence and systematic explanation of the multi-channel transition period of the 1980s and 1990s. This context building prepares us to be able to talk about and theorize a post-network era and remains unchanged and unchallenged by the last seven years. The second revised edition maintains that contribution here and also expands on what can be said of the post-network era as it has come into greater relief. The structure and organization of the book remain largely intact; substantive adjustments are noted below.

It is the curse of a project like this to be inevitably out of date. During the time lag of the production process alone, new developments will occur. There are even developments that transpired before I submitted the manuscript that I elected to exclude because their consequences and likelihood of lasting impact, though potentially substantial, were not yet clear (Aereo; Chromecast). Here, I've focused more on frameworks for understanding developments than on cataloging what different companies are doing in 2014 because of the inevitable change. For example, YouTube seems to have a new monetization strategy about every nine months, so instead of detailing the strategy in place at the moment of manuscript submission, I focus more on the industrial differences of advertising and subscription economic models and the consequences they've produced for content in other media. I find my voice bolder here than in the first edition, but I'm a conservative prognosticator by nature. I'm more enamored with the consequences of new technologies, regulations, or economic strategies than with crystal ball gazing, and when teaching the book, am interested in encouraging my students to think about new developments in this way as well. It remains an obvious class assignment to have students investigate what has happened since the book was written; and I'll have done my job well if I've given them enough tools here to make arguments about why, how, and to whom those changes matter.

When I first proposed the revision, I suspected I would need to add a great deal to the first version, but instead found a considerable amount I wished to cut. To me, the first edition now read as though the author felt she needed to prove her mettle on the topic by offering a detailed and exhaustive recounting of the vast experiments in each industrial process—and that was probably the case. Here I rest on the reputation of my past work and do less to chronicle each new thing. Maybe I've omitted something that will prove important in time, but the last seven years have offered countless lessons about television use, distribution, and financing, and I chose to focus more on areas now offering some evidence and consistency, rather than the most current thing. I don't doubt that a few years hence I'll either be back at this or we'll deem that a new book is needed and that this one has become a historical text. The frameworks the book uses for thinking about television are the parts that will live on regardless of further redevelopment of industry practices and changes in the companies that dominate them.

The introduction is now more succinct, and the detailed discussion of the different eras has been moved into the first chapter. In its place, I provide a new section that argues that we need to begin speaking of television content with more specificity when we consider the post-network era. I posit the categories of prized content, live sports and contests, and linear viewing as three such categorizations. I use these different ways of experiencing content throughout the book to illustrate how industrial adjustments affect each differently.

Except for the addition of the era explanations, chapter 1 remains the most unchanged. This book has been successful in reaching a variety of audiences, and this is the chapter that most marks it as a scholarly endeavor—which makes it more and less interesting to different audiences. In some ways the distinction of phenomenal television that I offered in 2007 is now made more precise through discussion of prized content and live sports and contests, which are distinctions defined as much by viewer behavior as by content. Given that only a narrow group of early adopters experiences television in the ways made possible by post-network distribution technologies, it is difficult to think much further about the uses of television in the post-network era beyond what was possible in 2007.

Adjustments to chapter 2, which focuses on technology, are mostly of the "update" variety. The development and fast penetration of smartphones required significant adjustment in the discussion of mobile uses of television, although with the exception of sports programming, these devices are used more for "portable" television than for "mobile" (live) television in the United States. I've also added two tables charting key moments in the shift to non-linear viewing and significant milestones in digitally distributed U.S. sports.

Chapter 3 includes new discussion of funding from outside the industry, new funding and distribution models such as the 10-90 sitcom, and the production mechanisms of series originally produced for online distribution. Other sections are updated; most notably, I include a discussion of the emerging use of social media in promotion. The conclusion to this chapter is also new.

I've reframed chapter 4's discussion of shifting windows. Strategies that seemed important in 2007 have been revealed to be but transitional practices in the broader evolution to nonlinear viewing norms. An abbreviated discussion of reallocation, repurposing, and DVD sales remains—as these were important strategies in the trajectory of change—and I've added extensive new discussion of video on demand and broadband streaming. Addressing the emergence of Netflix was a much-needed component, not only because of what Netflix does, but also because its competition pushed established entities to adopt endeavors such as TV Everywhere far more quickly than would have occurred otherwise. The discussion of distribution to the home is more streamlined, as the arrival of the DOCSIS 3.0 standard largely eliminated the distinction of IPTV that existed in 2006. I've updated the competitive and regulatory framework to acknowledge the role—albeit limited—of telco service providers and Google Fiber, the evolution of net neutrality as a regulatory issue, cable and Internet bundling, and the fact and fiction of "over-the-top" access as it had developed by 2014. Because of the increased technical language and jargon, I have added a table defining different delivery technologies early in the chapter; I've also added a time line chronicling the technological developments in the evolution of digital television distribution.

I have adjusted the framing of chapter 5 from *advertising* to *financing* to better address the variety of subscription and transaction financing structures being used by nonlinear television. The chapter now opens by addressing key differences between advertiser-supported and viewer-supported financing and the necessity of keeping these distinctions clear when we consider how content is received. In many ways, HBO and Netflix are more alike because they are non–advertiser-supported subscription services than different because one comes in through cable and the other over broadband—a distinction I suspect will be technologically nebulous the next time I revisit this book. A condensed discussion of advertiser-supported alternatives to the thirty-second advertisement remains, and the sponsorship section now addresses sports programming more specifically. I've expanded the discussion of efforts to save the thirty-second ad to include discussion of pre-roll video ads that support Hulu and YouTube. I also explore how new dynamic ad insertion technologies now make it possible for advertisers to monetize advertisements in VOD (video on demand) to an extent likely to expand this nonlinear form of distribution.

The exploration of audience measurement in chapter 6 has been updated to match the reframing of the previous chapter away from an exclusive focus on advertiser-supported television. I've reorganized the chapter to make the steady progression of new measurement norms developed throughout the last decade more systematic, and updated data based on new interviews with executives at Nielsen and follow-up interviews with some who were quoted in the first edition.

Chapter 7 is still built upon the same case studies as the first edition. When I considered replacing them with newer shows, I ultimately determined that I would lose more than I gained. The original case studies are extraordinary "firsts" that have given rise to so many similar successors that there is not much that is distinctive about newer shows; these case studies have become the new normal. The opportunity to take an even longer view of the original case studies and the more expansive distribution some have experienced also strengthens them through added richness. I have moved the discussion of *The Shield*'s early advertising struggles from the advertising chapter and incorporated it into the case study, and also moved the discussion using *Arrested Development* as an ideal text for a post-network distribution method to its case study. I've addressed the "progeny" of the different cases in the conclusion to each case study, including addressing *Off to War* as a precursor to YouTube original content.

The conclusion now looks ahead more than the previous version did, as I imagine what television might be like at the point when the post-network era has truly arrived. I extend my personal use anecdotes to identify the extensive ways my own television use has changed in the last seven years, and return to the discussion of the distinctions of prized content, live sports and contests, and linear television.

Without doubt, the most satisfying part of returning to this project was coming to find that the "end of television" discourse had finally abated and that considerable evidence existed suggesting that the moving pictures and sounds we've long characterized as "television" are enjoying expanded storytelling possibilities and integration into the lives of those who consume them. I know that this terrain will continue to evolve, and I am not interested in tying the future of television to its continued distribution by broadcast network or cable wire and channel. The passing of these technologies does not mark the death of television, but the passing of inferior distribution systems that were the best technologies available in another era. Exploring how new distribution technologies enable new ways of financing and producing the moving pictures and sounds we know as television provides much to consider in the years ahead.

ACKNOWLEDGMENTS

Second Edition

As much as I'd like to claim that I managed the task of a revised, second edition on my own, this was decidedly not the case. As *The Television Will Be Revolutionized* reached its fifth birthday, I thought I'd do a blog post updating it. That developed into a new introduction. After breakfast with Eric Zinner, it became a revised, second edition. Thanks for giving me the opportunity to really extend the life of the book, Eric.

A fabulous cohort of students in my graduate Analyzing Media Industries class had what was either the curse or the blessing of applied learning: they helped provide feedback on what was most needed for the update and identified preliminary data sources. The perspectives of voices outside my head were important for pushing me toward the considerable revision contained within. Special thanks to Derek Granitz and Amanda Cote for struggling with the more challenging chapters, and extra special thanks to my indomitable research assistant, Kitior Ngu, whose exceptional diligence is graciously matched with patience.

The scope of revisions required new research. I'm very thankful for the time and insight provided by Evan Shapiro, Jon Weiser, Jon Mandel, Brendan Kitts, Frances Croke Page, and Brian Fuhrer. The additional research was funded by a grant from the University of Michigan Associate Professor Support Fund, for which I'm also thankful. And a note of appreciation to the Cable Center's Barco Library for permission to use the late 1980s cable program guide image.

I offer personal thanks to those who reached out with kind words after reading the first edition. Even a second edition can be a long and mostly solitary process, and it was a tremendous motivator to know that the work had proven helpful, been appreciated, and reached a broad audience. The scope of the revision and the speed with which it came together were enabled by the patience of Wes, Sayre, and Calla and our it-takes-a-village mates, Beth and Jay Ellis, who are largely responsible for any work and life balance we've achieved. Finally, I offer sincere thanks to family and friends who have made life rich outside my fascination with television.

First Edition

The examination of the operation of cultural industries is a less common pursuit among media studies scholars; perhaps one reason is that this type of research poses particular challenges. Executive offices and the day-to-day operation of cultural industries are not easy for critically minded academics to access, but over the last five years I've attended a wide range of industry events and forums that offered meaningful glimpses into these worlds and informed this research in crucial ways. This research is built upon four weeks of participant observation of media buying, planning, and research departments and immersion in a number of industry conferences, including the Academy of Television Arts and Sciences Faculty Seminar, November 2002; the National Association of Television Program Executives (NATPE) Conference and Faculty Seminar, 2004–2007; the Future of Television Seminar sponsored by Television Week, September 2004; the International Radio and Television Society Foundation Faculty/Industry Seminar, November 2004; the International Consumer Electronics Show, January 2006; the National Cable and Telecommunications Association National Show, April 2006; and the Future of Television Forum, November 2006. Visiting these industry meetings and extensive reading of trade press provided more information about the industry than I could meaningfully report. The precise sources of all of the anecdotes, cases, and analysis in the following chapters are not always explicitly acknowledged, but my understanding of industry operations and struggles derives primarily from these sources. Immersing myself in the space of these industry events helped me understand the paradigm of thought that dominated the industry at various points in this adjustment, as everything from formal conference presentations to casual conversations overheard in hallways and ballrooms contributed to my sense of industry concerns and perspectives.

Many organizations, individuals, and funding sources enabled my research in crucial ways. I am incredibly grateful to the National Association of Television Program Executives Educational Foundation, expertly managed by Greg Pitts, for the various ways a Faculty Development Grant, Faculty Fellowship, and the organization's educator's rate and programming provided firsthand access to many of the executives making decisions about the industrial changes chronicled here. The Faculty Development Grant, and generous hosting by Mediacom, also offered invaluable perspective on the upfront buying process.

The Advertising Education Foundation's Visiting Professor Program allowed me to spend two weeks observing the operation of the media buyer

Universal McCann, and a schedule carefully arranged by Charlotte Hatfield exposed me to the many dimensions of buying, planning, and research, information that was exceptionally helpful in composing the advertising chapter. Thanks also to Sharon Hudson for her work on this great program and all those in the industry who support it.

I was honored by the International Radio and Television Society Foundation in November 2004 as the Coltrin Professor of the Year as a result of a case study exercise I wrote to explore the issues examined in this book with my students. In addition to providing a fine honor, IRTS constructed a number of excellent panels of industry executives who spoke to many of the central issues and provided valuable information and perspective. I am also grateful to the faculty who joined me in New York and participated in the case study. Thanks to IRTS, Joyce Tudryn, Stephen H. Coltrin, and all those who support IRTS for these opportunities.

Funding and support from the Denison University Research Foundation, NATPE Educational Foundation, and a University of Michigan Rackham Faculty Grant and Fellowship all supported various aspects of travel and industry conference fees upon which my research heavily relied. Course release in the winter of 2006 and a summer stipend allowed my attendance at a marathon of industry conferences and enabled focused and fast work, which has aided in the timely contribution of this book.

Many working in the industry offered insight in formal and informal interviews and responded to e-mail queries. The detail of description I offer here would have been impossible without their generous explanations. Thanks to Laura Albers, Pamela Gibbons, Todd Gordon, Heather Kadin, Deb Kerins, Michele Krumper, Jon Mandel, Mitch Oscar, Rob Owen, Frances Page, Brent Renaud, Shawn Ryan, Andy Stabile, Stacy Sullivan, and Susan Whiting for their time and insights.

Introduction

As I was dashing through an airport in November 2001, the cover of *Technology Review* displayed on a newsstand rack caught my eye. Its cover story was titled "The Future of TV," and the inside pages provided a smart look at likely coming developments.[1] Even by the end of 2001, which was long before viewers or television executives truly imagined the reality of downloading television shows to pocket-sized devices or streaming video online, it was apparent that the box that had sat in our homes for half a century was on the verge of significant change. The future that the author, Mark Fischetti, foresaw in the article depicted the television world that would be available to early adopters by the mid-2000s fairly accurately (by "2000s," I mean the first decade of the twenty-first century, not the century in its entirety). His focus, though, was on the living room television set, and his vision did not anticipate the portability of computing that would develop over the late 2000s to break down distinctions between television and "computer" screens, or that mobile phones would so quickly become pocket computers and portable televisions. But right there in his third paragraph is the sentiment that television and consumer electronics executives uttered incessantly beginning in 2006 as the mantra of the television future: "whatever show you want, whenever you want, on whatever screen you want."

Even though Fischetti presciently predicted the substantial adjustments in how we view television, where we view it, how we pay for it, and how the industry would remain viable and vital, many other headlines in the intervening years predicted a far more dire situation. Reports and articles bore ominous titles like "The End of Television as We Know It" (IBM Business Consulting Services), "The Death of Television" (*Slate*), "Why TV Will Never Be the Same" (*Business Week*), and "How Old Media Can Survive in a New World" (*Wall Street Journal*).[2] By 2007, a *Wired* article better captured the emerging contradictions with the title "The TV Is Dead. Long Live the TV."[3] Predicting the coming death of television became a new beat for many of the nation's technology and culture writers in the mid-2000s. When television contrarily persisted, the naysayers turned instead to the dominant cable delivery model, announced the imminent demise of the cable industry, and suggested that legions of viewers would soon cancel cable subscriptions. Sounding the death knell for cable, prognosticators proposed that viewers would go "over the top" (OTT) of their cable boxes to access favorite shows through Internet delivery of content by using services such as Netflix, Hulu, iTunes, or a wide range of authorized and unauthorized web-based sources; Max Fisher's *Atlantic* article "Cable TV Is Doomed" is indicative of the new apocalyptic theme.[4] But despite such claims and endless fawning over the latest gadget or gizmo that would usher in the demise of television or cable, both persisted. Showtime's CEO, Matt Blank, wryly joked at the 2013 Cable Show that industry journalists' favorite topics were companies with no revenues and no earnings, followed by those with some revenues and still no earnings; "old" television companies like his that were flush with both proved of little interest.

The journalists weren't alone in their uncertainty about the future of television or even the definition of television, as new ways to use television and new forms of content confounded even those who used the device every day. In 2004—before much legal or illegal streaming of video online occurred— the longtime broadcast television executive Rich Frank told a Las Vegas ballroom full of television executives about a recent visit with his young grandson. He asked the boy which network was his favorite, expecting to hear a broadcast network or perhaps Nickelodeon in response. But without a moment's hesitation the boy replied, "TiVo." By 2013, a child might instead answer "PBS.org" or "the videos on daddy's phone." If the period from 2000 through 2010 led audiences to imagine that television would become something different than it had been during the preceding half century, the period from 2010 through 2014 introduced and normalized aspects of the future of television, such as the presumption that "television" is not only viewed on a television set. By that time, the industry slowly but meaningfully expanded viewers' ability to watch "whatever show you want, whenever you want, on whatever screen you want."

We may continue to watch television, but the new technologies available to us require new rituals of use. Not so long ago, television use typically involved walking into a room, turning on the set, and either turning to specific content or channel surfing. Today, viewers with digital video recorders (DVRs) may elect to circumvent scheduling constraints and commercials, while others download or stream the latest episodes of their favorite shows, either within or outside the conventional setting of the living room. And this doesn't even begin to touch upon the vast array of content created outside the television industry that appears on video aggregators such as YouTube or social networking sites.

As a result of these changing technologies and modes of viewing, television use has become increasingly complicated, deliberate, and individualized. Television as we knew it—understood as a mass medium offering programs that reached a broad, heterogeneous audience and spoke to the culture as a whole—is no longer the norm in the United States, though most certainly neither is going "over the top." But despite what many initially thought, changes in what we can do with television, what we expect from it, and how we use it have not been hastening the demise of the medium; instead, they are revolutionizing television.

To explore this revolution, this book offers a detailed and extensive behind-the-screen exploration of the substantial changes occurring in television technology, program creation, distribution, and television economics, why these practices have changed, and how these changes are profoundly affecting everyone from television viewers to those who study and work in the industry. It examines a wide range of industrial practices common in U.S. television and assesses their recent evolution in order to explain how and why the images and stories we watch on television find their way to us as they do in the twenty-first century. These changes are so revolutionary that they suggest the nascent development of a new era of television, the effects of which we have only begun to detect.

What Is Television Today?

Television is not just a simple technology or appliance—like a toaster—that has sat in our homes for more than sixty years. Rather, it functions as both a technology and a tool for cultural storytelling. We know it as a sort of "window on the world" or a "cultural hearth" that has gathered our families, told us stories, and offered glimpses of a world outside our daily experience. It brought the nation together to view Lucy's antics, gave us mouthpieces to discuss our uncertainties about social change through Archie and Meathead, and provided a common gathering place through which a geographically vast nation could share in watching national triumphs and tragedies. A certain understanding of what television was and could be developed during our early years with the medium and resulted from the specific industrial practices that organized television production processes for much of its history. Alterations in the production process—the practices involved in the creation and circulation of television—including how producers make television programs, how studios finance them, and how audiences access them, have created new ways of using television that now challenge our basic understanding of the medium. Changes in television have forced the production process to evolve during the past twenty years so that the assorted ways we now use television are mirrored in and enabled by greater variation in the ways television is made, financed, and distributed.

We might rarely consider the business of television, but production practices inordinately affect the stories, images, and ideas that project into our homes. The industrial transformation of U.S. television has begun to modify what the industry creates. Industrial processes are normally nearly unalterable and support deeply entrenched structures of power that determine what stories can be told and which viewers matter most. But beginning in the mid-1980s, the U.S. television industry began reinventing itself and its industrial practices to compete in the digital era by breaking from customary norms of program acquisition, financing, and advertiser support that in many cases had been in place since the mid-1950s. This period of transition created great instability in the relationships among producers and consumers, networks and advertisers, and technology companies and content creators, which in turn initiated uncommon opportunities to deviate from the "conventional wisdom" or "industry lore" that ruled television operations. Industry workers faced a changing competitive environment triggered by the development of new and converging technologies that expanded ways to watch and receive television; they also found audiences willing to explore the innovative opportunities these new technologies provided.

Rather than enhancing existing business models, industrial practices, and viewing norms, recent technological innovations have engendered new ones—but it is not just new technologies that have revolutionized the television industry. Adjustments in how studios finance, make, and distribute shows as well as in how and where viewers watch them occurred simultaneously. None of these developments suggested that television would play a diminished role in the lives of the nation that spends the most time engaging its programming, but the evolving institutional, economic, and technological adjustments of the industry have significant implications for the role of television in society.

The industry remains in the throes of rapid and radical change in 2014 as the television transformation moves from a few early adopters to a more general and mass audience. As new uses become dominant and shared by more viewers, television's role in culture continues to evolve. Understanding these related changes is of crucial interest to all who watch television and think about how television communicates ideas, to those who study media, and to those who are trying to keep abreast of their rapidly changing businesses and remain up-to-date with new commercial processes.

Despite changing industrial practices, television remains a ubiquitous media form and a technology widely owned and used in the United States and many similarly industrialized nations. Yet the vast expansion in the number of networks and channels streaming through our televisions and the varied ways we can now access content has diminished the degree to which societies encounter television viewing as a shared event. Although once the norm, society-wide viewing of particular programs is now an uncommon experience. New technologies have both liberated the place-based and domestic nature of television use and freed viewers to control when and where they view programs. Related shifts in distribution possibilities that allow us to watch television on computer screens, tablets, and mobile phones have multiplied previously standard models for financing shows and profiting from them, thereby creating a vast expansion in economically viable content. Viewers face more content choices, more options in how and when to view programs, and more alternatives for paying for their programming. Increasingly, they have even come to enjoy the opportunity to create it themselves.

Although television maintains the technological affordances of a mass medium that, in principle, remains capable of serving as the cultural hearth around which a society shares media events—as we did in cases such as the Kennedy assassination or the *Challenger* explosion—it increasingly exists as an electronic newsstand through which a diverse and segmented society pursues deliberately targeted interests. The U.S. television audience now can rarely be categorized as a mass audience and is instead more accurately understood as a collection of niche audiences. Television has been reconfigured in recent decades as a medium that most commonly addresses fragmented and specialized audience groups, but no technology emerged to replace its previous norm as a messenger to a mass and heterogeneous audience. The development of broadband distribution substantially affected the circulation of ideas and enabled dissemination to even international audiences, yet the Internet allows us to attend to even more diverse content and provides little commonality in experience.

Television's transition to a narrowcast medium—one targeted to distinct and isolated subsections of the audience—along with adjustments within the broader media culture in which it exists, significantly altered its industrial logic and has required a fundamental reassessment of how it operates as a cultural institution. For the last sixty years, we have thought about television in certain ways because of how television has been, but the truth is that television has not operated in the way we have assumed for some time now. Few of the norms of television that prevailed from the 1950s into the 1980s remain in place, and such norms were themselves the results of specific industrial, technological, and cultural contexts long since passed. In particular, the presumption that television inherently functions as a mass medium continues to hold great sway, but the mass audiences once characteristic of television were, as the media scholar Michael Curtin notes, an aberration resulting from Fordist principles of "mass production, mass marketing and mass consumption."[5] Consequently, previous norms did not suggest the "proper" functioning of the television industry any more than did subsequent norms; rather, they resulted from a specific industrial, technological, and cultural context no more innate than those that would develop later.

Understanding the transitions occurring in U.S. television at this time is a curious matter relative to conventional approaches to exploring technology and culture. Historically, technological innovation primarily has been a story of replacement, in which a new technology emerged and subsumed the role of the previous technology. This indeed was the case of the transition from radio to television, as television neatly adopted many of the social and cultural functions of radio and added pictures to correspond with the sounds of the previous medium. The supplanted medium did not fade away, but repositioned itself and redefined its primary attributes to serve more of a complementary than competitive function. But it is not a new competitor that now threatens television; it is the medium itself and those who try to retain practices now clearly suboptimal.

The changes in television that have taken place over the past two decades—whether the gross abundance of channel and program options we now select among or our increasing ability to control when and where we watch—are extraordinary and on the scale of the transition from one medium to another, as in the case of the shift from radio to television. And it is not just television that has changed. The field of media in which television is integrated also has evolved profoundly—most directly as a result of digital innovation. The audience's experiences with computing and the emergence of the mobile phone as a sophisticated portable screen technology better thought of as a "pocket computer" than a "phone" are now as important to understanding television as the legacy behaviors of domestic viewing. Various industrial, technological, and cultural forces have begun to radically redefine television, yet paradoxically, it persists as an entity that most people still understand and identify as "TV."

This book explores this redefinition of television specifically in the United States, although these changes are also redefining the experience with television in similar ways in many countries around the world. From its beginning, broadcasting has been "ideally suited" technologically to transgress national borders and constructs such as nation-states; however, the early imposition of strict national control and substantially divergent national experiences prevailed over attributes innate to the technology.[6] Many different countries experienced similar transitions in their industrial composition, production processes, and use of this thing called television at the same time as the United States, but precise situations diverge enough to make it difficult to speak in transnational generalities and lead to my focus on only the U.S. experience of this transition. As Graeme Turner and Jinna Tay tellingly assessed in 2009, "'What is television?' very much depends on where you are."[7] The specific form of the redefinition—as it emerges from a rupture in dominant industrial practices—is particular to each nation, yet similarly industrialized countries are experiencing the transition to digital transmission, the expansion of choice in channel and content options, the increasing conglomeration of the industry among a few global behemoths, and the drive for increased control over when, where, and how audiences view "television programs." The development of an increasingly global cultural economy also has led the fate and fortune of the U.S. television industry to be determined beyond national confines.

Situating Television circa 2014

During its first forty years, U.S. television remained fairly static in its industrial practices. It maintained modes of production, a standard picture quality, and conventions of genre and schedule, all of which led to a common and regular experience for audiences and lulled those who think about television into certain assumptions. Moments of adjustment occurred, particularly at the end of the 1950s when the "magazine" style of advertising began to take over and networks gained control of their schedules from advertising agencies and sponsors, but once established, the medium remained relatively unchanged until the mid-1980s. First, the "network era" (from approximately 1952 through the mid-1980s) governed industry operations and allowed for a certain experience with television that characterizes much of the medium's history. The norms of the network era have persisted in the minds of many as distinctive of television, despite the significant changes that have developed over the past twenty years. I identify the period of the mid-1980s through the mid-2000s as that of the "multi-channel transition." During these years, various developments such as the growing availability of cable service and new cable channels, videocassette recorders (VCRs), and remote controls changed our experience with television, but did so very gradually, in a manner that allowed the industry to continue to operate in much the same way as it did in the network era.

Signs of a subsequent period, a "post-network era," began to emerge in the early 2000s. Many changes from the norms of the multi-channel transition are readily identifiable, but it remains too early to know the ultimate characteristics and conventions of the post-network era. What separates the post-network era from the multi-channel transition is that the changes in competitive norms and operation of the industry become too pronounced for many of the old practices to be preserved; different industrial practices are becoming dominant and replacing those of the previous eras.

These demarcations in time, which are intentionally general, recognize that all production processes do not shift simultaneously and that people adopt new technologies and ways of using them at varied paces. By the end of 2005, adjustments in how people could access programming—particularly through DVR use and purchase of full seasons on DVD—enabled a small group of early adopters to experience television outside the linear schedules of network programmers in a manner characteristic of a preliminary post-network era.[8] By 2014, a greater range of viewers were engaging television in places other than the living room screen, but such Internet, mobile phone, and even DVR time-shifted viewing accounted for only a small fraction of overall viewing. As an illustration, Nielsen data from the second quarter of 2013 indicated that the aggregate of time per month spent watching time-shifted TV (12 hours, 35 minutes), using a DVD/Blu-ray device (5 hours, 10 minutes), using a game console (6 hours, 27 minutes), using the Internet on a computer (not specifically for viewing video content) (27 hours, 21 minutes), watching video on the Internet (6 hours, 28 minutes), or watching a video on a mobile phone (5 hours, 45 minutes) amounted to only 63 hours and 46 minutes, a good bit less than half of the 146 hours, 37 minutes viewers still spent watching "traditional" TV.[9] Indeed, much about the nuances of shifts

Table I.1. *Characteristics of Production Components in Each Period*

Production component	Network era	Multi-channel transition	Post-network era
Technology	Television	VCR Remote control Analog cable	DVR VOD Portable devices Mobile phones Tablets Digital cable
Creation	Deficit financing	Fin-syn rules Surge of independents End of fin-syn Conglomeration and coproduction	Multiple financing norms Variation in cost structure and aftermarket value Opportunities for amateur production Challenging promotion environment
Distribution	Bottleneck Definite windows Exclusivity	Cable increases possible outlets	Nonlinear access TV Everywhere Netflix streaming MVPD app availability Erosion of time between windows and exclusivity
Advertising	30-second ads	Subscription Experimentation with alternatives to 30-second ads	Coexistence of multiple ad models—30-second ads, placement, integration, branded entertainment, sponsorship Multiple-user supported models—transactional and subscription
Audience measurement	Audimeters Diaries Sampling	People Meters Sampling	Cross-platform Census measurement Digital program ratings Online campaign ratings

in behavior is lost in aggregate averages, and change of this scale is necessarily gradual and profoundly varied when more individualized behaviors are considered. Even as I made final edits to this manuscript in early 2014, it remained impossible to assert that a majority of the audience had entered the post-network era or that all industrial processes had "completed" the transition, but the eventual dominance of post-network conditions appeared to be inevitable.

The characteristics of the three phases of television, explored in more detail in chapter 1, are summarized in table I.1.

And So, the Television Will Be Revolutionized

The world as we knew it is over.
—Les Moonves, president, CBS Television, 2003

The 50-year-old economic model of this business is kind of history now.
—Gail Berman, president of entertainment, FOX, 2003[10]

These bold pronouncements by two of the U.S. television industry's most powerful executives only begin to suggest the scale of the transitions that took place as the multi-channel transition yielded to new industrial norms characteristic of a post-network era. Television executives commonly traffic in hyperbolic statements, but the assertions by Moonves and Berman did not overstate the case. Here they reflected on the substantial challenges to conventional production processes as a result of scheduling and financing the comparatively cheap but widely viewed unscripted ("reality") television series that flooded onto network schedules in the early 2000s. Yet the issues brought to the fore by the success of unscripted formats offered only an indication of the broader forces that threatened to revise decades-old business models and industrial practices.

A confluence of industrial, technological, and cultural shifts conspired to alter institutional norms in a manner that fundamentally redefined the medium and the business of television. The U.S. television industry was a multifaceted and mature industry by the early years of the twenty-first century, when Moonves and Berman made these claims. As post-network adjustments became unavoidable, many executives expressed a sense that the sky was falling—and indeed, the scale of changes affecting all segments of the industry gave reasonable cause for this outlook. A single or simple cause did not initiate this comprehensive industrial reconfiguration, so there was no one to blame and no way to stop it.

An important harbinger of the inevitability of the post-network era occurred in mid-2004, when the rhetoric of industry leaders shifted from advocating efforts to prevent change to accepting the present and coming industrial adjustment. This acceptance marked a transition from corporate strategies that sought to erect walls around content and retard the availability of more personalized applications of television technology to efforts to enable content from traditional providers to travel beyond the linear network platform.[11] In his detailed history of the invention of media technologies, Brian Winston illustrates how existing industries have repeatedly suppressed the radical potential of new technologies in an effort to prevent them from disrupting established economic interests. Unsurprisingly, the patterns Winston identifies also appear in the television industry, in which "supervening social necessities" such as a desire for greater control over television content led inventors to create technologies that provided markedly new capabilities (such as the DVR), while those with business interests threatened by the new inventions sought to curtail and constrain user access.[12] Nonetheless, many of the conventional practices and even the industry's basic business model proved suboptimal in this new context and resulted in crises throughout all components of the production process. Considerable uncertainty persists about the new norms for programming and how power and control will be reallocated within the industry.

New technological capabilities and consumers' response to them forced the moguls of the network era to imagine their businesses anew and face fresh competitors who had a vision of a new era. As suggested by the duration of the multi-channel transition, this industrial reconfiguration often produced unanticipated outcomes and developed haphazardly. Much of the sense of crisis within the industry resulted from the inability of powerful companies to anticipate the breadth of change and to develop new business models in response. Those who dominated the network era sensed their businesses to be simultaneously under attack on multiple fronts, which often led to efforts to stifle change or deny the substance of the threats to conventional ways of doing business.[13] Entrenched network-era business entities consequently did not lead the transition to the post-network era; rather, mavericks such as TiVo, Apple, Google, and Netflix identified businesses that connected with viewers' desires and forced industrial evolution.

Contrary to the persistent headlines, television is not on the verge of death or in any way dying. Although indications of all kinds of change abound, there is little to suggest that the central box through which we view will be called anything other than television in this lifetime. Adjustments throughout the television industry will not turn us into "screen potatoes" or lead us

to engage in "monitor studies." We have processed and will continue to process coming changes through our existing understandings of *television*. We
will continue to call the increasingly large boxes that serve as the focal point
of our entertainment spaces *television*—regardless of how many devices we
need to connect to them in order to have the experience we desire or whether
they are giant boxes, flat sheets of glass mounted on walls, or some technology yet only imagined. We are even likely to conceptualize almost all video
that conforms to the conventions we've come to associate with television as
television, even if we stream it years after its production, if we watch it on
personal-size mobile phone screens, or if it is produced for entities never distributed as television, such as Netflix or Amazon. The U.S. television industry may be evolving, the experience of television viewing may be evolving,
but our intuitive sense of this thing we call television remains intact. A revolution is on its way, but it will not overthrow television; the growing accessibility and manipulability of video will expand its sovereignty and embed it
ever more deeply into our cultural experience.

The adjustments characteristic of a still largely imagined post-network
era will be far more profound than the changes evident so far. One thing
revealed by the current conditions of the nascent post-network era is that
television content no longer can be considered uniformly. Since the early
2000s, the broad constellation of television programming has fractured into
at least three distinct entities that are fundamentally different in ways very
meaningful to the commercial underpinnings of the industry.

One type of content enabled by the post-network era is what I distinguish
as "prized content." Prized content describes programming that *people seek
out and specifically desire*. It is not a matter of watching "what is on"; prized
content is deliberately pursued. Prized content also compels *some* audience
members to follow news of its development, to read endless chatter on blogs
and news sites, to seek out missed episodes, control viewing, and even pay
for this most valued content. Prized content is determined by the audience
member—what I prize may not be prized by you—though there may be
features more likely to make content prized by larger or specific audiences.
Prized content is a post-network-era phenomenon that emerges in defiance
of the technological affordances of mid-twentieth-century broadcasting,
which created the norm of a linear content flow that provided specific content at certain network-determined times and that has served as the dominant organization for television. Though these norms remain entrenched
and still persist in 2014, post-network-era affordances of digital distribution
enable prized content to be viewed in more deliberate ways, though also in
accord with the traditional linear scheduling. The opportunity to experience

television independently from an externally determined flow fractures the monolithic television experience to create this category of prized content.

Emergent distribution technologies have enabled a television practice that allows greater selection—perhaps parallel to the transition in filmgoing in the mid-1940s, when audiences began seeking out particular films rather than continuing the rote behavior of going to the theater each week and viewing whatever movie was playing. Many observers reference examples of what I consider prized content in declarations of a "new golden age" of television or in "Best of" lists developed in the last decade.[14] A sampling of content of the last fifteen years that was prized by a significant audience might include *The Sopranos, Mad Men, The Wire, Lost, West Wing, Friday Night Lights, Breaking Bad,* and *Downton Abbey,* among many others. Notably, that significant audience may only be two to three million viewers, far from the mark of contemporary mass hits that are watched by many more, but that may not inspire the same passion as prized content. Audiences with different tastes might include *Real Housewives, Jersey Shore,* or *Duck Dynasty* as similarly compelling cases of prized content, underscoring how *prized content is not an aesthetic or evaluative distinction* assessed based on features of the show, but is distinguished by how audiences desire to experience it.

Prized content is so compelling that it suffers from interruption, be it the interruption of commercial pods or the interruption of a week's passage between conventional "airings." The media scholar Jason Jacobs has also identified disruption—and digital television's ability to eliminate and reduce it—as a defining distinction of what I'd categorize as the network and post-network eras.[15] Preliminary data about the use of video on demand and DVR playback by genre reveal that the far greatest use of these devices is to view dramas, which affirms the idea that many viewers particularly desire a different experience with this narrative form than traditional television experience has allowed.[16] The desire for control over pace of viewing and the opportunity to re-view enables—and perhaps even makes superior—nontraditional economic models such as direct, transactional payment. Rather than model existing norms of viewer behavior, engagement with prized content might be more comparable to how audiences read a novel.

A second distinct type of content is that of "live sports and contests." Indeed, live sporting events are far from new—they can be found among the earliest broadcasts—but as the break from the multi-channel transition has become more profound, the exceptionality of live sporting events has become inescapable. Live sports, as well as live televised contests such as *American Idol* or *Dancing with the Stars,* resist all of the ways the technologies and distribution opportunities of the post-network era enable audiences to disrupt prized content from residual viewing norms and economic strategies. As they do with prized content, audiences place high value on watching particular contests as specifically *sought-after* content, but full enjoyment of this content features *exceptional time sensitivity* that necessitates live or near-live viewing. The formats of most contests naturally allow for action breaks, which has made sports programming resistant to the commercial-skipping and illegal-downloading technologies that have imperiled the economics of other programming forms.[17] Sports and contests thus remain optimal for the traditional mechanisms of television advertising and the economics that support it, and also offer seemingly endless opportunities for sponsorship and branding, further expanding their economic value.

Sports programming has been a frequent topic in discussions of the future of television precisely because the increasing fees demanded by rights holders and eventually passed on to television viewers—whether or not they *view* sports—have grown so significantly as to threaten the equilibrium of programming costs. The investment house Sanford C. Bernstein & Company released research in 2013 illustrating that live sports accounts for 20 percent of viewing by cable subscribers but 50 percent of the cost of their subscriptions.[18] The journalist Derek Thompson captured the dilemma of costs and audience demand for televised sports well: "Without live sports," he asserts, "the TV business could fall apart; and because of live sports, the TV business could fall apart."[19] The value of live televised sports has increased because so little other programming continues to unite comparatively large audiences who watch at an appointed time and remain captive through the commercials. When we talk of the future of "television," we must do so in a way that acknowledges that the features that distinguish prized content and live sports and contests prepare them for very different industrial norms.

I distinguish the final type of programming as "linear content," though most recognize this as plain old television. Not long ago, all television was linear, and much of what is viewed still is. Linear content is *what people watch when they watch "what is on,"* or it might be distinguished by the notion of "I'm going to watch television" as opposed to "I'm going to watch *Sons of Anarchy*." Like sports, linear content is viewed live, but likely with much less intention than most sports. The motivation for viewing is not watching particular content, but a desire for companionship, distraction, or entertainment that may or may not make the content the viewer's focus. Linear television might be the television viewed when you sit down in the evening to see what's on; it is the morning talk show that airs as you ready for work, and the evening news that plays as you prepare dinner. By definition, linear content is not time-shifted, so the established model of advertising remains effective, if "effective" is a term that could ever really describe the economic benefits of airing commercial messages in content that viewers attend to only casually.

I offer these categories of television to illustrate the need to speak of particular types of television content and make content-specific claims when postulating coming economic models. These three categories don't quite contain all viewing, and I'm sure we can imagine many instances that present features of multiple categories. My point is to begin to speak of television viewing with greater specificity, because viewers' increased ability to manage viewing differently has significant implications throughout television's industrial norms. Creating terminology that acknowledges the different attributes that are enabled by technological and distribution affordances of the post-network era aid in crafting a more sophisticated conversation about television's present and future. Though disruptions to conventional practices occur—such as iTunes sales of single episodes beginning in 2005 or Netflix's rich subscription-based on-demand offerings beginning in 2010—acknowledging the range of content now characteristic of television helps make clear that it is a variety of practices and norms that are imperiled, rather than television per se.

Key aspects of the post-network revolution include the enabling of new types of programming such as prized content and the establishment of profound distinctions in the experiences and economic possibilities among existing programming types such as live sports and contests and linear content. Another emerging aspect of the revolution can be found in the growing mechanisms for organizing and packaging content, whether by emergent *aggregators* such as Netflix or Hulu, emergent *devices* such as Roku and Boxee, or emergent *applications*, whether those enabling live streaming of channels over computers or devices such as gaming systems that enable accessing television content through Internet connection. We remain at a most nascent stage of what I suspect will be a massive disruption of norms of television delivery, and it is too soon to predict common viewing behaviors in the future. Nonlinear viewing—that is, viewing not at an externally appointed time—whether by DVR, video on demand (VOD), DVDs, or streaming—has become a primary way of engaging television for some viewers, though these behaviors remain irregular or completely unused by many more. Nonlinear viewing calls into question the continued need for previous ways of organizing television, such as the "channel," and these early years of preliminary post-network formation have featured the addition of new channel-like distribution and aggregation "middlemen" such as Netflix, Hulu, and YouTube that are each trying to reorganize the content experience. Adding more middlemen, at the same time that channels become increasingly superfluous, seems a short-lived disruption, though one likely to aid a longer-term paradigm shift.

The following pages update understandings about television's industrial practices from which others might build analyses of the substantial adjustments occurring within other media systems and their societies of reception. The book also contributes to the necessary rethinking of "old" media in new contexts. The deterioration of the foundational business model upon which the commercial television industry long has operated suggests that a substantive change is occurring. Examining the industry at an embryonic moment of norm creation sheds light on how power is transferred during periods of institutional uncertainty and reveals how new possibilities can develop from emerging industrial norms. There is a similarity between the industrial moment considered here and that examined in Todd Gitlin's 1983 book *Inside Prime Time*.[20] Both books chronicle the consequences of industrial practices of the television industry at the close of an era. Gitlin, however, captured this moment unintentionally, while this work is reflexively aware of the transitory status of the practices it explores.

Perhaps paradoxically, I take a particular type of television—"prime-time programming"—as the book's focus. Despite significant industry changes, as I completed the book, prime-time programming remained the most viewed and most widely discussed form of "television," though its high costs did not make it the most lucrative. The post-network era threatens to eliminate time-based hierarchies, but the distinctive status of prime time is determined as much by its budgets and production practices as by the time of day in which it has traditionally "aired." Changing industrial norms bore consequences for all programming. Adjustments in production components also affected affiliate and independent stations in significant and particular ways, but the breadth of these matters prevents me from addressing them here. Although the affiliates represent a large part of the television industry, the consequences of post-network shifts affected these stations in substantially different ways depending, among other things, on whether the station was owned and operated by a network, whether it was located in a large or small market, and the network with which the station was affiliated.[21]

The next chapter develops the distinctions of the network era, multi-channel transition, and post-network era with greater detail and briefly steps away from the book's main focus on how shifts in industrial practices and business norms affect programming to meditate on some of the more abstract and bigger issues—some might say theories—called into question by these institutional adjustments. Concerns about how television operates as a cultural institution, the adaptation of tools used to understand it, and the development of new ones aid us in thinking about intersections of television and culture that may not be the primary concern of those who work in the industry.

Such questions and concerns are nonetheless of crucial importance to the rest of us who live in this world of fragmented audiences and wonder about the effects of the erosion of the assumptions we have long shared about television.

Each aspect of production examined in chapters 2 through 6 changed on a different timetable in the course of the multi-channel transition. By 2005, profoundly different technological capabilities and distribution methods had emerged, though these new possibilities were pushed further by 2010, once broadband distribution of full-length professional content suggested greater change. Though indications of post-network technologies and distribution norms may now be evident, other production components remain insubstantially adjusted. Thus, each of these chapters focuses on a particular production component—technology, creation, distribution, financing, and audience measurement—and explores the process of transition, what practices have changed, and their consequences with regard to how television functions as a cultural institution.

With a focus on technology, chapter 2 explores how new devices have made television more multifaceted and allowed more varied uses than were common during the network era. By 2005, new television technologies enabled three distinct capabilities—convenience, mobility, and theatricality—that led to different expectations and uses of television and created a diversified experience of the medium in contrast to the uniform one common in the network era. Technologies including DVRs, VOD, portable television, and high-definition television—among many others—produce complicated consequences for the societies that adopt them as viewers gain greater control over their entertainment experience, yet become tethered by an increasing range of devices that demand their attention and financial support. The technological field expanded in 2010 as television delivered by Internet technologies expanded the mobility and portability of television and freed the medium from its staid domestic norm to be experienced on an array of screens.

Chapter 3 explores the practices involved in the making of television, particularly the institutional adjustments studios and networks made during and after the implementation of the financial interest and syndication rules, as well as the effects of these adjustments on the content the industry produces. Studios have responded to changing economic models by battling with creative guilds and unions to maintain new revenue streams, shifting production out of union-dominated Los Angeles, and creating vertically integrated production and distribution entities. Changing competitive practices among networks have resulted in significant adjustments in the types of shows the industry produces and expanded the range of profitable storytelling. The chapter thus examines how redefined production norms have created opportunities for different types of programming and required new promotion techniques.

Some of the most phenomenal adjustments in the television industry result from viewers' expanded ability to control the flow of television and to move it out of the home. Whereas a distribution "bottleneck" characterized the network era and much of the multi-channel transition, the bottleneck broke open in late 2005 and has expanded to offer nearly limitless possibilities for viewers to access programming. Chapter 4 explores how viewers gained access to television in an increasing array of outlets that featured differentiated business models. New distribution methods made once unprofitable programming forms viable and decreased the risk of unconventional programming, opening creative opportunities in the industry and contributing to the fundamental changes in the production processes discussed throughout the book.

Chapter 5 examines the shifting strategies for financing television, particularly the emergence of alternatives to the advertiser-supported model that dominated previous eras. In the last decade, U.S. television has experienced notable expansions in subscription and other direct-pay economic models, as well as a diversifying array of strategies for advertiser support. Here again distinctions among prized content, live sports and contests, and linear viewing reveal divergent futures for financing, and the emergence of more diversified strategies is symptomatic of the conditions of a multifaceted post-network era that relies upon multiple, coexisting financing strategies.

Following the examination of financing and advertising, chapter 6 explores the often-unconsidered role of audience measurement, which proved particularly contentious in the late years of the multi-channel transition and as the post-network era developed. The industry leader, Nielsen Media Research, endeavored to introduce technological upgrades that reallocated advertising dollars, while new distribution methods and advertising strategies required impartial measurement for validation. The existing paradigm of audience measurement proved increasingly insufficient for the variation characteristic of post-network television. This chapter considers the crucial role of audience measurement and developments during the tumultuous early 2000s, as well as the consequences adjustments in this sector might bring to the production of television in the future.

While chapters 2 through 6 include many examples that apply somewhat abstract industrial practices to specific shows and circumstances, chapter 7 takes a detailed look at how technology, creation, distribution, financing, and audience measurement intersect in five very different programs. Each of the five cases explored here owes its existence or success to production practices uncharacteristic of the network era and tells a particular and distinctive story about production processes at the end of the multi-channel transition. These shows, *Sex and the City*, *Survivor*, *The Shield*, *Arrested Development*, and *Off to War*, illustrate how changes in multiple practices interconnected to expand the range of stories that could be profitably told on U.S. television, as well as point to some of the implications of this expanded storytelling field for the industry and culture. I also address more recent shows that continue the paths charted by these early shows, but still lack as full a history, such as *Girls* and *Video Game High School* or the fortunes of the "stars" produced thus far by YouTube.

The perspective here involves looking ahead, not to predict, but to prepare for a new era of television experience and criticism. The precise forms that the technologies and uses of television will take are not definite, but substantial industrial ruptures are already apparent, and the need for practical information and conceptual models to rethink the medium is evident. The following pages may consequently both serve as a eulogy to the television we have experienced to this point and prepare our understanding of the medium yet to come.

1

Understanding Television at the Beginning of the Post-Network Era

Before continuing further, we must develop the distinctions between the U.S. network era, the multi-channel transition, and the post-network era in greater detail. This chapter opens with a concise recounting of the major developments of sixty-five years of U.S. television that explains the norms that produced a particular experience of television in each era. The second half of the chapter steps back from the details of industrial operation that otherwise motivate the story told in these pages to reflect on how television has been thought of and the role it has been perceived to play in society. The adjustments in industrial norms have produced corresponding changes in television content and in viewers' experience of television. The second half of the chapter discusses past consequences and conjectures about future developments in television's transition from a medium designed first for a mass audience and then for niche audiences.

The Network Era

The series of fits and starts through which U.S. television developed complicates the determination of a clear beginning of the network era. Early television unquestionably evolved from the network organization of radio. This provides a compelling argument for dating the network era to the first television broadcasts of the late 1940s, if not to the days of radio. Alternatively, the industrial conditions of early television enabled substantial local production and innovation, which made these early years uncharacteristic of what developed in the early through mid-1950s and became the network-era norm. Dating the network era as beginning in 1952 takes into account the passage of the channel allocation freeze, the period from 1948 to 1952 during which national television licensing was halted as the FCC organized its practice of frequency distribution, color television standard adoption, and other institutional aspects that regularized the network experience for much of the country.[1]

Television certainly began as a network-organized medium, but many of the industrial practices and modes of organization that eventually defined the network era were not established immediately. By the early 1960s, network-era conventions were more fully in operation: the television set (and for some, an antenna) provided the extent of necessary technology; competition was primarily limited to programming supplied to local affiliates by three national networks that dictated production terms with studios; the networks offered the only outlets for high-budget original programming; thirty-second advertisements—the majority of which were sold in packages before the beginning of the season—supplied the dominant form of economic support and were premised upon rudimentary information about audience size.

The network era of U.S. television was the provenance of three substantial networks—NBC, CBS, and ABC—which were operated by relatively non-conglomerated corporations based in the business center of New York. These networks were organized hierarchically with many layers of managers, and each was administered by a figurehead with whom the identity and vision of the network could be associated.[2] Established first in radio, the networks spoke to the country en masse and played a significant role in articulating postwar American identity.[3] Networking was economically necessary because of the cost of production and the need to amortize costs across national audiences. Networking used economies of scale to recoup the tremendous costs of creating television programming by producing one show, distributing it to audiences nationwide, and selling advertising that would reach that massive national audience. Gathering mass audiences through a system of national network affiliates enabled networks to afford "network-quality" programming with which independent and educational stations could not compete.

The financial relationships between networks and the production companies that supplied most television programming have changed throughout the history of television, but the dominant practices of the network era were established by the mid-1960s. Film studios and independent television producers had only three potential buyers of their content and were thus compelled to abide by practices established by the networks. In many cases the networks forced producers to shoulder significant risk while offering limited reward through a system in which the producers financed the complete cost of production and received license fees (payments from the networks) that were often 20 percent less than costs. Studios also sold the programs to international buyers or in syndication to affiliates after the program had aired on a network, but the networks typically demanded a percentage of these revenues, as detailed in chapter 3.

The conventions of advertising and program creation were multifaceted in television's early years. As was the case in radio, much early television featured a single sponsor for each program—as in the *Texaco Star Theater*—rather than the purchase of commercials by multiple corporations, a practice that later became standard. Only in the late 1950s and early 1960s did the networks eliminate the single-sponsorship format, thereby wresting substantial control of their schedule away from advertising agencies and sponsors. The earlier norm eroded precipitously in part as a result of the quiz show scandals, which revealed advertisers' willingness to mislead audiences, but as explored in chapter 5, this erosion also had much to do with adjustments in how the networks sought to operate, as well as differences in the demands of television relative to radio. After the elimination of the single-sponsorship format, networks earned revenue from various advertisers who paid for thirty-second commercials embedded at regular intervals within programs in the manner that is still common today. Advertisers made their purchases based on network guarantees of reaching a certain audience, although many of the methods used to determine the size and composition of the audience were very limited.

Viewers had few ways to use their televisions during the network era in comparison with the television we know in the twenty-first century. Most viewers selected from fewer than a handful of options and primarily chose from three nationally distributed networks, inconsistently dispersed independent stations, and the isolated and underfunded educational television stations that became a slow-growing and still underfunded public broadcasting system. As the name implies, in the network era, U.S. "television" meant the networks ABC, CBS, and NBC.

In addition to lacking choice, network-era viewers possessed little control over the medium. Channel surfing via remote control—an activity taken for granted by contemporary viewers—did not become an option for most until the beginning of the multi-channel transition—although, as established, there were few options to surf among. Viewers possessed no recourse against network schedules, and time shifting remained beyond the realm of possibility. If the PTA bake sale was scheduled for Thursday night, that week's visit with *The Waltons* could not be rescheduled or delayed.

In the network era, television was predominantly a nonportable, domestic medium, with most homes owning just one set. Even by 1970, only 32.2 percent of homes had more than one television.[4] The communication scholar James Webster describes television in this era as an "old medium," in which programming was uniform, uncorrelated with channels, and universally available.[5] Such basic characteristics of technological use and accessibility contributed to the programming strategies of the era in important ways. Network programmers knew that the whole family commonly viewed television together, and they consequently selected programs and designed a schedule likely to be acceptable to, although perhaps not most favored by, the widest range of viewers—a strategy the CBS vice president of programming Paul Klein described as "least objectionable programming."[6] This was the era of *broad*casting, in which networks selected programs that would reach a heterogeneous mass culture, but still directed their address to the white middle class. This mandate was integral to the business design of the networks and led to a competitive strategy in which they did not attempt to significantly differentiate their programming or clearly brand themselves with distinctive identities, as is common today.

The network era featured very specific terms of engagement for the audience regardless of the broader distinctions in how the industry created that programming or how the business of television operated. Viewers grew accustomed to arbitrary norms of practice—many of which were established in radio—such as a limited range of genres, certain types of programming scheduled at particular times of day, the television "season," and reruns. These unexceptional network-era conventions appeared "natural" and "just how television is" to such a degree that altering these norms seemed unimaginable. However, adjustments in the television industry during the multi-channel transition revealed the arbitrary quality of these practices and enabled critics, industry workers, and entrepreneurs to envision radically different possibilities for television.

As the arrival of technologies that provided television viewers with unprecedented choice and control initiated an end to the network era, the multi-channel transition profoundly altered the television experience. To be sure, many network-era practices remained dominant throughout the multi-channel transition, but during the twenty-year period that began in the mid-1980s and extended through the early years of the twenty-first century, these practices were challenged to such a degree that their preeminent status eroded.

The Multi-Channel Transition

Beginning in the 1980s, the television industry experienced two decades of gradual change. New technologies, including the remote control, videocassette recorder, and analog cable systems, expanded viewers' choice and control; producers adjusted to government regulations that forced the networks to relinquish some of their control over the terms of program creation;[7] nascent cable channels and new broadcast networks added to viewers' content choices and eroded the dominance of ABC, CBS, and NBC; subscription channels launched and introduced an advertising-free form of television programming; and methods for measuring audiences grew increasingly sophisticated with the deployment of Nielsen's People Meter. As in the network era, this constellation of industrial norms led to a particular viewer experience of television and enabled a certain range of programming. Many of these industrial practices are explored in greater depth in chapters 2 through 6, which focus on new norms emerging in production processes, including technology, program creation, distribution, financing, and audience measurement, and how these norms adjusted the type of programming the industry creates. In introducing this distinction between the multi-channel transition and the post-network era here, we must first establish the difference in viewers' experience of television. The subsequent chapters then detail the modifications in industrial practices that introduced these changes for viewers.

The common television experience was altered primarily as a result of expanded choice and control introduced during the multi-channel transition. As competition arising from the creation of new broadcast networks, such as FOX (1986), the WB (1995), and UPN (1995), expanded broadcast viewing options, a rapidly growing array of cable channels also drew viewers away from broadcast networks. The combined broadcast share—the percentage of those watching television who watched broadcast networks—declined from 90 to 64 during the 1980s, and that percentage was shared by six broadcast competitors instead of three.[8] Broadcast networks (ABC, CBS, FOX, NBC, the WB, and UPN) collected an average of only 58 percent of those watching television at the conclusion of the 1999–2000 season, and only 46 percent by the end of the 2004–2005 season.[9] Alternative distribution systems such as cable and satellite enabled a new abundance of viewing options, and 56 percent of television households subscribed to them by 1990—a figure that grew to 85 percent of households by 2004.[10]

The development of new technology that increased consumer control also facilitated viewers' break from the network-era television experience. Audiences first found this control in the form of the remote control devices (RCDs) that became standard in the 1980s. The dissemination of VCR technology further enabled them to select when to view content and to build personal libraries. For many, the availability of cable, remote control devices, and VCRs resulted in significant change all at once. The diffusion of these technologies was complexly interrelated. Viewers did not need to purchase a new remote-equipped set to gain use of an RCD. Many who acquired cable boxes and VCRs first accessed RCDs with these devices, while the new range of channels offered by cable and the control capabilities of VCRs expanded viewers' need for remotes.[11]

Substantial changes within the walls of the home also altered how audiences used television during the multi-channel transition. Because of the limited options of the network era, programs were widely viewed throughout the culture, but the explosion of content providers throughout the multi-channel transition enabled viewers to increasingly isolate themselves in enclaves of specific interests. As Webster explains, "new media" provide programming that is diverse and is correlated with channels, and they make content differentially available.[12] For example, although many cable channels can be acquired nationwide, the varying carriage agreements and packaging of the channels by locally organized cable systems create different availability based on geography and subscription tier.

Webster argues that this programming multiplicity results in audience fragmentation and polarization.[13] While much of the concern within the industry about audience fragmentation focuses on the consequences of smaller audiences for the commercial financing system that supports U.S. television, cultural critics are now considering how the polarization of media audiences contributes to cultural fissures such as those that emerged around social issues in the 2000 and 2004 elections as well as the continued sense that different sites of television news perpetuated divergent realities of national issues. Here, polarization refers to the ability of different groups of viewers to consume substantially different programming and ideas, rather than simply to the dispersal of audiences. New technologies contribute to this polarization in various ways; for example, control technologies, which enable audiences to view the same programs at different times, decrease the likelihood of viewers sharing content during a given period, while the new surplus of channels spreads the audience across an expansive range of programming.[14] Moreover, viewers' ability to use recording technologies to develop self-determined programming schedules also diminished the already languishing notion of television as an initiator of watercooler conversation—a notion once enforced through the mandate of simultaneous viewing.

The emergence of so many new networks and channels changed the competitive dynamics of the industry and the type of programming likely to be produced. Instead of needing to design programming likely to be least objectionable to the entire family, broadcast networks—and particularly cable channels—increasingly developed programming that might be most satisfying to specific audience members. At first, this niche targeting remained fairly general, with channels such as CNN seeking out those interested in news, ESPN attending to the sports audience, and MTV aiming at youth culture. As the number of cable channels grew, however, this targeting became more and more narrow. For example, by the early 2000s, three different cable channels specifically pursued women (Lifetime, Oxygen, and WE), yet developed clearly differentiated programming that might be "most satisfying" to women with divergent interests. These more narrowly targeted cable channels increased the range of stories that could be supported by an advertising-based medium. By the mid- to late 1990s, some cable channels built enough revenue to support the production of "broadcast-quality" original series such as *La Femme Nikita* (USA) and *Any Day Now* (Lifetime), and their particular economic arrangements allowed them to schedule series with themes and content unlikely to be found on broadcast networks.[15] Their niche audience strategy and the supplementary income they gained from the fees paid by cable providers led cable channels to develop shows that have much more specific target audiences than those of broadcast series. The ability of cable channels to succeed with smaller audiences has made broadcasters' mission difficult; once given an option, viewers have found the more precisely targeted content offered by many cable channels more satisfying than broadcasters' least objectionable fare. Yet the expanse of cable channels has spread audiences thinly, and most cable channels remain constrained by their much smaller audiences and related lower advertising prices.

Indications of a Post-Network Era

The choice and control that viewers gained during the multi-channel transition only continued to expand as evidence of a post-network era emerged. Others (myself included) have previously used "post-network" to indicate the era in which cable channels created additional options for viewers—similar to the way I use the phrase "multi-channel transition" here. The term "post-network" is best reserved, however, as an indicator of more comprehensive changes in the medium's use, one I suspect will ultimately take a nonlinear form. Here, "post-network" acknowledges the break from a dominant network-era experience, in which viewers lacked much control over when and where to view and chose among a limited selection of externally determined linear viewing options—in other words, programs available at a certain time on a certain channel. Such constraints are not part of the post-network television experience in which viewers now increasingly select what, when, and where to view from abundant options. Though I once intended the post-network distinction to simply indicate the erosion of network and channel control over how and when viewers watch particular programs, it has grown more feasible to imagine a post-network era devoid of networks or channels as the distinctive industrial entities they've served as thus far. In the early years of the post-network era, networks and channels have remained important sites of program aggregation, operating with distinctive identities that help viewers find content of interest. The gradual transition away from previous norms of mass audiences, new ways of accessing and paying for content, and the economic crises to which they contribute have decreased the profit margins of some sectors of the industry. Though multi-channel video programming distributors (MVPDs) such as Comcast, Charter, or DirecTV have passed costs for the ever-growing array of content to their subscribers, evidence of a breaking point for incremental cost increases has presented itself and reveals the inefficiency of bundling collections of unwanted channels into cable packages, and even of bundling unwanted programs into channels. Though this inefficiency could be tolerated in an era in which the license fees for channels were more modest, the fast growth in program costs—described by a cable industry financial analyst in 2013 as "multiples of the rate of inflation"—may yield a more radical departure from existing norms of packaging content, especially for those who predominantly view prized content and live sports and contests.[16]

Chapters 2 through 6 provide detailed considerations of the new industrial conditions that suggest the gradual establishment of a post-network era. These conditions include emerging technologies that enable far greater control over when and where viewers watch programming; multiple options for financing television production that develop and expand the range of commercially viable programming; greater opportunities for amateur production that have arisen with and been augmented by a revolution in distribution that exponentially increases the ease of sharing video; various advertising strategies including product placement and integration that have come to coexist with the decreasingly dominant thirty-second ad, as well as new economic models built on subscription and transaction payment; and advances in digital technologies that further expand knowledge about audience viewing behaviors, allow for pay-per-click advertisement pricing, and create opportunities to supplement sampling methods with census data about use. Once again, adjustments in the production process change the use of television as viewers gain additional control capabilities and access to content variation. Additionally, other new technologies have expanded portable and mobile television use and have removed television from its domestic confines.

Unlike the fairly uniform experience of watching television in the network era, by the end of the multi-channel transition, there was no singular behavior or mode of viewing, and this variability only continues to increase as the post-network era develops. For example, research on early DVR adopters found that they sometimes engaged television through the previously dominant model of watching television live. However, at other times and with other types of programming they also exhibited an emergent behavior of using the device not only to seek out and record certain content but also to pause, skip, or otherwise self-determine how to view it. Control technologies have effectively added to viewers' choice in experiencing television, as they have enabled far more differentiated and individualized uses of the medium.

Two key non–television-related factors also figure significantly in creating the changes in audience behaviors that characterize the post-network era: computing and generational shifts. The diffusion of personal computers relates to changing uses of television in notable ways. During the multi-channel transition, when viewers began to experience television as one of many "screen" technologies in the home, the initial contrast between the experience of using computers and watching television led users to differentiate between screen media according to whether they required us to push or pull content, lean back or lean forward, and pursue leisure or work. Subsequently, however, digital technologies have come to dismantle these early differentiations and tendencies of use and have allowed for the previously unimagined integration of television and computers—of all types—in the post-network era. This integration occurred concomitantly first with the growth in home computer ownership, which rose from 11 percent in 1985 to 30 percent in 1995, 67 percent by 2005, and 80 percent by 2013.[17] Second, the rapid diffusion of smartphones—devices in 65 percent of homes owning televisions by 2013—further disrupted expectations of devices and the capabilities of screens even since preliminary suggestions of the post-network era emerged. Mobile devices, however, are not necessarily used for mobile viewing: 2013 research revealed that 64 percent of smartphone TV/video use occurs in the home, suggesting that most use is simply for the convenience of viewing more places within the home.[18] Smartphones have quickly accustomed their users to expect to be able to engage in any "computer" need from any location, which has encouraged expectations of convenient access to television content.

The technological experience of personal computing and smartphone use is important beyond the growing convergence of media in the latter part of the multi-channel transition period because of the new technological aptitudes and expectations embodied in their use. The presumption that technologies "do" something useful and that we "do" something with them has played a significant role in adjusting network-era behavior with regard to television. The media theorist Dan Harries refers to the blending of old media *viewing* and new media *using* as "viewsing."[19] Thinking about such activities as being merged, rather than as being distinct, takes important steps beyond the binaries between computer and television technologies commonly assumed in the past and addresses the multiple modes of viewing and using that audiences began to exhibit by the end of the multi-channel transition. The distinction of "computing" has also become as unstable as "television," as computing has moved beyond the PC and become integrated into a wide range of mobile devices used constantly and in an array of contexts in a manner that makes the notion of computing as an activity done at a desk increasingly residual.

Related generational differences have also played a key role in changing uses of television.[20] Many of the distinctions such as broadcast versus cable—let alone between television and computer—that have structured understandings of television are meaningless to those born after 1980. Most members of this generation (dubbed "Millennials" or "digital natives") never knew a world without cable, were introduced to the Internet before graduating from high school, and carried mobile phones with them from the time they were first allowed out in the world on their own.[21] The older edge of this generation provoked a new economic model in the recording industry through rampant illegal downloading, while their younger peers may own elaborate music collections that lack any physical form such as CDs.

Acculturated with a range of communication technologies from birth, this generation moves fluidly and fluently among technologies. Anne Sweeney, then cochair of Disney media networks and president of the Disney-ABC television group, cited research in 2006 indicating that 40 percent of Millennials went home each evening and used five to eight technologies (many simultaneously), while 40 percent of their Boomer parents returned home and only watched television.[22] Similarly, a 2006 report by IBM Business Consulting Services emphasized the "bimodality" of television consumers in coming years. It predicted a "generational chasm" between the "massive passives" (mainly Boomers who retained network-era television behaviors), "gadgetiers" (members of the middling Generation X who were not acculturated with new technologies from birth but were more willing to experiment with them), and "kool kids" (the Millennials).[23] Younger generations, who have approached television and technology in general with very different expectations than their predecessors, have also introduced new norms of use. For example, the television scholar Jason Mittell reflects on the significance of the arrival of a DVR in his home at the same time as his first child, and notes that when she came to ask, "What is on television?" the question referred to what shows might be stored on the hard drive, as she had no sense of the limited access to scheduled programming assumed by most others.[24] The widespread availability of control technologies provides a different experience for younger generations, who may never associate networks with television viewing in the same manner as their predecessors. As the generation that came of age using television to watch videos and DVDs and to play video games becomes employed in the industry, it will enable even greater reimagining of television content and use.

At a summit entitled "The Future of Television" sponsored by the trade publication *Television Week* in September 2004, all but one of the panelists used evidence drawn from observations of their children's approach to television as justification for their arguments about the new directions of the medium. In addition to their children not operating with a model of television organized by networks and linear schedules, the executives noted, with awe, the mediated multitasking that defined their children's television use. Research that continues to show growth in all media use supports these anecdotes. For example, as of 2007, time spent viewing television had not diminished despite continued expansion in time spent using the Internet; instead, multiple media have come to be simultaneously used. By 2012, the industry had termed this "second screen" use and found that 85 percent of tablet or smartphone users use these devices while watching television at least once a day, mostly to check e-mail or social media or send text messages.[25] Generations who are growing up with smartphones and tablets are accustomed to using multiple technologies to achieve a desired end, whether to access information, find entertainment, or communicate with friends. Such comfort in moving across technologies—or what those in the industry refer to as "media agnosticism"—has been crucial to the adoption of devices for watching television and ways of doing so that further facilitate the shift to the post-network era.

In sum, while features of a post-network era have come to be more apparent, such an era will be fully in place only when choice is no longer limited to program schedules and the majority of viewers use the opportunities offered by new technologies and industrial practices. Post-network television is primarily nonlinear rather than linear, and it could not be established until dominant network-era practices became so outmoded that the industry developed new practices in their place. The gradual adjustment in how viewers use television, and corresponding gradual shifts in production practices, have taken more than two decades to transpire, which is why I distinguish this intermediate period as the multi-channel transition. During this time, viewers experienced a marked increase in choice and achieved limited control over the viewing experience. But the post-network era allows them to choose among programs produced in any decade, by amateurs and professionals, and to watch this programming on demand on main "living room" sets, computer screens, or portable devices.

Implications of a Post-Network Era

I used to start each semester by surveying my classes in search of a show we all shared in common to draw examples from throughout the term. This was a pretty easy feat in my first few years of teaching in the early 2000s. Usually I found a show on my first (*Friends*) or second (*ER*) try. By 2007, I gave up on such unanimity. Instead I now gather a sense of what different factions of students might be watching, as it has been a while since I taught a class in which we had all seen the same show at least once (yes, even *American Idol*, *Jersey Shore*, or *Real Housewives*). This development illustrates an important consequence of the choice in viewing provided by the post-network era. The hundreds of channels offering programming by the end of the multi-channel transition significantly fragmented the audience. Then, by the end of 2010, viewers could readily access hundreds of television shows from any era on DVD or online, and an amateur video clip could reach as large an audience as a network show. Although only early adopters may have been viewing television in these new ways by this time, these developments suggest additional coming fragmentation.

   One of the first amateur videos to reach a mass audience appeared in April 2006. By the time "The Evolution of Dance," a humorous six-minute amateur performance of the progression of popular dance styles from the 1950s through the present, had been on YouTube for four months, it had been played at least thirty million times. That figure grew to over 210 million seven years later.[26] Site users re-posted the video multiple times, and in at least three different languages, taking advantage of one of YouTube's technological strengths—the ease with which videos can be linked to other sites—but making it difficult to sum up how many times it had been viewed across these multiple postings and on other sites. As a point of comparison, the most-watched television show of the preceding 2005–2006 television season—*American Idol*'s Tuesday-night performance episodes—averaged 31.2 million viewers each week. FOX's blockbuster hit included judges paid roughly $30 million a year, and the network earned $700,000 for a thirty-second advertisement, in addition to the at least $25 million per season paid by each of the three series sponsors.[27] In contrast, "The Evolution of Dance" featured the negligible production values of a video camera set up in the audience of a comedy club and was originally posted by the video's creator and dancer, Judson Laipply. Laipply did not profit directly from the millions of viewers (this was before YouTube enabled profit participation in its advertisements), although stories about the video's popularity appeared on the *Today Show*, *Good Morning America*, and *Inside Edition* and drew attention to his work as a public speaker and "inspirational comedian." YouTube benefited from the high traffic to the site that may have initially clicked through some of the banner advertisements and later from pre-roll advertisements—the video advertisements that play before selected clips—and the video's 2006 debut aided growing cultural awareness of the site. When I queried my classes in the fall of 2006 about their familiarity with the video, some had seen it—although fewer than I had expected and by no means as many as had seen various television shows. When I asked coworkers (faculty and staff over the age of thirty), most responded by asking what YouTube was—until a few months later, when Google's $1.65 billion purchase of the site drew much attention.

By the end of the decade, video sharing and YouTube use had grown considerably: By 2013, YouTube reported one billion unique visitors to the site each month.[28] Laipply posted "Evolution of Dance 2" in January 2009, but lacking novelty, drew only twenty million viewers in four years. There have been other so-called viral or spreadable hits such as "David after Dentist" and "Charlie Bit My Finger," though neither has been viewed nearly as many times as the YouTube topper as of 2013, Psy's "Gangnam Style" video, which had been streamed 1.8 billion times. Most web video circulates through distinct taste communities, and though nearly 2 billion views seems broad, with the possibility of an international audience of 2.5 billion Internet users on computers and 1 billion worldwide smartphone-enabled users, and that these are not unique viewers, even "Gangnam Style" can be seen as a niche phenomenon.[29]

The changes in how we view, experience, and use television made evident by these anecdotes have massive implications for how we think about television and its role in culture. The increased fractionalization of the audience among shows, channels, and distribution devices has diminished the ability of an individual television network or television show to reinforce a certain set of beliefs to a broad audience in the manner we long believed to occur. Although television can still function as a mass medium, in most cases it does so by aggregating a collection of niche audiences. The narrowcasting that became common to television during the multi-channel transition has thus required adjustments in theories about the mass nature of the medium, while the exponential expansion in viewers' choice and control since the network era has necessitated an even more substantive reassessment of television. Taking up these issues, this chapter provides an overview of some of the central ideas that have governed the study of television and culture as well as some preliminary tools for making sense of television in the post-network era.

Defining Television

The industrial changes that developed during the multi-channel transition made uncertain the object called "television" as new forms and ways of using the device required us to reconsider how we determine "what is television?" The term "television" has been broadly used to refer to a singular technology—a box with a screen—though it has enabled a range of experiences since the network era. But television is more than just a technology—more than a composite of wires, metal, and glass. It possesses an essence that is bound up in its context, in how the screen is most commonly used, in where it is located, in what streams through it, and in how most use it, despite the possibility for broad variation in all the factors. It is primarily this essence— derived from existing use—that distinguishes a television from a computer monitor, particularly in the context of contemporary technological convergence and the manufacturing of digital "televisions" that have no tuning capability—that is, the ability to receive signals over the air.

Lisa Gitelman argues for a definition of media as "socially realized structures of communication, where structures include both technological forms and their associated protocols."[30] Protocols include "normative rules and default conditions," such as the greeting "Hello," monthly billing cycles, and a system of wires and cable for the U.S. phone service. Understanding that the protocols of television contribute to distinguishing the medium helps us rectify some of the inadequacy of defining the medium only in terms of the piece of equipment and to address how the technology becomes a television when it receives signals via broadcast, cable, or satellite transmission. A television is not just a machine, but also the set of behaviors and practices associated with its use. The media scholar James Bennett has helped in the retheorization of television's scope through his distinction of "digital television":

Television as digital media must be understood as a non–site-specific, hybrid cultural and technological form that spreads across multiple platforms as diverse as mobile phones, games consoles, iPods, and on-line video services such as YouTube, Hulu, Joost, and the BBC's iPlayer, as well as computer-based mediaplayers such as Microsoft's Windows Media Player and Apple TV.[31]

Like Bennett, I find it useful to allow "television" to expand the narrow confines of its network-era operation. This new stage—what Bennett terms digital television and I distinguish as post-network television—doesn't mark an end of television, but the beginning of a new era.

I approach television with the presumption that our cultural understanding of this medium does indeed conceive of it as more than a monitor, piece of hardware, or gateway to programming, and that television is less defined by how the content gets to us and what we view it on than by the set of experiences and practices we've long associated with the activity of viewing. All of these technical attributes unquestionably contribute to how a culture uses and understands television, yet inherited meanings, expectations, and habits also circumscribe it in particular ways. New technologies and industrial practices have introduced radical changes in technological aspects of television, its use, and its consequent cultural significance, but various aspects of sociocultural experience still define television in our minds in specific and meaningful ways, particularly for those generations who knew television in the network era.

Television may not be dying, but changes in its content and how and where we view have complicated how we think about and understand its role in the culture. The transition of radio in the 1940s provides an illustrative parallel. As television first entered homes, radio had to fundamentally redefine itself—both in its programming and in the ways and places that listeners used it. Before television, radio was primarily a domestic-bound technology that played particular programs on a known schedule; after television usurped the captive home audience, radio became a portable medium and shifted to emphasize ongoing music or talk formats. Nonetheless, after television, the technology remained commonly understood as "radio" despite the substantial difference in the medium and adjustments to its role as a cultural institution. In truth, "video" provides the more accurate term for the cross-platform circulation of "television" content, but as the television experience has encompassed new capabilities and spread to additional screens in recent years, established cultural understandings have shifted accordingly so that we still continue to comprehend different experiences as watching "television."

In introducing a collection of essays that consider various aspects of the wide-ranging transitions that occurred by the beginning of the twenty-first century, Lynn Spigel reflects on the title of the anthology—*Television after TV: Essays on a Medium in Transition.* "Indeed," she notes, "if TV refers to the technologies, industrial formations, government policies, and practices of looking that were associated with the medium in its classical public service and three-network age, it appears we are now entering a new phase of television—the phase that comes after 'TV.'"[32] Although the title of the collection is eye-catching and provocative, it suggests a far more absolute rupture than that which occurred; it is also arbitrary in affording the norms of the network era such eminence as determinant of the medium. Still, attention to transition and uncertainty about the present status and likely future of television evident in the anthology and its title were not uncommon by late in the multi-channel transition. The title of another important article queries, "What is the 'television' of television studies?"—a question that similarly asserts concern about ambiguity regarding the fundamental attributes of television.[33] Those who write about television have never adequately addressed which of the "technologies, industrial formations, government policies, and practices of looking," to borrow from Spigel, might particularly establish the ontological boundaries of the medium—the things that make television "television." We err in allowing those norms established first to "determine" the medium; they are as arbitrary as any subsequent formation.[34]

Thinking about Network-Era Television

Scholars in fields as diverse as literature, film studies, political science, sociology, psychology, and communication developed different ways of thinking about television and its role in culture. Those in the area of "media studies" have attended most closely to the ways programs, audiences, industries, and sociocultural contexts intertwine in the creation and circulation of television, and their ideas are most relevant here. Scholars of media studies—and critical television studies in particular—have developed detailed theories and empirical studies that examine the multifaceted nature of cultural production common to television. But in the network era, there was no need for esoteric discussions of what television is, as it was assumed to be a simple technology whose variation spanned little more than screen size and color or black and white.[35] Much television theory continues to presume network-era norms in explaining the cultural and institutional functions of television, and draws from distinctive national experiences with the medium. This book and the conditions of the post-network era call many of these assumptions into question.

Foundational understandings of television view it as a—if not *the*—central communicative and cultural force in society. Its centrality derived from its *availability* and *ubiquity*; as early as 1960 more than 87 percent of U.S. households had televisions, and the technology increasingly was available in spaces outside the home, such as taverns and hospitals.[36] The accessibility of television was in many ways enabled by the low cost of acquiring its programming. Either as a result of advertising support in the United States or public funding in most other countries, viewing television programs did not require the same type of per-use fee associated with most other entertainment and informational media such as films, newspapers, and magazines. To be sure, commercial media "cost" societies in ways obscured by simple presumptions that proclaimed that network-era television was "free"; nonetheless, it was reasonable to assert that television's low barriers to access greatly contributed to its cultural importance in the network era.

During that time, the medium gained its status as a primary cultural institution precisely because network-era programming could and did reach such vast audiences. Television derived its significance from its capacity to broadly share information and ideas and facilitate an "electronic public sphere" of sorts.[37] Its stories and ideas reached a mass audience that some have argued enabled television programs to negotiate contradictory and contested social ideas, while others have proposed that this reach allowed television to enforce a dominant way of thinking.[38] Significantly, both perspectives ascribed importance to television because of its pervasiveness.[39] Viewers' lack of control over the medium and the limited choice at this time aided its ability to function as both forum and ideological enforcer. Network-era norms imposed the synchronousness of linear viewing, and television earned its status as an instigator of "watercooler conversation" by providing shared content for discussion. Coworkers and neighbors chose from the same limited range of programs each night, and thus were likely to have viewed the same program.

Assessments of television that consider how it contributes to the sharing and negotiation of ideas understand it to operate as a "cultural institution"—that is, as a social conduit that participates in communicating values and ideas within a culture by telling stories and conveying information that reflects, challenges, and responds to shared debates and concerns. Educational systems, clubs and societal orders, and religious organizations are also cultural institutions, although we may more readily identify and accept the influence of these sites on how we know and understand the world around us.[40] At the same time television functions as a cultural institution, however, it is also a "cultural industry." That is, in a context such as the United States, the television industry operates as a commercial enterprise that primarily seeks to maximize profits, while nonetheless producing programs that are important creative and cultural forms that communicate social values and beliefs. Industry workers may primarily make decisions based on what types of programming they perceive to be most profitable, yet these decisions still have important cultural implications for what stories are told, by whom, and how society comes to understand the worlds that television presents. Remembering the commercial mandate of television—again, particularly in the United States—is imperative: in the cultural industry of television, business and culture operate concurrently and are inextricable in every aspect.

Studies that explain the economic and industrial norms of television in the network era are particularly relevant to the focus here upon television as a cultural industry. Until recently, few attempted to bridge the chasm between humanities-inflected theories about the operation of media in culture and political economy research that emphasizes economics and industrial operations.[41] This history of avoidance, and at times hostility, between approaches is increasingly being corrected by theories and methods that deliberately merge aspects of culture and economics or explore quotidian industrial processes to better understand the agents, organizations, and processes involved in cultural production—as I attempt here.[42]

As is the case of dominant cultural theories about television, most political economy work assumes television to be a mass medium and attributes much of its importance to this characteristic. The notion of mass media and the scale of such businesses are important to political economy approaches examining the assemblage and distribution of labor and capital, while the mass audience was crucial to cultural approaches because of the necessity for programs to be widely shared within the culture. In both cases, the breadth of the audience reached by network-era programming allowed television to circulate ideas in a way that asserted and reinforced existing power structures and dominant ways of thinking within a society.

In many cases, the changed industrial context has not negated the value of theoretical tools provided by these perspectives, but some require reconsideration and adjustment. For example, Horace Newcomb and Paul Hirsch's argument that television programs provide a cultural forum to negotiate ideas within society makes sense insofar as television continued to facilitate this cultural role after the network era on certain occasions; however, broad and heterogeneous audiences now rarely share individual programs in the manner that could be assumed in the network era. Television might continue to provide a cultural forum for those who tune in to a particular show, but it has become increasingly unlikely that television functions as a space for the negotiation of contested beliefs among diverse groups simply because audiences are now more narrow and specialized.[43]

Other theories, such as Raymond Williams's network-era theory of "flow," require more significant revision.[44] Williams used the idea of flow to comment on the nature of the steady stream of programming through the set and the manner in which narrative, advertisements, and promotions all intermixed. The continuous infiltration of control devices into television use has greatly disrupted flow as a fundamental characteristic of the medium, at least in terms of television flow being determined by someone other than the individual viewer.

Television's transition from its network-era norm as a mass medium toward its post-network-era function as an aggregator of a broad range of niche and on-demand viewing audiences has required significant adjustments to industrial assumptions about the medium. For example, in his 1989 book *The Capitalization of Cultural Production*, Bernard Miège located television among media industries that operate under a "flow" model (this use of the term differs substantially from that of Williams) and rely on "home and family listening," "an undifferentiated, indirect mass market," the "instant" obsolescence of content, and the use of a programming grid that creates daily interaction and cultivates viewer loyalty, all of which eroded during the multi-channel transition. By the mid-2000s, the market characteristics of U.S. television had come instead to resemble those of his "publishing" model, which features a "segmented mass market" and the "dialectic of the 'hit and catalogue,'" along with the purchase of individualized objects—in this case, particular episodes of television shows.[45]

Noting that "'television' now functions as a bookstore, a news stand, or a library," Newcomb has departed from the "cultural forum" concept he and Hirsch offered in 1983 and conceived of the medium similarly to Miège's publishing model.[46] Television has adopted multiple possible revenue streams in ways that mirror the bookstore (DVD sell-through, iTunes downloading), magazine subscription (premium cable networks such as HBO), and the subscription library (MVPD on demand, Netflix). Each of these possible transactions of capital for content created new and distinct relationships between the economic model, programming, and how these forms of television might function as a cultural institution. And, as Newcomb notes, these alternative transaction or publishing models thrive on specialty, distinction, and niche taste, all of which unmistakably distinguish the practices of the multi-channel transition and post-network era from network-era norms that privileged the opposite characteristics.

Post-network-era practices have led the television audience not only to fracture among different channels and devices, but also to splinter temporally. The control over the television experience that various technologies offer has ruptured the norm of simultaneity in television experience and enabled audiences to capture television on their own terms. Moreover, as the *New York Observer* columnist Tom Scocca notes, the ephemerality once characteristic of the medium has also come to be less prominent; for example, the video experiences offered by YouTube allow for archiving images so they may be called up at will.[47] New devices have provided tools to capture television and consequently have produced a norm of asynchronous viewing that has altered the interaction of the culture with the medium in crucial ways. Television devices remain ubiquitous and accessible in the post-network era, but the ubiquity of specific content has been eliminated as broad audiences have come to share little programming in common and less frequently view it simultaneously.

The nature of post-network television will likely be profoundly different than that of the network era, but the contradictory affordances of these changes make assessing the relative quality of either era difficult. The uncertain future often instigates a nostalgia for past norms that imagines the past differently than it was experienced. As *Time*'s James Poniewozik opines,

> The irony of the nostalgia for TV's "golden age" is that it romanticizes the very things people used to condemn. Mass media were once homogenizing; now we miss how they unified us. Cultural critics once said TV appealed to the lowest common denominator; now cable's ambitious niche shows cater to elitists. Some even romanticize commercials—commercials!—as making TV for the masses possible.[48]

And it is not only the experience of viewers and insights of scholars that are changed, the adjustments to the U.S. television industry chronicled here provided as extraordinary a shift for those who work in it. The diversification in economic models, changing industrial relationships, and challenges to regulatory practices posed by new technologies all required revisiting many of the foundational industrial assumptions of television and how it operated.

Theorizing Niche Media: Identifying Phenomenal Television

Regardless of whether we have truly reached the post-network era, the U.S. television industry and its norms of operation have changed significantly. The most noteworthy adjustment, already evident by 2005, was the erosion of television's regular operation as a mass medium. Although it has continued to play this role in isolated moments, television is no longer organized in this way and has not been since the mid-1990s. By then, it was already apparent that we needed to reassess television and see it as a medium that primarily reaches niche audiences. Continued transition in television's core economic models would only further adjust the type of programming that could be profitably produced and television's operation as a cultural institution.

No mass medium arose to supplant television in the wake of its industrial change, and it might be that mass media as they existed in the twentieth century were remnants of a particular set of industrial and economic relations from another era.[49] Niche-focused media long have played an important role in society by communicating cultural beliefs, albeit to narrower groups than mass media. Women's magazines provide an illustrative example, as ample critical scholarship has explored how this media form that targets a specific audience consistently reproduced certain discourses of beauty, identity, and female behavior.[50] Niche media are identified as important voices to specific communities, but have received less critical attention than mechanisms of mass messaging.

Theorizing the cultural significance of niche media might begin by exploring those industries that have operated in this organization for some time, and the magazine industry—with its era of mass distribution earlier in the century—may provide the most relevant point of comparison. In considering the process through which this industry transitioned from mass market publications with titles such as *Life*, *Look*, and the *Saturday Evening Post* to more narrowly targeted magazines, Joseph Turow argues that demand from advertisers to reach ever more specific audiences fueled the fragmentation.[51] While acknowledging the economic value and efficiency targeting provides to advertisers, he raises a cautionary flag about such fragmentation and rightly notes the dangers for ideals of democracy and community that result from what develop into "gated informational communities."[52] The redefinition of television in the course of the multi-channel transition as a medium that supports fragmented audiences and polarized content consequently has exacerbated the cultural trends and outcomes that Turow identified in the magazine industry.

Television's new abundant offerings make it difficult to determine a proper frame through which to examine programming and assess its significance. We are accustomed to moral panics and activism that develop from concern about the vast reach of mediated messages. Thinking about television in the age of narrowcasting requires that we take into account the substantial variation now encompassed by its programming. "Successful" television programs might now gather audiences that range from tens of thousands to tens of millions, while channels might be accessible in anywhere from three million to one hundred million homes. Some programs stream into the home without any viewer payment, others require a subscription for a channel of programming (HBO), and viewers now can buy particular programs on DVD or as single-show downloads. With such ample variation in the availability and ubiquity of television programming, we need more specific models for understanding television's operation in the culture, ones that will enable us to differentially assess its significance.

Toward this end, I propose "phenomenal television" as a particular category of programming that retains the social importance attributed to television's earlier operation as a cultural forum despite the changes of the post-network era. In the network era, television content derived its relevance simply from being on the air, which necessarily meant that it was widely viewed because of the vast and substantive audiences programs had to draw to survive. Often popular shows were particularly important sites of analysis because broad viewership on a mass medium denoted a certain scope of influence. In a narrowcast environment, content must do more than appear "on television" to distinguish itself as having cultural relevance, since now much that appears on television might be seen by just a few viewers. For example, the particular economic model of advertiser-supported cable networks allows them to produce shows viewed by 1 percent of the available audience and for these shows to still be considered hits. Network-era theories might still apply to some programming produced in this narrowcast environment, and phenomenal television denotes such programming. Although the task of determining relevance and distinction is more difficult in the post-network era, phenomenal television does have identifiable attributes, as specified below.

Themes, topics, and discourses that appear in multiple and varied outlets indicate a form of phenomenal television. The criterion here is not purely quantitative—that is, a topic that appears in seven shows is not necessarily "more" phenomenal than one appearing in six. Rather, multiplicity might indicate a society-wide negotiation of an issue or a crisis in existing understandings in the same manner it did in the network era. Trans-show or trans-network themes derive importance in a narrowcast environment because such scope indicates content that has achieved or is likely to achieve uncommon audience breadth despite fragmentation and polarization. Ideas appearing in multiple shows—particularly different types of shows—might indicate concerns relevant to the broader society rather than distinct subcultures.[53] For example, in the year after the September 11, 2001, attacks on New York and Washington, DC, multiple narratives exploring fictional renditions of the aftermath appeared across at least twelve shows on seven networks.[54] Cultural critics could not look to just one of these shows as indicative of cultural sentiment on the subject, or even just that of television; instead, the niche media environment required a more holistic evaluation of the multiplicity of stories that likely reached varied audiences. This attribute responds to the way that individual programs and episodes rarely have the cultural significance previously common because of the fragmentation of audiences, although when thematically similar content is viewed and considered in aggregate, television has the potential to operate much as it did in the network era.

Attention to institutional factors such as *what network or type of network airs a show relative to the network's common audience* derives increased importance after the network era and plays a role in determining phenomenal television. Despite all being forms of television, broadcast, basic cable, and subscription cable have different regulatory and economic processes that contribute to their norms of operation and the possible programs they can create. These outlets also vary amply in audience size, and this too is a factor we must address in considering the reach and importance of a program or theme. Many programs—particularly those on premium and basic cable—reached narrow audiences throughout the multi-channel transition, but too often particular audience conditions were not addressed in framing analyses of or concerns about programs. Additionally, factors such as whether viewers watch content as part of linear schedules or on demand have come to further distinguish contemporary television programming as more viewers incorporate new control devices into their regular viewing habits. In the network era, we could assume a broad and heterogeneous audience who viewed linear schedules of network-planned programs. Now we cannot presume that the audience represents the culture at large; instead, it embodies only a distinct segment or component thereof. Assessing the type of network providing programming offers significant insight into the audience of a particular program.

Programs that achieve *watercooler status* earn a certain degree of importance due to their ability to break through the cluttered media space, but this alone does not indicate phenomenal television. We must also explore how and why a program achieves this prominence. A watercooler show that is supported by a particularly large promotion budget might be less meaningful than a show that captures the zeitgeist of the moment or gains its attention from the way that it resonates with a cultural sentiment or a struggle percolating below the surface of mainstream discourse. Phenomenal television can "go under the radar" and circulate out of sight or beyond the awareness of most of society, but examinations of such television must attend to how and why such shows are important. In the network era, watercooler shows were often those that were somehow boundary-defying, but few boundaries remain, and merely airing on television has become less indicative of social significance than was once the case.

Incongruity suggests another feature of phenomenal television, which has a tendency to break into unexpected gated communities. For example, incongruity might exist in cases where the ideology of a story conflicts with the dominant perspective anticipated to be shared by the audience of that network. The ability for like to speak only to like is one of the greatest consequences of narrowcast media because it decreases the probability of incongruity and disables the type of negotiation theorized to be central to the ability of network-era television to operate as a cultural forum. In many ways, the significance of a show such as *All in the Family* resulted from the heterogeneous audience that had their views alternatively challenged and reinforced by the differing perspectives articulated by Archie and Meathead. Similarly, a show such as *The Cosby Show* was particularly important because its depictions of upper-middle-class black life reached both black and white homes in a segregated society accustomed to representations of African Americans as being poverty-stricken or criminals. It remains significant to have a dramatic series focused on the lives and sexuality of a group of gay men (*Looking*) or lesbian women (*The L Word*), but these shows aired on a subscription channel that built an identity as the destination for gay and gay-friendly people, which made the content of these shows far more congruous than if they had aired elsewhere. Incongruous moments, such as the sophisticated negotiation and deconstruction of patriarchal masculinity provided by *Playmakers* and aired on ESPN or the critical exploration of the abuses of the Taliban against women on the WB family drama *7th Heaven*—which notably aired before September 11, 2001—expose audiences to ideas they may not normally self-select. The incongruity of these shows relative to what the audiences of these channels and their programs might expect can defy the tendency of narrowcasting to perpetuate gated media communities.

Programming affirmed by *hierarchies of artistic value and social importance*—those programs imbued with what Pierre Bourdieu terms "cultural capital"—indicate another distinction of phenomenal television. I do not wish to suggest that what I term "phenomenal television" is categorically "better" than other television, in the manner that "quality" television has been inconsistently used. Rather, what I am proposing is that television programming of specific aspiration and accomplishment—whether this be an ambitious period drama, a rigorous piece of investigative journalism, or a pointed political satire—might also distinguish itself as phenomenal because of its particular effort to enrich or expand cultural dialogue or thinking and to maximize the creative potential of the medium.

This delineation of characteristics of phenomenal television is not intended to suggest that programs that do not meet any of these criteria are unimportant. Rather, it marks a preliminary effort to develop a richer vocabulary and to build multifaceted theory in response to the growing multiplicity of television and its operation as a niche medium. The conditions of the post-network era require reconsidering everything we once knew about television and more clearly differentiating among its many forms. Size of audience is a significant consideration, but there are also features that distinguish programs in terms of content and in ways that are important to assess. The idea of phenomenal television provides a way to adjust our assumptions about television while keeping its increasingly niche operation in mind.

In many cases, the presumptions of network-era theory remain relevant in thinking about the cultural role of niche media and require only slight modification. For example, in 1978 John Fiske and John Hartley described the "bardic" role of television, noting how programs could "articulate the main lines of the established cultural consensus about the nature of reality."[55] Such a premise remains relevant in a narrowcast environment, but with the difference that television articulates the main lines of cultural consensus for the particular network and its typical audience member rather than for society in general. A so-called boundary-defying program such as *The Shield*, which explores the psyche and actions of a corrupt detective, may seem too far outside the accepted reality of the television audience on the whole, but the ambiguity of right and wrong it represents does appeal to a specific group of viewers who accept the complexity of human action and the arbitrariness of the justice system.

A category such as phenomenal television is just an initial tool for understanding the role of niche media in society; much more thinking in this area is certainly needed. Theories of niche media can in most cases reasonably assume certain characteristics of the audience—as niche media succeed because of their ability to tap into certain affinities that bring audience members together. But even though television programming of the multi-channel transition and post-network era increasingly targeted niche audiences, the breadth of content transmitted through the medium remained accessible to many beyond those targeted audiences. Those who watch niche content, but for whom it was not intended, might be viewed as "cultural interlopers"— as when teens' parents watch MTV or liberals view Fox News; although not all niche programming is equally susceptible to such practices. Industrial and economic factors such as how media are paid for vary the likelihood of interlopers across different types of television and in comparison with other niche media such as magazines. For example, subscriptions that provide access to a package of cable channels readily allow cultural interloping; subscriptions to specific programs, as in the case of pay-per-view, do not. Television watching is also often a shared activity in households, which increases the probability of cohabitants exposing others to television content not geared toward them.

Such possibilities for cultural interloping may further change as post-network distinctions solidify. Television purchased on a transactional basis, such as the pay-per-episode model available on iTunes, may be less likely to reach interlopers because of the added fees required to access this content. By contrast, subscriptions to channels might better facilitate interloping—as in the case of a viewer who subscribes to HBO for the movies, but samples *Looking* because it has no added cost. Important similarities and differences might be identified with other media such as magazines that have a fee per use and tend to be consumed in solitude, but can often be picked up in places like waiting rooms and read for free. These discrepancies and variations suggest the degree to which a one-theory-fits-all-media—and even a one-theory-fits-one-medium—framework is inadequate for theorizing niche media and post-network-era television. Similarities among media might exist, but specific contexts remain crucial in assessing the particularities of varied media.

The Persistence of Television as a Cultural Institution

The ubiquity that earned television much of its perceived significance has also been changing as a result of post-network reconfigurations. As the possibilities for portable and mobile television explored in the next chapter indicate, television is everywhere it has ever been and in many more places. Paradoxically, though, individual "pieces" of television (shows, episodes) are shared by fewer and fewer viewers. Together, these developments further the need to consider specific contexts and factors that are far narrower than a simple construction "television" allows. For example, in March 2006, two University of Chicago professors released a study widely reported in newspapers across the country that found that children who watched television were not substantially harmed by the behavior.[56] Such reports—with varied findings—appear yearly (even monthly) from researchers in many different fields. With rare exception these studies talk about the effect of "television," as though there were no differences in the experience of it, no differences in what is watched or how. Certainly, effects studies are not the only form of research to suffer from such unspecified generalizations concerning television, but the point is that variations in the medium that emerged throughout the multi-channel transition indicated how untenable these generalizations had become, if they were ever meaningful.

Instead of utilizing uniform assumptions and explanations of television, we might diversify our thinking by establishing "modes of television" that group similar functions of the medium. Indeed, for all the differences in viewing, every instance is not so distinctive as to be fundamentally unlike any other. Establishing some frequent modes of television use aids in distinguishing characteristics in a great many of television's iterations. At least four distinctive modes of television function existed by 2005: television as an electronic public sphere; television as a subcultural forum; television as a window onto other worlds; and television as a self-determined gated community. These persisted as general norms relevant to thinking about television's role in culture a decade later; but by 2014, the industrial distinctions among prized content, live sports and contests, and linear television seemed more indicative of how those in the television industry were distinguishing among television contexts.

Television as an electronic public sphere identifies the operation of television in the network era as it was explained by Horace Newcomb and Paul Hirsch's cultural forum model, Todd Gitlin's delineation of television's ideological processes, or John Fiske and John Hartley's notion of the medium's bardic role.[57] Drawing primarily on Newcomb and Hirsch, we might say that television operates as an electronic public sphere when it reaches a vast and heterogeneous audience and offers a shared experience or content that derives its importance from the scope of its reach, its ability to provide a space for the negotiation of ideological positions, and as a process-based system of representation and discourse. Now, however, television decreasingly operates in this way. When it does, it usually does so on unplanned occasions, except for a few remaining events such as the Super Bowl. At the same time, though, it is helpful to see the electronic public sphere as existing on a continuum. For example, in comparison with the network-era reach of television—when top shows were watched by 40 to 50 percent of television households—popular contemporary shows such as *American Idol* have a narrower scope—only 19.8 million out of a universe of 114 million homes watched it most weeks when it was at its peak.[58] But even with only an average of 17 percent of U.S. television households watching the show, it was among the most widely viewed regular programs in a given year.

Television operates as *subcultural forum* when it reproduces a similar experience as the electronic public sphere, but among more narrow groups that share particular cultural affinities or tastes. MTV is likely to be the best example, in that the network provides the lingua franca for adolescents and operates as "must-see TV" in order for teens to achieve cultural competence. The key difference between the electronic public sphere and a subcultural forum (note the embedded "cultural forum" in the terminology) is that the latter is characteristic of television that reaches smaller and more like-minded audiences. For example, Fox News provides a version of daily news and events that serves viewers who choose to watch a news outlet with its particular sensibility. Importantly, when television operates as a subcultural forum, it is often integrated with the use of other media that similarly reflect subcultural tastes and sensibilities. Viewers incorporate a television network or set of programs into a broader set of media, reproducing particular silos of specific worldviews. Broadband distribution of television and aggregators such as YouTube channels now serve far more narrow subcultures than were possible with the television available throughout the multi-channel transition.

Post-network television also can function as a *window onto other worlds*. In some ways this is a corollary to its function as a subcultural forum, as the ubiquity and availability of television make it a convenient means for exposing oneself to programming targeted to a different audience—or to interlope. Television makes it easy to be a casual anthropologist and travel in worlds very different from one's own, although by no means are all those worlds equally available. Viewers engage in television as a window onto other worlds when, as cultural interlopers, they view niche media not targeted to them. Parents trying to understand teen culture can gain glimpses into it on MTV—although understanding how teens receive the content or how any intended audience makes meaning of programming is another matter entirely. In leaving my own silo of information and taste culture, I have explored the excessive and regressive masculinity offered by Spike or the stories of masculinity in crisis aired on the FX shows *Rescue Me* and *Nip/Tuck*. Ever-expanding cable systems make available ever more worlds, including networks and content originating from outside the United States, although, again, all worlds and perspectives are far from equally available. Perhaps one of the most telling aspects of post-millennium U.S. culture emerges from the uncommon use of television as a window onto other worlds relative to television as a subcultural forum.

Finally, a fourth mode of television as a *self-determined gated community* has emerged particularly as a result of increasing flexibility in distribution and opportunities for viewers to access programming on demand. Here, television's cultural role is even more specific than when it functions as a subcultural forum. This mode encompasses particular uses and personalized organizations of television, as well as individuals' pursuit of specific content, including that which may be amateur-created. The operation of this more nascent mode can be observed in the videos submitted to aggregators such as YouTube or in those attached to social networking sites. Here self-created television becomes a forum of expression and a way for viewers to communicate—most likely with established peers—which they do by sharing their television. As television and web viewing become more integrated and convenient, viewers will also share recommendations, links, and viewing lineups that contribute to the personalization of television. Self-determined viewing behaviors include deliberately shifting among the variety of modes of use noted here and creating specialized viewing communities.

Certainly other modes may already exist that I have not included. The expanding fan cultures facilitated by social media perhaps suggest another distinct mode of television that might be labeled "television as cult conduit." The point I wish to highlight is the variety and differences in just the few functions given here. Each mode features varied characteristics and leads to very different cultural outcomes—television "means" differently in each of these modes—and does so in ways that previous explorations of television have not considered. I do not intend these four modes to account for all of television viewing; rather, I hope that identifying them will encourage others to consider television in terms of specific contexts and uses, rather than thinking about television-related phenomena as characteristic of television at large.

Delimiting the different ways television functions leads us to foreground the multifaceted nature of television and the growing diversity of uses viewers may adopt. You use the television that you flip on in the background while making dinner differently from the way you use the set to record a show you reserve for a time when all other distractions can be avoided—and that content consequently has different meaning and importance. Likewise, the television you view on a portable device on a daily train commute or the videos you search out online also indicate still other relationships between content and use. Each of these examples illustrates fundamental distinctions among use, content, and audience. In each case, the viewer may be watching television, but to understand the behavior and its cultural function, we need to develop more precise frameworks to explain differentiation among types of television content—such as phenomenal television—and why viewers watch in particular ways.

Key Ideas for Thinking about Television's Revolution

Finally, I turn to the key ideas and definitions particularly important for the reconsideration of television offered in the remaining pages. Most viewers remain unaware of the business of television, such as the intricate processes involved in deciding what shows to produce, selling them to networks, and finding advertisers, but understanding the business of television and how it is changing is crucial to comprehending why the industry produces certain shows and how to intervene in this system. Those who have sought changes in the cultural output of television—shifts in depictions of nondominant ethnic groups and women, for example—have been most successful when they have illustrated that their goal was a matter of "good business" for the industry, as has been the case for many social initiatives.[59] The production of expressive forms like television shows is a challenging business, and no matter the extent of market research, many of the tools other sectors of business use to understand what their consumers want and to predict success are ineffective in the creation of cultural forms.

I closely examine many components of production that figure centrally in the creation of U.S. television programming and focus exclusively on the commercial sector despite the existence of a small public broadcasting system. As I noted in the introduction, rather than thinking of production as just the making of a show, I define production as *all of the activities involved in the creation and circulation of television programming*. I organize this broad conception of production into five "production components"—technology, creation, distribution, financing, and audience research—and explore each in subsequent chapters. I do not intend any prioritization in the production components catalogued here. Sometimes technologies and distribution practices enhanced preceding developments intentionally, while other adjustments occurred independently.

Although I distinguish these five components as different activities, they must be understood as interrelated processes connected by multidirectional influences. Thus, for example, changes in advertising can introduce adjustments in how producers create programs, while changes in the creation of programs can likewise affect how advertisers are integrated in programming, as well as how much advertiser support networks or studios need. Moreover, the relations among the five components are constantly in flux. During the multi-channel transition, when adjustments in distribution capabilities affected economic models, the altered economic models then enabled certain creative norms—all of which affected the type and range of programming likely to be produced. Such an approach to production differs considerably from ideas about industry operation that assume power and influence operate in a one-way, top-down, hierarchical manner and that allow factors such as ownership structures a more deterministic role in the creation of expressive forms and day-to-day industry operations.

Just as production encompasses multiple components, production also exists as one "cultural process" in what some have termed the "circuit of culture" and others a "circuit of media study."[60] These circuit-based models or frameworks for studying media such as television provide a sophisticated conceptualization of the relationship between the creation of culture and the imperatives of commercial industries. Processes and factors other than production, such as reception, sociohistorical context, and particular cultural artifacts, are interconnected, with each affecting and being affected by the others. The production of television involves the negotiation of many different interests and requires a complicated model to adequately address the intersections of varied commercial and regulatory interests that also mediate in the creation of cultural forms.

Because of my focus on production, other parts of the circuit of culture receive minimal attention despite their relevance to the changes that mark the emergence of the post-network era. Often, these other cultural processes serve as structuring forces that significantly affect the conditions of production. For example, regulatory actions dating to the 1920s continue to determine the fundamental characteristics of the competitive terrain upon which the television industry operates. However, I attend little to the details of many of these broad, structuring regulatory actions, except in the instances when they particularly affect specific production practices, because they remain consistent throughout the history of broadcasting.

Here, though, I must emphasize the significance of the deregulatory policy that allowed expansive consolidation and conglomeration throughout media industries that the government began implementing at the beginning of the multi-channel transition, even though it is not a topic examined extensively in the book. This policy produced considerable regulatory consequences despite the reduced regulatory influence that the term "*de*-regulation" might suggest. Most notably, deregulation significantly changed what type of owner predominated throughout the television industry. Ownership of the roughly 1,400 television stations nationwide was substantially consolidated by the networks and a few station groups, while conglomerates also gathered broadcast networks, cable channels, production facilities, and even distribution routes such as cable and satellite providers into common ownership. New media entities were often integrated into these vast media conglomerates—as in the case of the AOL/Time Warner merger—although in many cases the architects of the new media age (Yahoo!, Microsoft, Google, Apple, Facebook, Amazon) and the consumer electronics industry (Sony, Samsung, Panasonic, Apple) remained separate from conglomerates dominating television content (News Corp, Viacom, Time Warner, and Disney) and many other legacy media.

Regulators had the perfect opportunity to intervene in broadcast norms during the digital transition mandated by the Telecommunications Act of

1996. The forced transition to digital transmission could have allowed Congress and the Federal Communications Commission to revisit the vaguely defined mandate that stations operate in the "public interest, convenience, and necessity" in exchange for the opportunity to use public airwaves to secure billions of dollars in profits, but regulators largely ignored this opportunity. Although regulatory rhetoric might have proclaimed that deregulation would lead to competition, most of the actions of the FCC since the end of the network era have been strongly influenced by the powerful industries the agency was created to regulate.[61] For the most part, the changes in industry operation chronicled here did not result from the competition that deregulation was supposed to inspire; instead, they came largely from the actions of companies outside the FCC's purview (consumer electronics and computing).

The degree to which the medium and the industry redefined themselves with remarkably little re-regulatory input introduced notable challenges by the late multi-channel transition. The interventions made by the regulatory sector—seen most distinctly in the fin-syn rules (explained in chapter 3), shifting cable policies, and deregulation of ownership—had massive implications for the industry's operation and in structuring the norms of production. At the same time, the relative swiftness with which production components responded to changes in various production practices decreased the relevance of the lumbering regulatory sector in establishing the regulatory conditions appropriate to emerging post-network norms. Regulators could radically adjust the playing field for the industry at any time—as the intermittent threat of mandating à la carte cable service suggested throughout the mid-2000s—but they seemed unlikely to deviate from the "market-driven" logic underscoring their decisions for the previous two decades—that is, except in the case of content regulation. The developments of the multi-channel transition merited sweeping regulatory action that revisited broadcasting's regulatory foundations; however, by 2013, regulators had established no clear principles that reflected the substantial industrial adjustments occurring.

Each of the following five chapters focuses on a different aspect of production in order to explain the broad changes that have taken place over the past twenty years, the new norms being established in the post-network era, and why these changes in how television is made affect both the types of programs that are produced and the role of these stories in the culture. The next chapter looks at the changing technologies viewers have used to watch television and how new devices have enabled viewers unprecedented control over how, where, and what they view—and increasingly, to even make their content.

2

Television Outside the Box

The Technological Revolution of Television

Never before have the balance sheets, strategies, constituent relationships and very existence of media conglomerates been shaped so
radically by technology and changing consumer habits. Never before
has so much revenue been put on the line, and never before has there
existed the potential for so much content, distribution, packaging
and pricing to be placed beyond the reach of the media giants.
—*Hollywood Reporter*, 2005[1]

TV has evolved in the past, but the current digital revolution shock
is unprecedented. And, just as in earlier periods of fecundity, TV
production, distribution, and consumption are all being redefined
and refreshed by outsiders, from Apple's Steve Jobs to the new amateur producers peopling YouTube or Blip.tv.
—*Wired*, 2007[2]

The first epigraph, taken from an uncharacteristically forward-looking think
piece by one of the industry's key trade publications, captures the uncertainty
and anticipation of the industry as early as 2005. Industry workers knew
that technological change was approaching. Many had seen the diverse platforms and applications debuting at electronics shows long before they would
reach the living rooms of middle America, but no one could be certain of
how audiences would use the new technologies, how quickly they might
adopt them, which devices would prove essential, or what might be the next
"killer application." To make things more challenging, constant technological
development made the earth seemingly shift beneath the feet of those trying to adapt. By the time a mid-2000s strategy making use of then-emerging
DVD and video-on-demand (VOD) distribution opportunities was in place,
unanticipated possibilities in downloading to portable devices made many of
those preliminary efforts seem woefully inadequate.

Technologies involved in the digital transition enabled profound adjustments in how viewers used television, and these newly enabled capabilities

necessitated modifications in many other industrial practices. Emerging devices considerably enhanced viewers' ability to control television first in, then out of the home. The increased control over how, when, and where to view provided by digital video recorders (DVRs), DVDs, electronic programming guides (EPGs), digital cable boxes, laptops, smartphones, and tablets expanded *convenient* uses of television. These devices enhanced the comparatively limited capabilities first afforded by analog technologies such as VCRs and allowed viewers far more flexibility in when and where they watched television.

Live broadcast has long been perceived as an inherent technological attribute of television—perhaps because audiences lacked technological tools to control their viewing throughout the network era. As new technologies emerged, the industry initially perceived liveness to be important and endeavored to identify technological solutions that would enable reception of live television almost anywhere—what I categorize as *mobile television*. However, as U.S. viewers came to use various television technologies, the ability to control viewing, which included using tools to both time-shift and place-shift, emerged to have far greater desirability than the ability to view live television outside the home.

In discussing technological advancements, I differentiate between *portable* and *mobile* television use, a distinction not commonly observed in industry discussions. In my parlance, viewers use mobile television when they access *live* television outside the home, as opposed to the *time- and place-shifting* characteristic of portable television. These distinctions can seem confusing because many devices allow both portable and mobile television use: Laptops, tablets, and smartphones can be used to download content and store it for nonlinear viewing or to stream live video. While viewers may rarely think of these distinctions, they are important because the liveness of mobile viewing preserves the advertising model particular to linear viewing in a manner quite different from the mostly nonlinear use of portable television. Portability fits clearly within the realm of expanding convenient uses of television, while the desire for immediacy characterizes mobile television. In sum, this chapter distinguishes between *mobile* television technology use, which accounts for out-of-the-home *live* viewing, and *portable* television technology use, in which viewers take once domestic-bound content anywhere for viewing at any time.

Additionally, technological advancements in audio and visual quality—many of which resulted from the digital transmission of television signals—expanded the *theatricality* of television until the distinction intended by the word (to signify the feeling of watching a film at the theater) became insignificant. The emergence of high-definition sets as replacements for the long inferior NTSC television standard particularly contributed a technological revolution in the quality of the television experience. Digital transmission alone allowed some enhancement of television's audio and visual fidelity, but the high-definition images in particular appeared as crisp as reality and offered the detail available on film.

Each of these attributes of post-network technologies—convenience, mobility, and theatricality—redefined the medium from its network-era norm. Their significance results from the considerably revised and varied uses of television that consequently have emerged and that contrast with the unstoppable flow of linear programming, the domestic confinement, and the staid aesthetic quality of the network era. Rather than these technological assassins causing the death of television, as many writing about television in the mid-2000s claimed, the unprecedented shift of programming onto tiny mobile phone screens, office computers, and a wide range of portable devices ultimately reasserted the medium's significance. But the new technological capabilities required adjustments in television distribution and business models in order to make content available on the new screens, which provided a challenging task given the inconsistent interests of rights holders such as content creators (studios) and distributors (traditional channels and networks). Studios sought to maintain tight control over content so as not to disrupt the traditional revenue streams and the long-term value of copyright ownership. Networks desired strategies likely to drive viewers to other content on the network or to other network locations—like network-owned websites—from which the distributor might earn additional value. Both studios and networks sought to maintain their established status in the television industry while new broadband distributors and consumer electronics developers endeavored to rethink many of the business's established "rules."

The various post-network technologies produced complicated consequences for the societies that adopted them. Viewers gained greater control over their entertainment experience, yet became attached to an increasing range of devices that demanded their attention and financial support. Many viewers willingly embraced devices that allowed them greater authority in determining when, where, and what they would view, although as fees and services enabling new conveniences proliferated, they also struggled with the burden of the many costs previously borne by advertisers. In many cases, the "conventional wisdom" forecasting that the new technologies would have negative consequences for established industry players proved faulty; technologically empowered viewers used devices to watch *more* television and provided the industry with unexpected new revenue streams at the same time they eroded old ones. The emergence of these technologies consequently resulted in contentious negotiations within and between factions as view-ers assessed what capabilities were worth the cost, the consumer electronics industry endeavored to embed its products in the daily use of as many as possible, legacy media (such as broadcast networks) evolved their business models, and legacy and new media services (cable providers and Internet aggregators like YouTube and Hulu) developed mechanisms to make new technologies useful and programming accessible.

Network-Era and Multi-Channel Transition Technologies

The lack of technological variation during the network era enforced a fairly uniform television experience for viewers. Television sets that received very few signals over the air functionally defined the technological experience in the network era, while the use of antennas and CATV added complex-ity and limitations for some viewers. These devices, however, tended only to enable viewers in rural or mountainous areas the same access to the medium enjoyed by their urban brethren. Either way, viewers had no technological control over television and little choice among content. Certainly, the transi-tion to color television was significant, and many of the technologies that began to revolutionize television use during the multi-channel transition were introduced to early adopters while all other characteristics of the net-work era remained firmly intact. For most, however, a single television in the home without remote control or VCR characterized the network-era techno-logical experience with television. This uniformity of use aided the industry's production processes because it enabled the industry to assume certain view-ing conditions and rely on viewers to watch network-determined schedules.

Technological developments of the multi-channel transition introduced profound changes for users, first by enhancing choice and control with ana-log technologies such as cable network distribution, VCRs, and remote con-trol devices. Experiments with remote controls began in the early days of radio and continued through its refinement and into the television era.[3] The industry sold as many as 134,000 remote-equipped televisions as early as 1965. Despite these early starts, Bruce Klopfenstein argues that 1984 to 1988 marked the period of most rapid overall remote control diffusion, due to the simultaneous distribution of cable and VCR remote controls in concert with those of television sets.[4]

The VCR is one of the first technologies to trouble our understandings of "television." The distribution of the VCR as an affordable technology, which achieved mass diffusion at the same time as the remote control, sig-nificantly expanded viewers' relationship with and control over television entertainment. Nearly 50 percent of U.S. homes owned VCRs by 1987;[5] this figure increased to 65.4 percent by 1990, 88.1 percent at the end of the century, and peaked at 98.4 percent in 2003, after which VCR rates declined and DVDs (and later DVRs) began replacing the technology.[6] The recording devices allowed viewers to negate programmers' strategies through time shifting and introduced new competitors such as the home video purchase and rental market. In addition to enhancing viewers' television capabilities by allowing them to record and review television shows, the VCR also enabled the television set to function entirely independently of the networks' linear program schedules.

Another key characteristic of the early multi-channel transition resulted from the arrival of cable, which introduced profound changes in both technology and distribution. As a technology, cable substantially altered viewers' experience with its introduction of a vast array of channels. In 1988, 50 percent of U.S. households subscribed to cable, which was the subscription base analysts believed necessary for cable operators to provide a large enough audience to achieve profitability.[7] This subscription level marked an increase from just 19.9 percent in 1980, grew to 56.4 percent by 1990, and reached 68 percent in 2000. By 2000, nearly ten million additional households received programming via direct broadcast satellite (DBS—services such as DirecTV or Dish Network).[8] In the mid-2000s, "telcos"—companies traditionally known for providing phone service, such as Verizon and AT&T—began competing in some markets, offering packages of channels and then-state-of-the-art Internet service. By 2014, 10 percent of homes received television content from a telco. Since the mid-2000s, cable, satellite, and telco penetration grew to roughly 90 percent of U.S. television homes.[9]

Broadcasters maintained many of their network-era programming practices throughout the multi-channel transition even though the audience that regularly viewed them decreased in scope. The increase in program outlets significantly shifted the size and composition of the audience watching the Big Three networks, but it required decades for this change to reach an economic crisis point. In some ways, a paradox of remaining the "most mass" programming outlet reaffirmed the status of broadcasters and allowed them to remain disproportionately dominant throughout much of the multi-channel transition despite their slipping share of the audience. Cable channels drew audiences, but the multiplicity of cable channels was significant only in aggregate; any one channel drew a small fraction of the audience still reached by a broadcaster, and other than content-specific channels such as CNN, MTV, and ESPN, cable channels created very limited original programming during the multi-channel transition. The broadcast networks achieved some cost cutting by scheduling more programs from cheaper genres such as newsmagazines and early "reality" shows, but the broadcast networks were able to maintain many of their core practices throughout the multi-channel transition. Broadcasters' continued dominance and cable channels' limited encroachment enabled a conciliatory coexistence that ruptured once cable channels began producing "broadcast-quality" series in the early 2000s and deviated from broadcast norms of season length and scheduling patterns.

Analog technologies enabled limited control and choice during the multi-channel transition, but the arrival of digital television technologies at the end of the century vastly reconfigured technological capabilities and introduced characteristics of a post-network era. The shift from an analog technology such as the VCR to the DVR and DVD may seem insignificant in terms of the similar capabilities each provides, but the arrival of digital technologies profoundly changed television.

Digital Media and the Post-Network Era

Welcome to the age of fast-food TV: nuggets of news and entertainment that can be consumed on cellphones, video game consoles and digital music players. Whether the programming is downloaded via iTunes software or over a cellular network, the trend is changing where—and how—TV watchers are tuning in.
—Meg James, *Los Angeles Times*, 2005[10]

The digital revolution produced two types of consequences for television: interoperability and efficiency. These capabilities adjusted viewers' experience as the common language of ones and zeros shared in the digital transmission, reception, and home recording of television advanced the medium considerably. It provided the technological opportunity to converge televisions, computers, and other home technologies, and also allowed more efficient signal transmission and storage. Digital transmission further expanded choice, as broadcasters and cable providers were able to relay more information in their broadcast spectrum and cable wire by using a digital signal and eventually Internet protocols. Digital technologies enabled broadcasters to offer multiple "channels" in the six megahertz of spectrum previously required to transmit one analog channel—and the number of channels continued to grow with better compression technologies. Likewise, cable providers expanded channel offerings and added on-demand services (VOD) once they were able to more efficiently compress their signals. The compression technologies allowed digital cable to increase channel offerings with additional niche channels that sought increasingly precise tastes; for example, the general sports-caster ESPN eventually competed in a sector of sports channels for various regions (MSG, Big Ten Network) and sports (NFL Network, Golf Channel, World Fishing Network), as an indication of the expanding fragmentation. The consequences of choice were widely experienced by the early 2000s and receive limited examination here, where I focus instead on the newer developments of convenience, mobility, and theatricality.

The State of Technology Adoption

Viewers' use of television expanded considerably as they adopted the tech-nologies developed and deployed throughout the multi-channel transi-tion. Many of the technological shifts introduced incremental change to the industry in a manner that did not substantially challenge existing industrial practices, while other technologies instituted such considerable modifica-tions that they contributed to adjustments throughout the production pro-cess. The technologies launched during the multi-channel transition were neither uniform in character nor deployed in an organized or coherent man-ner. Rather, many different sectors, such as computer and consumer elec-tronics industries, governmental regulators, cable and satellite providers, and broadcasters, had varied stakes and visions for the role of these technologies in the future of the U.S. television industry.

All of the domestic technologies explored here were widely available by mid-2005, although penetration rates were still low for some of the more significant devices such as DVRs and high-definition (HD) televisions. A snapshot of technological diffusion and use collected in the spring of 2005 and then the fall of 2013 reveals the varied emergent and integrated status of different technologies (see table 2.1). The years between 2005 and 2013 are marked by a 40 percent gain in homes with DVRs, a nearly 40 percent increase in homes with broadband connection, and the emergence of "sec-ond screen" technologies such as tablets as well as the introduction of smart-phone technology. These devices, along with VOD capability, feature tech-nological affordances that enable viewers who desire a nonlinear television experience the ability to organize their viewing free from network schedules.

In the spring of 2005, 82 percent of homes with a television reported own-ing two or more sets, and nearly half (45 percent) owned a television with a screen larger than thirty inches.[11] Number of sets became a decreasingly relevant statistic as the screens upon which television might be viewed mul-tiplied well beyond the population of a given home. Ten percent of homes

Table 2.1. A Snapshot of Television Technology Diffusion, 2005 and 2013

	Spring 2005 (%)	Fall 2013 (%)
Sets		
Homes with television	100	95.8
Homes with 3 or more sets	45	67
Signals		
Homes with cable	67	54
Homes with digital satellite	20	31
Homes with telco signal	n/a	10
Total multi-channel	87	92
Homes with only over-the-air reception	14	~9
Channels and Devices		
Homes receiving 40 or more channels	82	unknown
Homes receiving 100 or more channels	40	58 (in 2008)[a]
Homes with a VCR	87	55
Homes with a DVD player	76	83
Homes with a video game system	39	56
Homes with high-definition TV	9[b]	83 capable
Homes with a digital video recorder	7[b]	49
Homes with a computer	67[b]	80 (Internet connected)[c]
Homes with a mobile phone	72[b]	87
Smartphones (of mobile phone population)	n/a	65
Tablet	n/a	29
E-book reader	n/a	26[d]
Homes online	59[b]	72[c, d]
Homes with broadband connection	28[b]	70[c]

Sources: Unless otherwise indicated, data from Nielsen Media Research. Fall 2013 data from "An Era of Growth: The Cross-Platform Report," March 2014, or "The Digital Consumer," February 2014; 2005 data pulled from Npower for 15 May 2005 and based on total U.S. homes with a television.

[a]Nielsen Media Research, "Average U.S. Home Now Receives a Record 118.6 TV Channels," 6 June 2008, http://www.nielsen.com/us/en/press-room/2008/average_u_s__home.html.

[b]Knowledge Networks Statistical Research, "The Home Technology Monitor: Spring 2005 Ownership and Trend Report" (Crawford, NJ: Knowledge Networks SRI, 2005). The survey uses a telephone sampling method, so the data are figured from a base of telephone households with one or more working television sets.

[c]Nielsen reports 8 percent more homes with Internet-connected computers than does Pew; I'm not sure which report is more in error.

[d]Pew Internet and American Life Project, "Three Technology Revolutions," http://www.pewinternet.org/Trend-Data-(Adults)/Device-Ownership.aspx.

owned a set larger than fifty inches—a figure growing about 4 percent per year—while 26 percent reported having a home theater or Surround Sound audio system.[12] Such audio technology reached this rate in 2000 and maintained considerable consistency, suggesting a likely adoption plateau.[13] By 2005, only 9 percent of homes owned a high-definition set, although that was nearly twice as many as two years earlier.[14] Substantially greater penetration of HD technology was achieved by 2013; however, not all homes that owned HD sets received HD content because of the varying availability of HD packages from cable and satellite services and general confusion on the part of set owners. As recently as October 2012, a Nielsen study revealed that only 29 percent of prime-time broadcast and 25 percent of prime-time cable programming was viewed in HD.[15] Such data reveal how limited HD viewing remained despite the much greater presence of HD-capable sets.

By 2005, a multi-channel norm was clearly in place, with 87 percent of television households receiving signals from a non–over-the-air source, 67 percent of which subscribed to cable and 20 percent to satellite. Cable maintained its dominant status in 2013, though it had lost share to the telco providers—AT&T and Verizon—that began competing in selected markets with a fiber-to-the-home product. Only 19 percent of television households subscribed to digital cable in 2005, although it was available to 85 percent of wired cable homes.[16] By 2012, 81 percent of cable households subscribed to digital cable.[17] Digital cable allowed for advanced and two-way signal transmission between cable providers and homes that enabled more robust video-on-demand offerings and eventually, DVRs using cloud-based storage. Eighty-two percent of television homes received more than forty channels, and 40 percent received more than one hundred in 2005, a figure that increased to 58 percent by 2008. Arguably counting channels was never a good indicator of the range of content people might view. Nielsen reported in 2014 that "the average U.S. TV home now receives 189 TV channels" but watches an average of only 17 channels.[18] Once broadband-distributed video became widely available around 2010, counting channels became an even less informative indicator.

The penetration of convenience devices provides one of the most marked shifts between 2005 and 2013. VCR ownership continued to decline in 2005, and was down to 87 percent. By contrast, and unsurprisingly, ownership of DVD players continued to increase, with 76 percent of homes owning a DVD player in 2005. Despite the omnipresence of DVRs in industry discussions by 2005, only 7 percent of homes had the device, which marked an increase from 4 percent in 2004.[19] DVR penetration began to grow much more rapidly once the devices were integrated into the set-top boxes provided by the companies providing cable service, typically for a monthly fee. (Henceforth, the companies that provide video service—whether by cable, fiber, or satellite—will be noted as MVPDs, the common industry acronym that stands for multi-channel video programming distributor.) Twenty-six percent of homes reported they had access to video-on-demand (VOD) services in 2005, but only 11 percent reported viewing a free or pay VOD program in the previous month; the title availability at this point was most limited and emphasized theatrical film content.[20] The transition to digital cable led to greater availability of the service between 2007 and 2011. Only 37 percent of homes had set-top boxes that enabled VOD in 2008, but they were in 60 percent of households by 2013.[21] Notable expansion in VOD content availability and then use occurred around 2011; the measurement service Rentrak reported that free television VOD increased more than 40 percent in 2012 from the previous year.[22] In 2005, 39 percent of television households owned a video game system that attached to the television. This distribution level had been steady since 2000, but began increasing as the devices added Internet connections that expanded the range of activities gaming systems could be used for and newer-generation systems such as the Wii introduced controllers allowing motion control gaming, which also expanded the gaming audience.[23]

In terms of the broader home technology space in 2005, 67 percent of homes had a computer, while 23 percent owned two or more, and patterns of growth in home computer ownership suggested that demand had nearly reached equilibrium.[24] Eighty-eight percent of computer households (59 percent of all households) used the computer to go online, a use level that remained steady since 2001. Fifty-two percent of online households connected through a regular telephone line, while nearly half (28 percent of all households) used a broadband high-speed method, with nearly even distribution between cable modems and DSL service.[25] Significant growth in access to high-speed broadband connections occurred by 2013, with 70 percent of homes connected to the Internet through a broadband connection and just two percent through dial-up, though significantly, 28 percent of homes remained without Internet access.[26]

This substantial gain in Internet speeds resulting from the shift to broadband, as well as the emergence of smartphones and compression technologies that enabled video to be accessed over 3G then 4G mobile data networks, provided the technological basis for the most significant adjustments in television technology. Seventy-two percent of homes owned a mobile phone by 2005, and 41 percent owned two or more.[27] Though 31 percent of households had an Internet-capable mobile phone, or what were then called personal data assistants (PDAs), only 11 percent used the devices to access the

Internet, and just 5 percent of mobile phone homes owned phones capable of receiving television-like video, and even fewer used this feature.[28] By 2013, 65 percent of the 235 million U.S. mobile phone subscribers had a smartphone easily capable of accessing and screening video content, though this population spent only an average of 1 hour, 23 minutes per month, or 2.7 minutes per day, using their phone in this way.[29]

Considering television technology now requires looking beyond long-standard figures such as the number of sets and VCR penetration. As these data illustrate, Internet access and smartphone availability are just as important technological pieces to understanding post-network-era access to television content. And though notable increases are evident in the 2013 figures—so much so as to call a new era of television distribution into existence—it is important to reiterate that 28 percent of U.S. homes still had no Internet access.

Likewise, even by 2005, choice—measured by subscription to non–over-the-air providers and the number of channels available—seemed to have reached useful capacity. There always might be room for more—endlessly so, thanks to broadband-distributed video—but the average number of channels that audiences viewed suggested that few had interest in the expansion of linear channels. Nielsen estimated that despite exponential growth in availability, the number of channels viewed by a household tended to increase only slightly. A household with 31 to 40 channels viewed an average of 10.2, while those with 51 to 90 viewed just over 15. The number of channels viewed remained at 17 from 2008 through 2014, despite an increase in average channel availability from 129 to 189.[30] The call for à la carte cable packaging that would allow viewers to select only the channels they desired had begun by the early 2000s, though it would take the perceived disruption of cord cutting a decade later to yield serious industry consideration.

On one hand, there was good evidence that a post-network era was emerging and that the "state" of television needed different forms of evaluation than those that had marked its quick rise to ubiquity. But there were also signs of network-era persistence: Even by 2012, aggregate Nielsen data indicated that despite nearly 50 percent DVR penetration and expanding VOD offerings, time spent viewing time-shifted content accounted for just under 8 percent of time spent viewing.[31] Again, that is an aggregate figure, and perhaps the significance of time-shifted viewing gets lost in the use of background television; among some sectors of the audience, time-shifted viewing accounted for well over 20 percent of prime-time viewing. The stakes of such moments of transition are so considerable that selective release of data could affect perceptions just as significantly as not recognizing the new questions and methods that need to be explored.

Digital Control Yields Convenience

Digital technologies allow for such a new array of television uses that it is difficult to sort out the variety of technological affordances provided by different devices and how those then map onto or deviate from how audiences actually use them. The typical audience member may focus only on what different technologies do, and thus organize them differently than I do here, as I am also concerned with how the capabilities of devices also require adjustments in other production practices. This section examines the new conveniences offered by technologies, beginning with technologies that make home viewing more convenient, then those that enable portable (non-live) viewing out of the home, and finally a consideration of how the convenience of breaking from the linear schedule occurs across domestic and portable technologies.

Viewers first gained the convenience of defying networks' schedules with the VCR, which established a modest beginning that since has been expanded by DVRs and digital devices that integrate Internet and television to vastly expand consumer control. The first technologies made television more convenient by allowing viewers greater control over when they would view, though continued to bind that viewing to domestic sets. By the time these domestic control technologies began reaching audiences larger than early adopters—around 2010—a second expansion in convenience emerged as the technology, infrastructure, and distribution strategies converged to meaningfully establish portable television. These technologies (laptops, tablets, and smartphones) and broadband-delivered aggregators (such as Netflix, YouTube, and Hulu) expanded the nonlinear viewing made possible by the first technologies by enabling viewers to access this content anywhere they could receive a broadcast signal, access a wireless Internet connection, or even receive a mobile phone signal—which, for those with ample financial resources, meant virtually anywhere in the United States.

The first digital control device, the DVR, initially appeared to offer little additional capability than the VCR. Yet its efficiency and ease of use made its contribution significant. While programming a VCR was perceived as so difficult that a joke about the flashing 12:00 VCR clock became ubiquitous, DVRs featured one-step recording capabilities from their introduction in 1999. For some DVR users, time shifting became the default mode of viewing for most programming—particularly in prime time—a difference suggestive of a shift from mere control to convenience. Even early-generation devices featured on-screen menus and programming schedules far easier to navigate and quicker to load than those offered by digital cable systems over a decade later. The remote capabilities available with some machines that enabled viewers to program the DVR from out of the home by accessing it via computer or mobile phone further illustrated the convergence of digital technologies and expanded control.

The ease of recording common to DVRs and their tape- and disc-free, hard-drive–based archive made them a significant threat to the conventional practices of the television industry. VCR users could—and unquestionably did—"zip" through commercials in recorded material, but VCR use tended to be restricted to more isolated occasions of particular shows; based on Nielsen data, MAGNA Global estimated that VCR recording accounted for only 6 percent of the average prime-time audience in 2005.[32] Industry analysts marveled at the level of satisfaction earned by DVR technologies, as adopters recounted that their DVR "changed their lives" and professed "love" for the machine. Like many skeptics, I saw the DVR as an insubstantial advance from the VCR, until I used one. I quickly joined the converted as my whole approach to viewing television changed radically once I could easily control so many aspects of the experience. By the 2007-2008 season, as DVR penetration rates reached above 25 percent, networks and advertisers agreed to begin buying and selling advertising based on the "C3" report Nielsen had developed that included live viewing plus DVR viewers who watched recorded programs and viewed commercials within three days of recording.[33] DVR penetration continued to grow, and networks agitated for a move to a "C7" measure that would count playback done within a week. Yet, though nearly 50 percent of homes had DVRs in 2013, Nielsen reported that homes with DVRs watched only eleven hours, thirty minutes of recorded programming each month—a mere 8 percent of average total viewing.[34]

The introduction of the DVR affected television in wide-ranging ways. DVRs were many viewers' first experience with nonlinear viewing, a way of viewing that growing VOD offerings and online streaming from Netflix or Hulu would expand. The DVR also provided the first clear technological threat to the conventional advertising model of thirty-second commercials embedded in programming, and fear of this technology led to adjustments in advertising strategies and program financing models.

Yet by 2014, just as the DVR had infiltrated nearly half of television homes, its future began to appear uncertain. Distribution technologies such as VOD and online streaming that developed after the initial DVR diffusion suggest that its role in the transition to the post-network era may be more as a technology that provides a bridge to the post-network era, rather than as a technology of the post-network era. Subsequent, arguably complementary, technologies made the effort of selecting and recording content that DVRs required—even if largely mechanized through "season pass" type settings—seem somewhat burdensome. With viewer control embedded in its name, VOD expanded viewers' control over their television experience and is a technology characteristic of the industry's shift beyond a mere multi-channel transition toward a more full-fledged, post-network era as VOD requires no linear television experience.[35] VOD technologies provide a range of services akin to DVRs: both devices enable viewers to pause, stop, and rewind or fast-forward through programs. But the key distinction between them lies in where the technologies store content. DVRs pull content from the twenty-four-hour linear stream of programming that networks transmit and store the recordings on a device in the home (though next-generation devices have cloud-based storage, which makes this distinction negligible). VOD technologies store content on a server maintained by cable providers, and viewers access this programming bank at will, choosing among the offerings of the provider. VOD functionality can also be controlled by the MVPD, most commonly by disabling fast-forward capability.

The slow pace at which MVPDs introduced VOD offerings and developed a robust programming supply is far more a matter of the complicated rights allowances required in a business of many middlemen and advertising protocols than of technological capability. The development of VOD libraries required extensive negotiation between content creators, cable channels, and MVPDs in order to identify a financial model that would serve all three entities and still be desirable to viewers.[36] Throughout the mid-2000s, MVPDs rebuilt their infrastructure and offered VOD as part of top-tier digital subscription packages. By 2005, free VOD was available to 26 percent of cable subscribers, but primarily allowed them to view only "extras" and "bonus footage" rather than full episodes.[37] Some cable services experimented with subscription video on demand (SVOD), but the model of paying specifically for the on-demand content was less popular (unsurprisingly) than the "free" access, which cable services included as a "value-added" perquisite to encourage digital cable subscription. Subscription services such as HBO did include on-demand access to much of its content as part of subscription comparatively early on, providing its subscriber base with the earliest access to full-length on-demand content.

The other impediment slowing VOD development resulted from finding a way to monetize content. Distributors lacked a motivation until VOD viewing could be "counted" in ratings. Not until around 2012 did technologies allowing "dynamic advertisement insertion"—or the capability to change the advertisements included in VOD streams over time or by subscriber zip code—develop. These systems could change the advertisements based on date or location of the viewer and provided an economic motivation for MVPDs to push VOD adoption.

In just the seven years between editions of this book, the VOD world has changed substantially—at least for subscribers of some cable systems. As concerns about viewers cutting cable subscriptions to access broadband-delivered programming through services such as Netflix magnified beginning in the fall of 2010, large MVPDs began creating libraries and "any device, anywhere" availability more comparable to the convenience and choice being offered by broadband-delivered programmers. By 2013, the cable giant Comcast, for example, offered subscribers unlimited free access to 30,000 titles, including episodes of 600 television series, through their Xfinity VOD television service, 270,000 titles on Xfinity.com, and 20,000 television shows and movies through the Xfinity app for iPad, iPhone, or iPod touch. This was part of the TV Everywhere initiative launched by Comcast and Time Warner in 2009 as a way to allow authenticated subscribers access to their "living room" content on a range of devices and eventually outside the home.

MVPDs aggressively overhauled their value proposition to audiences in the face of widespread cultural adulation of the alternative Netflix offered, but as of 2014, it remains difficult to claim VOD as a victory for the cable and telecommunication industry. These industries underutilized VOD capability for a long time and developed them only when a threat emerged. The comparison of the development of VOD in the U.S. market with the iPlayer, the British Broadcasting Corporation's on-demand application, reveals considerable insight into the implications of the public service mandate versus commercial mandate on innovation, particularly those operating with a functional monopoly, as was the case of the U.S. MVPDs. The BBC launched its iPlayer in 2007, which featured an interface more akin to Netflix's graphic interface than the text-heavy and awkward-to-navigate interfaces still offered by MVPDs in 2014. BBC's self-control of program rights and mandate to make programming accessible yielded far more immediate experimentation than evident in the United States, where publicly held companies are punished by the stock market for reinvesting in technological development, and filling libraries requires extended rights negotiations with a multiplicity of parties. The first MVPD to make an interface similar to the iPlayer available began rolling it out in mid-2012, but had not reached its full subscriber base a year later; availability seemed limited to those markets in which competition from a telco existed. Unsurprisingly, despite growing availability, research in 2012 revealed limited use of VOD, and the industry source *Variety* categorized the report as evidence that the cable industry had "fumbled badly with VOD," missing "a $6 billion business" and "paving the way for the emergence of over-the-top alternatives like Netflix."[38]

Despite its slow start, because VOD is an endeavor of the MVPDs—companies that connect 80 percent of homes to the Internet as well as cable—VOD has a structural advantage likely to secure its centrality to the post-network era.[39] Robust and consistent libraries will leave viewers with little need to record programs on their own or to seek additional middle-men to aggregate content, so long as VOD offerings don't include the bloated commercial pods characteristic of linear viewing. However, creating and maintaining robust and consistent libraries remains a significant challenge for MVPDs that own minimal content rights.

Convenience technologies—including the DVR, VOD, DVD, broadband-delivered program services such as Netflix (also known as SVOD, subscription video on demand), and mobile applications that can be used on devices such as phones and tablets—enabled viewers to more easily seek out specific content and view it on living room screens and in an ever-expanding variety of venues. These technologies increased viewers' ability to select not only when to watch, but also where, and provided the most expansive and varied adjustments in the technological capabilities of the medium. Convenience technologies encourage active selection, rather than passive viewing of the linear flow of whatever "comes on next" or "is on," and consequently lead viewers to focus much more on programs than on networks—all of which contributes to eroding conventional production practices in significant ways and to producing the distinctions among prized content, live sports and contests, and linear viewing highlighted in the introduction. The viewing behaviors these technologies enabled, in tandem with the vast choice among outlets that viewers could now access, were vital to the shift of television from what Bernard Miège theorized as a "flow" industry to something more like a "publishing" industry.[40] Convenience technologies also increased the deliberateness in viewers' use of television, which allowed for adjustments in how programs were created, funded, paid for, and distributed.

Matters of Space: The Convenience of Portable Television Devices

DVRs and VOD allowed viewers to capture television from the dictates of the networks' linear schedules, but on their own, these technologies still confined viewers to conventional "living room" viewing. Freeing viewers to watch content anywhere they desired required another set of technologies that allowed portability. Viewers first experimented with this possibility by watching television series sold on disks on portable DVD players, but rapid technological diffusion quickly made portable viewing much easier. By 2005, the more elegant solution of downloading programs to iPod players and devices also used for gaming, such as the PSP (PlayStation Portable), freed portable television from requiring a physical medium. TiVo-brand DVRs also expanded the convenience of the device through the TiVo ToGo application, which offered easy transfer of programs it recorded to laptops and portable media devices. All this would soon seem most insignificant, though, as broadband-delivered program providers unshackled television from its domestic confines and enabled viewing on laptops, smartphones, and tablets.

As of 2014, the dominant experience of television that developed through its network era and multi-channel transition made it seem like the conveniences of portability and mobility were distinctive technological affordances that required specific assessment, if for no other reason than their different relationship to existing economic models. Per my definition, mobile television is linear, while portable television is nonlinear. Mobile television consequently remains useful for conventional advertiser support, while portable television is chosen in a manner consistent with prized content and may be better monetized through transaction payment. The decades during which living room viewing dominated the experience of television remain paradigmatic in the minds of many in a way that rhetorically counterpoises mobile and portable viewing as some sort of threat, but this is simply our imagination of the future being tainted by knowledge of the past. The previous impossibility of mobile and portable will be forgotten as quickly as the place-based past of telephone calling has been, which will allow fluidity of viewing spaces to seem a "natural" use of television technology instead of the battle for supremacy that characterizes contemporary outlooks.

An important early volley in portable television came with Apple's October 2005 announcement of the sale of individual episodes for $1.99 on iTunes. Apple's announcement was important because it attached a particular economic value to an episode of television and the beginning of a repository of television shows available for purchase, which has since been expanded by Amazon and Google. Also important to sketching this history was the release of the *Saturday Night Live* short "Lazy Sunday" on YouTube in December 2005. This video, initially posted without authorization by the rights holder, NBC, attracted 1.2 million views in just ten days, offering a most preliminary suggestion of a slightly different application of broadband-distributed, nonlinear television. Though YouTube had been designed with the purpose of facilitating amateur sharing, the flurry of posting broadcast- and cable-originated content to YouTube indicated how the site might also serve as a repository that could help viewers manage the abundance of post-network video.

Though there were important developments in the next few years (see table 2.2), the real start to the revolution in portable television began in 2010. In January, Apple announced the iPad, which it released three months later. Though the technology alone is of limited use without applications, the tablet developed into a technology more preferred for portable viewing than the existing laptop and mobile phone screens. Applications and broadband-delivered program services began developing at this time as well. HBO GO launched in February 2010 and was widely hailed by users and the industry as a model for broadband distribution. Users appreciated its easy navigation and depth of content, while the industry appreciated the minimal disruption to existing models through the authentication system that allowed only linear HBO subscribers access. Then, in April, the sports giant ESPN rebranded its ESPN 360 service as ESPN3 and by October established an agreement with Time Warner Cable that allowed subscribers full access to ESPN on computers, then mobile phones in April 2011. Finally, in November 2010, Comcast offered its Xfinity TV app for the iPad, which was followed by Time Warner's TV Anywhere application in early 2011, and the efforts of other MVPDs to allow subscribers to access the content available on their living room set through other devices and in other locations through authentication soon followed.

The fall of 2010 marks the beginning of what I categorize as the "Netflix Surge": a period during which Netflix presented a much more disruptive—though short-lived—model for broadband-distributed, nonlinear television. In the fall of 2010, Netflix began offering a streaming-only service and, by many measures, provided top-tier content: it featured the content of the subscription network Starz and other licenses achieved at low rates before license holders realized the potential of the service. Netflix's quick subscriber gains—reported in frenetic blog-era trade press accounts—inspired and

Table 2.2. Key Developments in the Transition to Nonlinear Television

2005 June	YouTube launches.
2005 October	iTunes announces $1.99 downloads.
2008 January	Netflix offers unlimited streaming plan.
2009	Hulu becomes culturally relevant. It doubles its content library, adds Disney as a partner, and by October 2009, has over 855 million video views.
2010 February	HBO GO launches.
Late 2010–early 2011	Major strides in TV Everywhere: Comcast, Time Warner Cable, and Cablevision release authenticated apps for iPad.
2011–2012	Smartphone use expands from 30 to 56 percent of the mobile phone market.
	MVPD-provided VOD becomes increasingly robust. Begins offering most recent five episodes of original cable and many broadcast series.
2013 February	Netflix releases full season of House of Cards.
2013 May	Netflix releases full season of Arrested Development.
2013 October	Comcast announces package allowing viewers access to HBO with Internet and limited basic TV subscription.

fueled unreasonable anxiety about cord cutting, and soon the acronym OTT (over-the-top) began appearing to acknowledge concerns that cable subscribers would use Netflix and other broadband streaming and downloading distribution services as an alternative to cable subscriptions. Certainly, emerging Internet use figures, such as that nearly 25 percent of all evening Internet traffic was Netflix use, were noteworthy,[41] but given that Netflix traffic and all that streaming and downloading were, for the most part, traveling through "tubes" provided by the same MVPDs who also provided cable service, the level of anxiety seemed excessive.[42]

It should have been clear to anyone with a background in television economics and distribution that between the cost of renegotiating their content with the expanded subscriber base and the unlikely ability of circa 2010 Internet infrastructure to accommodate growth of Netflix beyond a niche, the Netflix Surge would flame out quickly. Netflix investors went on a wild ride in 2011, with stock prices rising to a high of $298.73 on July 13, 2011, and the service achieved a subscriber base of 24.59 million in the United States at the end of June 2011.[43] By the end of September 2011, the service made missteps, announcing a splitting of the by-mail and streaming service in what was effectively a doubling of cost to consumers and lost 800,000 U.S. subscribers, which was most significant for its deviation from high quarterly subscriber gains, and sent the stock price falling to $113.27.[44] But this doesn't diminish Netflix's contribution to inaugurating a post-network era, and by many measures the service had recovered by 2014. The Netflix Surge was crucial for offering more than a hypothetical thought experiment of how post-network television might operate. It had significant ramifications in pushing MVPDs to innovate to avoid disaggregation of content and presented viewers with a usable interface and the nonlinear tools of recommendation engines and queuing, which helped fuel viewer desire for better nonlinear alternatives.

Netflix and MVPD services' TV Everywhere initiatives are addressed in greater detail in chapter 4's exploration of changes in television distribution. Key to the discussion of the convenience of portable television here is the role Netflix played in enabling audiences to consider laptops and tablets "television screens." Netflix, perhaps more than any other entity, disrupted the long acculturated sense that television content should be viewed on a television set. Similar to the rabid TiVo fandom that permeated the popular culture of those of a privileged habitus in the early 2000s, Netflix again captured pent-up demand for a different kind of television experience, and for a few months, suggested a new world of television. The realities of television economics and the fact that Netflix—at this point a quintessential middleman—owned neither content beyond a handful of shows nor the connection into the home made apparent that Netflix was unlikely to overtake those who produced content or could deliver to audiences, but it could force revolution on those who did.

Matters of Time: Breaking from the Linear Schedule

By increasing asynchronous viewing, convenience technologies expanded the audience fragmentation and social polarization that preliminary choice and control technologies had already enabled during the multi-channel transition. Whether DVR owners who reschedule viewing on their own terms, viewers who wait several months to purchase full-season DVDs, or those who stream shows via VOD or Netflix, users of convenience technologies have come to select their own viewing conditions, including the crucial one of time. The resulting temporal fragmentation may seem comparably insignificant relative to other adjustments—such as the fragmentation of viewers among a multiplicity of channels—but it has had important implications in disabling the coterminous circulation of television within the culture, which significantly changed the way television operated as a conduit of cultural discussion. Beginning in 2004, feature articles in the popular press recounted the trend of audience members waiting until a full season of a series was available on DVD and then watching the full season at a self-determined pace.[45] Rachel Rebibo, a DVD owner who preferred this viewing experience, explained, "With a DVD player, I can set my schedule and turn it off anytime. It's my choice."[46] Another DVD viewer, Gord Lacy, offered, "I loved *West Wing*. I watched eight episodes in one night. I had only ever seen the pilot, and I'm Canadian watching a show about a U.S. President."[47] Those who turned to DVDs for control began changing the television viewing experience, and many who left the linear world for prized content, aimed never to return. By 2011, the ease of streaming full seasons of programs though Netflix or by accessing increasingly robust VOD caches of programming offered further tools to those willing to wait in order to obtain greater control of their viewing.

But until 2012, viewers made the personal choice to defer viewing until they could amass a stock that would permit favored pacing—distributors still released content in weekly intervals of individual episodes. Once it ventured into content creation, Netflix defied the model of weekly episode release and made available all of the episodes of the first "season" of its original series *Lilyhammer* simultaneously, a strategy it reproduced with much greater notice when releasing *House of Cards* in February 2013. Netflix suggested a possible future in which viewers would not have to wait for a linear, weekly delivery of content. The premiere of *House of Cards* generated extensive debate about the economic and cultural merits and consequences of this release strategy, and despite the voluminous commentary, reflection on the utter arbitrariness of the existing norm—given new delivery capabilities— went uncommented upon.

Some who waited to view complete seasons suggested that the primary benefit resulted from the elimination of commercials and the tedium of week-long waits between plot developments.[48] Willingness to engage television in this manner also indicated a very active selection process and personal planning of viewing uncharacteristic of the network era. As this behavior emerged, full seasons of DVDs typically were not available until a year after the original airing; many studios timed the DVD release to provide a promotional function and to refresh viewers before the start of a new season. This substantially disrupted the immediacy once considered characteristic of television and the capacity for its stories to function as the source of "watercooler" discussion for all viewers the next day at work. Those who waited for a complete season to be available found it difficult to avoid hearing about key plot developments from those who viewed original airings, which also suggested that viewers who elected to wait for full-season releases had a very particular relationship with the medium and its content. As early as 2005, the unexpected death of a central character in the series *Six Feet Under* provided an illustrative case. The surprise and significance of that creative development, which many television critics discussed and which made this plot twist difficult to avoid, led to a discussion of the propriety of including "spoilers" in reviews because so many audience members deferred their viewing—particularly non-HBO subscribers awaiting DVD release—and learning this key piece of information prematurely changed the nature of their viewing experience.[49] By the 2013 release of *House of Cards*, journalists writing about television had adopted many of the conventions long familiar to book reviewers and wrote of series in a manner that addressed key themes and character development but either did not assume that audiences viewed at the same pace or warned viewers before proceeding, though social media discussions were not so easily controlled.

The fact that only 20 to 25 percent of U.S. television households subscribe—and thus have access—to premium cable encouraged deferred full-season viewing for the large audience unable to access weekly. The first opportunity nonsubscribers had to see these shows was not through conventional weekly syndication, but through full seasons released on DVD or streamed. The viability of highly serialized, nonsubscription-originated series such as *24* and *Alias* also was enabled by this behavior, as viewers who preferred not to wait a week between each suspense-filled installment chose to wait for full-season access instead. The opportunity to compress viewing allowed better memory of meaningful details that might be forgotten if viewing was stretched over months and suggested the new potential viewing pleasures that might develop from the possibility of condensed viewing. Indeed, the category of prized content somewhat emerges in response to this technological affordance. The various ways television viewing became more convenient increased opportunities for viewers to access serialized content, which in turn encouraged greater production of ongoing stories and storytelling with a memory, what Horace Newcomb identifies as a "cumulative narrative."[50] Programs in the multi-channel transition experimented with storytelling strategies that pushed the episodic organization common throughout the network era, but with the exception of prime-time soap operas such as *Dallas* and *Dynasty* and the multi-episode plot arcs introduced—controversially—by *Hill Street Blues* and *St. Elsewhere*, networks' timidity in creating programs that relied on viewers to make a commitment to regular viewing largely prevented the development of programs with significant seriality. The technological and distribution opportunities that ushered in a post-network era enabled storytelling forms that matched the emerging ways to engage with content.

Before the attack on linear viewing that Netflix perpetrated with its *House of Cards* release, the service issued a much less widely considered attack through its basic queuing procedure. Though initially devised for the by-mail movie service, Netflix acculturated its subscribers to construct a ranked list—or in its terms, a queue—of the content viewers desired to have sent to them. A viewer might view and review this list intermittently and become accustomed to divorcing the activity of selecting viewing from the moment of actually viewing it. Though Netflix's advance into streaming largely eliminated the time gap between selection of content and its availability to viewers, the notion of the queue persisted in this nonlinear distribution system. With Netflix, viewers never faced the question of choosing among what was on at a particular time. Managing the abundance of offerings—mostly with a full season if not seasons' worth of episodes—made the queue a useful tool.

Again, the practice of queuing—though initiated by Netflix—does not embed Netflix in post-network television. Rather, the queuing that Netflix introduced provided its subscribers with a different paradigm for thinking about and organizing viewing behavior, and one that substantially challenges the long dominant, linear, "what's on" proposition. As new technologies and means of distribution allow viewers greater convenience, their behaviors adapt in accord with new affordances. A viewer who learns of a show from a friend no longer has to make the effort to watch at a particular time or schedule a recording of the current episode, but now often has access to those episodes aired previously, allowing her to place that show in queue to view after completing the one she is in the midst of viewing. Significantly, as MVPDs introduced more sophisticated user interfaces beginning in 2013, many included the option of creating a queue, as did applications such as HBO GO.

The distinctions among prized content, live sports and contests, and linear content are important to acknowledge in this discussion of convenience. The asynchronousness of delayed but controlled viewing can be important for prized content, but so detrimental to the experience in the case of live sports and contests as to make it rare, and it is by definition contrary to linear viewing. The "spoilers" of sports outcomes cannot be avoided, and much of the inherent pleasure in sport viewing derives from the certainty that there will be a winner determined by the end of the event and the unpredictable process of identifying that winner. Even the manufactured contests of reality competitions encourage synchronous viewing because participation in online fan discussion occurs as televised events unfold.

Despite concerned hand wringing over the loss of television as a forum for producing synchronous cultural attention, the ability to find missed bits of television on broadband aggregators paradoxically allows television to take on new cultural functions, and perhaps provides an "antidote" to the cultural fracturing produced by the flood of available content. As early as 2004, an estimated three million people viewed a contentious verbal exchange between Jon Stewart and Tucker Carlson on Carlson's cable show *Crossfire*, only 25 percent of which saw it during its original airing on CNN. Once news of the exchange circulated among fans and bloggers, they disseminated the clip online and linked it to their commentaries 1,880 times, and an estimated 2.2 million people viewed it (this was before the phenomenon of YouTube, where it was viewed more than 200,000 additional times over two years after the exchange).[51]

This was an early case of what became a much larger and regular phenomenon in which audiences could be assured of catching up on anything notable in a clip they could find online or that would be sent to them directly through social networks. The advent of YouTube and Hulu eliminated much of television's remaining ephemerality, and this made the cultural impact of bits of programming difficult to gauge based on the size of the original linear audience. Late-night talk and comedy shows seemed to make the most effective use of asynchronous possibilities, as the unexpected gaffes in guest interviews or uncommonly humorous bits were widely shared and viewed in

Figure 2.1. Electronic program guide (EPG), 1988.

following days and often effectively archived indefinitely so long as copyright was not enforced. In effect, then, convenience technologies have not uniformly contributed to audience fragmentation; rather, they have produced varied results that are related to the new ways of controlling content.

Navigating Convenience

In addition to devices that provide greater convenience in selecting when to view content both in and out of the home, another set of technologies has emerged that helps viewers find content of interest. Devices that enable viewers to "pull" content from television have revised the network-era function of the medium to "push" content organized into channels; the organization mechanisms used to find content can also be categorized as convenience technologies. DVRs and VOD have certainly made it easier for viewers to find programming of interest, but one of the most substantial adjustments in the convenient use of television came not from a new technology but from a refinement of an existing application. Negotiating the multi-channel environment became cumbersome as the number of channels expanded. One

Figure 2.2. Interactive programming guide (IPG), 2013.

need only look at the difference in a *TV Guide* circa 1980 versus 2004 to see the enormity of the shift cable introduced in viewing options. In the 1980s, the publication informed audiences of the prime-time offerings of five channels, while by 2004, it compressed listings for some seventy-six different channels into tiny print on small pages. In October 2005, as the explosion of channels, their inconsistent availability nationwide, and the development of more convenient electronic and interactive onscreen program guides eliminated the utility of the publication, *TV Guide* effectively surrendered to change and discontinued program listings and adjusted its format to report about shows and celebrities. *TV Guide* evolved and launched its own channel in February 1999 that offered a program listing on a screen shared with original programming focused on celebrity and program "news." By the end of 2012, the original programming had shifted into full screen in all but 17 percent of the homes reached by the channel, with plans to eliminate the program guide aspect entirely.[52]

The first innovation from the weekly *TV Guide* magazine or local newspaper listing was the electronic program guide (EPG), a dedicated channel featuring a constant scroll of the television schedule. The slow scroll and limited information about only the next two hours of programming restricted the utility of EPGs, especially as channel offerings expanded (see figure 2.1). Digital technologies from satellite providers, from digital cable, and embedded in DVRs enabled the significant advance of the interactive programming guide (IPG), a refinement of the EPG that allows viewers to search among channels in the present and future. IPGs provide the convenience of one-click recording (in the case of DVRs), and often include more information—such as episode title, plot description, and actors in addition to the series' name (see figure 2.2). IPGs enable viewers to search as far as two weeks in advance and more easily negotiate the growing multiplicity of networks, although different devices feature search functionalities with disparate capabilities. Artificial intelligence within the TiVo device provided widely hailed early advancement for its ability to identify content like that which it had been programmed to record and to maintain standing recording requests for content with certain key words.

The abundance of programming in this era created a substantial need for a mechanism to help find and gather content of interest from among the channels and programs already available. Few MVPD-provided set-top boxes provided these more sophisticated search capabilities until 2013, slowing the pace at which viewers adopted television viewing behaviors that emphasized precisely searching out content. The ideal IPG seemed something akin to "Google for television," and unsurprisingly viewers' experiences watching and searching for video online created new expectations of greater customization of their conventional television viewing environments. Thus, much of viewers' demand for greater television functionality has derived from their use of and acculturation with computers. The shift from EPGs to IPGs illustrates a fundamentally different conception of how to use television that developed between the network era and the computer age. The availability of devices that enable viewer-determined, convenient uses of television unquestionably has been critical to establishing the post-network-era characteristic of nonlinear television viewing. Vast choice among channels and the ability to record them that was typical of the multi-channel transition meant little to viewers without devices to help manage these options.

MVPDs slowly began introducing next-generation interfaces beginning in 2012, largely as a competitive response to broadband-delivered program providers and the sophistication of their IPGs. Because MVPDs mostly operate as regional monopolies, there was little other external market pressure to motivate development and innovation. The pleasure expressed by those who used interfaces such as those of Netflix and HBO GO revealed suppressed consumer frustration, and product development became a competitive necessity as entrepreneurial companies such as Roku and Boxee offered more intuitive and sophisticated search functionalities. Additionally, the perceived threats from evolving Google TV and Apple TV products eventually drove MVPDs to develop more innovative products. The next-generation Television Everywhere interfaces, which were gradually introduced beginning in 2012, reproduced the visual graphics, personalization features, and elaborate recommendation engines of entrepreneurial disruptors, though were in few homes in 2014.

Further, IPGs and EPGs emerge from an experience of television that is becoming more and more residual, at least in terms of prized content. As described above, the phenomenon of queuing provided a real paradigm shift from the linear viewing experience and acculturated at least the subculture of Netflix television streamers to think about developing personal viewing plans. Indeed, there is nothing natural about or inherent to the paradigm of linear viewing that persists now, but its comparative inconvenience will lead to its erosion as alternative experiences become more available and coming generations of viewers never know this norm. Personalized queues in combination with recommendation algorithms are valuable tools for navigating an environment of post-network programming abundance.

Although convenience technologies may not have caused changes in specific programs, they have substantially contributed to creating a realm in which the viewer, rather than the network, controls the viewing experience. Moreover, even as industry entities such as networks and studios still determine what programs are produced, many of these convenience technologies have come to support new forms of distribution that allow studios to profit from shows such as serials and cult favorites that offered only limited revenues in the network era. Whereas the conditions of production and distribution during the network era necessitated a narrow range of cultural goods, the post-network conditions of nonlinear use and wide-ranging distribution platforms have enabled the more varied production and use of television.

Mobile Television: Immediacy Unshackled from the Home

Television's ability to transmit live images distinguished it from other media in the network era. Many of the media events that persist in our cultural memory—the days after the Kennedy assassination, the wedding of Prince Charles and Lady Diana, and later, Diana's death—are connected to viewing primarily done on our domestic-bound sets. Mobile television devices defy network-era norms by untethering live television from a specific physical space and enabling it to function much like a hybrid of many preexisting technologies. The capability of mobile television reasserted the significance of television's immediacy, as moments in which viewers have desired live images have long transcended the times they were at home. Images of people flocked around television sets in offices and restaurants were common on September 11, 2001, particularly for those living in the eastern half of the country who were already at or on the way to work when the disaster began. Such images are likely to become less common in the future as events requiring immediate viewing will be watched on individual workplace computers, while coffee shops and other public venues may fill with people watching pocket-sized screen devices and congregating nevertheless.

Mobile uses of television have existed since the 1960s, but the ability to watch "live" television on a mobile phone or a laptop computer significantly expanded previous versions of this largely unused television attribute. Mobile television is more comparable to the contemporary radio, adding a visual dimension to a medium that became ubiquitously accessible after television took its place as the primary means for domestic entertainment. By the mid-2000s, mobile phones offered the main opportunity for mobile television. The far greater disruption arrived as the processing ability of computers and phones and the capacity of the nation's wired, wireless (cellular), and wifi networks grew robust enough to support the downloading of video content at tolerable speeds. The mobile "phone" became a portal for Internet access of video programming, though rarely was live material the desired content.

Whether truly live programming or merely that downloaded or streamed, a key component of post-network television is its ubiquitous availability. Experience to date suggests that small mobile screens serve substantially different cultural functions than the domesticated or even just place-based version of television when they are used to access television outside the home.[53] Mobile television technologies lead us to reimagine the contexts, meanings, and uses of viewing and trouble public/private dichotomies that we've long believed inherent to the medium. Moreover, while the dominant use of television as a domestic technology continues to structure cultural ways of knowing television, emergent uses, particularly by younger viewers, have begun challenging this framework.

Certainly the ability to view television outside the home is by no means new, and some technologies perpetuate uses of television similar to those most dominant in the medium's earliest days.[54] Mobile television—as delivered to mobile phone and computer screens—freed television from its domestic confines in the same manner as the earliest portable sets from the 1960s, but the cultural meaning and motivation of this mobility differ significantly forty years later. Lynn Spigel notes that the portable televisions of the

1960s "opened up a whole new set of cultural fantasies about television and the pleasure to be derived from watching TV—fantasies based on the imaginary possibilities of leaving, rather than staying, home."[55] But these sets had limited use, due to their size and need to access broadcast signals. In enabling viewers to take an array of television with them anywhere they can receive a mobile phone signal or access a wireless Internet connection, the new technologies erase nearly all spatial limitations of television as a medium.[56]

Various industries—consumer electronics, mobile phone service providers, and the television industry—have all eagerly considered how mobile television might yield new revenue. In the early 2000s, various proposals for countless ways to utilize this expanded capability emerged: streaming live shows, producing original vignettes for this smallest screen, and creating a wide array of other programming such as interactive gambling shows. Innovation was less a question of what could be done technologically and more one of coordinating technological capability with existing needs and uses desired by viewers. Even though one might be able to watch a live episode of a cinematically detailed series such as *Game of Thrones* on a screen the size of a postage stamp, did anyone really want to? The industry pursued multiple possibilities in hopes of being involved in whatever might emerge as the "killer app" of this new media form. Lucy Hood, president of Fox Mobile Entertainment, explained the perspective of those pushing these services: "What are the three things that you always have with you? Your money, your keys, and your cellphone. If we can deliver a fun entertainment experience on this device, that will make it a very powerful medium."[57]

What Hood did not anticipate was that there was no need to create special content for this device. Once effective content aggregators such as YouTube and Hulu established themselves, viewers sought content with mobile devices that they might view just as well on a laptop or an Internet-connected living room screen. Though regular live mobile television has caught on in other countries, in only two situations is there much of a desire for live, mobile television viewing in the United States: sports events and unexpected events. I'll focus on the second case first, as these are rare and as such do not drive technological use. Unexpected events include weather disasters and situations in which viewers urgently desire the most recent news. Before mobile television, viewers may have stayed home to remain abreast of developments or followed them by radio or online if maintaining access to a television was not an option. Very few events continue to offer updates that are meaningfully consumed live; U.S. cases such as the 2012 Sandy Hook Elementary School shootings and 2013 Boston Marathon bombing revealed that users could more efficiently follow events through a newsfeed app that would push new content to them instead of the ongoing recapping and speculating characteristic of live television coverage. Where video radar images are helpful in gauging emergency weather conditions, video images are less important in ongoing news and events in which developments are slowly released. Breaking news coverage still relies on linear assumptions of an audience waiting and watching in real time. There is no technological reason why video updates will not be sent—pushed to viewers in the future as a video equivalent of blog updates—but that is not how the technology is being used at this point.

The situation of live sports, however, differs. Unlike unexpected events, live sports events are both regular and planned. Despite this regularity, it is often impossible to schedule one's life around events, and the nature of contests gives live viewing particular salience regardless of tools that now allow viewers the convenience of recording and reviewing these events at a later time. By 2014, perhaps the best illustration of live mobile sports viewing in the United States could be found in CBS's streaming of all the games in the NCAA men's basketball tournament and in NBC's Olympics coverage. Both events occur in a contained period of time and throughout the day. Such events have always been difficult to schedule and view because of the relative scarcity of time for broadcasting; broadcasters had to choose among a wide array of simultaneous games and contests to "air" just one, and network-era television's embeddedness in the domestic space made it difficult to follow these events when away from home. Both NBC's Olympic endeavors and CBS's coverage of the Final Four tournament evolved throughout the early 2000s, as each sought to use digital distribution tools to expand the audience by making available previously unavailable content, while simultaneously maintaining the large "event" audience increasingly particular to such sports contests (see table 2.3).[58]

By the 2012 London Olympics and the 2013 NCAA men's basketball tournament, both broadcasters came close to delivering the pinnacle of these events. Viewers could watch all sixty-seven games of the NCAA basketball tournament live on mobile devices, all of which also aired on broadcast and cable networks. Likewise in 2012, NBC streamed all Olympic events for a total of 3,500 hours of content, including all 32 athletic competitions and 302 medal ceremonies. Though streaming quality depended somewhat on the mobile and wifi networks used, the experience required limited buffering and in many cases offered HD-quality images, if supported by the viewing device. Somewhat paradoxically, the mobile devices were often still optimal for prime-time, domestic viewing of games even when viewers were likely to be free of work and other out-of-home duties. These applications provided

Table 2.3. Key Developments in Digitally Distributed U.S. Sports on Television

2002 March	MLB.com begins streaming live video of Major League Baseball games to subscribers.
2003 March	CBS offers online streaming of NCAA men's basketball tournament for a $19.95 subscription, draws 20,000–25,000 subscribers.
2006 February	NBC streams the final event of the Turin Winter Olympics—the hockey final.
2006 March	CBS offers free streaming of NCAA men's basketball tournament. Receives nearly 200,000 advanced sign-ups.
2006 Summer	ESPN360 service begins offering live streaming of sports games, as well as on-demand content. TWC, Comcast, and Cox reject the service because ESPN demands carriage fees.
2007 March	CBS continues ad-supported streaming with registration, drawing 1.4 million unique viewers.
2007 September	ESPN360 is relaunched after some major adjustments to the programming strategy to emphasize live streaming.
2008 August	NBC Beijing Olympic coverage streams 2,200 hours of live events for twenty-five different sports; also offers 3,000 hours of on-demand content; and 1,400 linear hours on seven NBC channels
2009 March	CBS offers NCAA men's basketball tournament streaming in high-definition.
2010 March	CBS's NCAA men's basketball tournament provides over 11.7 million hours of live streaming.
2010 April	ESPN360 is rebranded as ESPN3.
2010 October	Time Warner Cable and ESPN enter into an agreement to allow ESPN subscribers access to a live simulcast of the entire ESPN lineup.
2011 March	NCAA men's basketball tournament available on iPad.
2011 April	Launch of WatchESPN service that allows viewers to watch any channel within the ESPN portfolio through an iPhone.
2012 May	WatchESPN becomes available on Comcast Xfinity.
2012 August	NBC streams all of the London Olympic games on the website NBCOlympics.com and also through a mobile app, NBC Olympics Live Extra.
2013 March	CBS's NCAA men's basketball tournament coverage, now called March Madness Live, has over 45 million live streams, an increase of 18.2 million streams from 2012.

access to those events and games that the broadcasters didn't deem the "most" important—or most likely to gather the largest audience—and were consequently not available by broadcast or cable channel.

Notably, distinctions emerge in the accessibility desired even within the realm of live sports and contests. Events such as the Olympics and Final Four tournament are relatively unusual, whereas many other sports competitions are built around schedules more likely to match up with the leisure time of the culture so that games are scheduled in the evening and on weekends. Baseball, the professional sport with most games played during "work hours," was one of the first to make use of a digital distribution system for

regular games as well. As early as 2002, MLB Advanced Video began streaming real-time video with a goal of reaching fans at work.[59]

Though sports have provided the only form of live mobile viewing that
audiences have significantly embraced, television has still been very much
untethered from the living room screen. It isn't television's "liveness" that
has driven portable and mobile uses, but its capacity as an audiovisual storyteller. Content described as "snack TV," short-form snippets of programming, have proven most desirable in this use—which at this point remains
constrained by the robustness of mobile and wifi networks and the pricing
of data plans. A key component of out-of-home television use was innovation of a portable device that didn't require a physical media form, evident
first in the video iPod, which was then integrated into subsequent generations of iPhones, smartphones, and then tablets. These devices made accessing content seamless, and users' main challenges were related to network
traffic and download speeds.

The wide range of capabilities of smartphones contributed to their quick
ubiquity. Use of the devices exploded in the United States in 2011–2012,
moving from 30 to 56 percent of the mobile phone market in just eighteen
months.[60] A key aspect of the smartphone revolution was the way it combined so many technologies in one and provided Internet access to homes
that might still be lacking broadband.[61] The real test of mobile television
use had yet to occur, though the number of television channels launching
apps that would simulcast the linear stream of programming grew considerably in 2013. Though it remains "early days" for mobile television, it is worth
restating that as of June 2013, 82 percent of tablet TV/video watching and
64 percent of that done on a smartphone was done *in* the home.[62] These figures suggest that these devices are more valuable to viewers for the control
of what and when to watch than to view outside the home. Coleen Fahey
Rush, executive vice president and chief research officer of Viacom networks,
noted in 2013 that "You can see mobile device usage growing, but it's not eat-
ing into TV-watching."[63]

As endeavors toward mobile television developed in 2005, the industry
imagined a "three-screen" world and sought to aid viewers in easily moving
content among a thirty-two-plus-inch "living room" screen, a seventeen- to
twenty-inch "computer" screen, and a two- to four-inch screen carried with
one at all times—the already omnipresent, though not yet smart, phone serving as a ready candidate. Some also forecast a fourth screen that would be
portable but not as ubiquitous—more in line with my distinction between
portable and mobile use, such as the portable DVD player or portable gaming devices. Though yet another screen—the tablet—has emerged in the

interim, the three-screen forecast endures; if anything, a two-screen future seems more likely than four. The consumer electronics industry will likely spend the next decade negotiating precisely what size screen viewers most want to carry with them and the most crucial functions of that device. Just as smartphones have offered considerably consolidated functionality, a similar evolution will come to household screens as the functions of televisions and computers merge.

Theatrical Television: Enhancing Television's Aesthetics

Current concepts of home, work, technology, and leisure differ significantly from those of the 1960s, when mobile television was first introduced. Though Spigel observes that the marketing of portable sets then related to the New Frontier rhetoric of the time, the adoption of cable and early home theater systems in the 1980s encouraged a return to the home that many trend and marketing specialists identified as "cocooning."[64] At the same time that television screens have become infinitesimally small and portable, domestic screens have expanded and offered unprecedented and compelling visual images. A curious dichotomy has developed between television desired for its out-of-home convenience, regardless of image and audio quality, and television that provides a rich sense experience and is homebound. This bifurcation is not entirely new; Spigel identified varied discourses of theatricality and mobility in the promotion of television sets as early as the 1960s.[65] Yet technologies in the twenty-first century make both applications far more compelling than in television's early years.

Calling technologies that enhance television's visual and audio fidelity "theatrical" may trouble some readers, and indeed the term is not completely satisfying to me either. Whereas it can call up the notion of the stage, what I refer to here is the cinematic experience in the theater. Because cinema predated television as a screen technology, its norms and capabilities have long served as the standard against which television has been measured—and consistently found inferior. In her book exploring the rise of the "home theater craze" and its effects on the film industry and film consumption, Barbara Klinger notes that the "Holy Grail" of home-based visual media has been achieving the "replication of theatrical cinema."[66] A number of the technological advances that she considers, such as high-fidelity audio systems, VCRs, and the DVD, were central to developing enhanced home theater environments throughout the multi-channel transition. But the single most important advance in the enhancement of television quality—high definition (HD)—did not begin to enter significant numbers of homes until later.

In the late 1990s and early 2000s, industry workers and the trade press did not treat enhancement technologies such as high definition as likely to introduce extensive change to television. High-definition sets increase the number of scan lines on the screen from 480, the previous standard, to either 720 in cheaper, substandard high-definition sets or 1080 in what became the U.S. industry standard (although both are "technically" considered high-definition). The consumer electronics industry continues to manufacture standard-definition sets with 480 lines, though mostly just very small sets, and by September 2012, 75 percent of U.S. homes had HD-capable sets.[67]

The Telecommunications Act of 1996 mandated that broadcasters begin switching to digital signal transmission, but providing HD service was not similarly required. It did, however, become technologically feasible as a result of the greater efficiency of digital signals. HD consequently developed in a highly haphazard way, mainly as a competitive strategy and point of differentiation. Some programs were produced only with 720 lines of resolution, while others produced the full 1080. The consumer electronics industry also made both 720 and 1080 sets available. To add further confusion, there is also the variation of whether the signal is "interlaced" (720i, 1080i) or "progressive" (720p, 1080p), with progressive scan providing the better image.

Perhaps the easiest way to understand the issue of HD resolution is to know that there is a clear and, I'd argue, stunning difference between 480 and 720. The difference in picture quality within HD—say the difference between 720p and 1080i—is much more in the eye of the beholder, and most likely to be noticed primarily on larger screen sizes and at closer distances.[68] More apparent is that almost all HD sets feature the rectangular 16:9 ratio common to film screens instead of the more square 4:3 ratio previously standard to television. These adjustments, which have produced markedly more vivid and lifelike images, have required shifts in production techniques and technologies. High definition has evened the playing field between television and film, which has long been considered superior for its finer image resolution, and by 2014, most filmmakers had shifted to digital video production.

During the transition to digital broadcast transmissions from 1996 to 2009, many prospective television buyers confused digital and HD because of the simultaneity of their introduction and the coterminous new availability of flat-screen technology and the 16:9 aspect ratio, neither of which correlated with digital or HD. Digitally transmitted images have better quality than analog, but do not provide nearly the enhancement of HD. While HD and digital are linked in the sense that the size of the HD signal requires digital transmission, the confusion for many viewers arises from the fact that not all digital sets were necessarily HD and that set capability did not guarantee that a viewer was receiving an HD signal.[69] By August 2006, industry research reported that only 36 percent of HD set owners had made the transition to the HD service that would allow them to receive HD signals on their sets.[70] And, as noted previously, as late as October 2012 Nielsen found that only 29 percent of prime-time viewing was viewed in HD, due to households continuing to subscribe to non-HD service from providers or the unavailability of chosen content in HD.[71]

To some degree, HD operated as an afterthought and an also-ran for many in the industry after Congress legislated the digital transition in the mid-1990s. The digital transition proceeded because of regulatory fiat, and the industry consequently did not experience a process of negotiation in this transition in the same manner as other technologies that required adjustments throughout multiple production components prior to their implementation. Legislating the transition timetable from analog to digital occupied many in Washington, and the shift certainly proved costly and cumbersome for broadcast station owners. Yet by 2006, journalists had spilled far more ink on how the DVR and new forms of distribution such as iTunes had revolutionized television than on HD. As a result, viewers have continued to exhibit only limited understanding of and interest in HD, which has slowed adoption and prevented HD from substantially affecting other aspects of production. The Nielsen finding of such limited HD viewing some six years later and after such widespread HD set purchase indicates that visual quality remains a most limited driver of viewer behavior.[72]

Because it seems that HD can operate within many conventional production norms, many in the industry have worried far more about other technological developments and their consequences for television's industrial norms. While HD increases production costs with little opportunity for correspondingly higher advertising rates, it does not do so to an extent that has caused existing economic models to collapse. Also, the dominant role regulators played in mandating the transition to digital independently of the market fundamentally divorced the attendant technological development of HD from other technological changes of the time. The possibility of a better image standard originally contributed to the necessary regulatory push to embark on the digital transition, but once broadcasters were forced into the digital realm, they became far more eager to explore the expanded revenue possibilities of "multiplexing" their signals—the term for using the spectrum to broadcast three or four standard-definition "channels"—or leasing their unneeded spectrum to others than in implementing HD.[73] But to those who own HD sets in the United States and receive HD service, there has been a radical adjustment in the visual experience of television. Extensive HD offerings now exist, with most MVPDs offering HD packages with 110 to 200 HD channels.

The U.S. digital transition was only a few months complete when the first 3D televisions were displayed at the annual Consumer Electronics Show in January 2009. Press coverage of 3D largely repeated the rhetoric and claims that were made of HD nearly fifteen years earlier. But the future for 3D became uncertain by 2014 after one of the few 3D providers, ESPN, announced it would cease operating its ESPN3D channel in June 2013, and the focus of the consumer electronics industry shifted to 4K/UltraHD. This new set and signal offer four times the resolution of 1080p HD, though not 3D. Both technologies had been adopted by a most negligible proportion of the population at the time of writing, and though adoptions will certainly grow, the lack of a regulatory mandate results in a substantial difference between the digital/HD and 3D or 4K/UltraHD transitions. Although the discourse surrounding these technologies may be similar, the paths of adoption may differ significantly.

Theatrical technologies affect how we think about the conditions in which viewers watch television and how and why they view. Technologies that enhance the visual quality of the television experience are distinctive from the ubiquity enabled by devices that make viewing more mobile or convenient. Theories reserved for cinematic viewing have become increasingly relevant to examining television as some audiences have constructed home theater environments that more precisely reproduce the cinematic experience, and changing production norms have encouraged refinement of craft over the expediency demanded by the pace of producing twenty-two episodes a year. Because of the comparatively inferior quality of network-era television images, few observers have analyzed the more formal characteristics of the television image—just as the substandard audio capabilities of early sets resulted in few assessments of the role of sound in television storytelling— but HD has opened up new aesthetic discussions concerning television.

The distinctions among prized content, live sports and contests, and linear viewing are meaningful in considerations of image quality as well. Sports viewing has been a significant driver of HD purchases (a 2013 survey by the Consumer Electronics Association found that one in five HD owners purchased an HD set for watching the Super Bowl), and part of what may lead viewers to distinguish certain content as prized may derive from its attention to and achievement in visual quality.[74] A better-quality viewing experience may not even register when viewers are engaging in linear television viewing with its less purposeful pursuit of particular content.

Production components have responded to the enhanced technological capability of the home theater environment to support the new possibilities in image and sound quality and used them to differentiate content in this increasingly crowded space.[75] For example, even before many of the convenience technologies disrupted previous norms, subscription services such as HBO and Showtime cultivated a production culture that prioritized aesthetic excellence and originality in a manner that distinguished their shows from those of conventional television—arguably a necessary distinction for content for which viewers must pay. Emphasizing modes of production that seek to maximize the artistic potential of television, these channels created content with budgets and production values once common only to films produced by major studios. These productions are best appreciated with HD and high-fidelity sound, and in the wake of the success of these subscription services, basic cable outlets such as FX and AMC have similarly established themselves as the purveyors of a distinctive type of television content. Thus, it is not just new technological capabilities that have led to programming of increasingly sophisticated visual and aural quality, but a nexus of industrial, cultural, and technological forces.

Another consequence of theatricality, when merged with convenience technologies, is the production of a new variation of event television. In the network era, event television meant the televising of "media events," or what Daniel Dayan and Elihu Katz distinguish as "mostly occasions of state—that are televised as they take place and transfix a nation or the world."[76] By this definition, media events derived their status from their vast reach and the attention they commanded—common attributes of television in the past. After the network era, such regular, planned media events became increasingly rare. While the Super Bowl does continue to be an annual event of this order, other sporting events such as the Olympics fragmented among multiple channels and devices in a manner that decreased the status of any single competition, although the opening and closing ceremonies may provide exceptions. In the meantime, events such as the annual Academy Awards telecast dwindled in significance, though the high-quality production values and wide range of performers of the Grammys strengthened the cultural relevance of its telecast. Indeed, live sports exhibit much of this distinction, but only because so little else can gather audiences on this scale, and even the live sports audience is not that significant in comparison with network-era norms.

Instead, media events have become more personalized, as viewers have come to confer "event" status on programs that they themselves make special—much of which is characteristic of prized content—often with the aid of technology. Using control devices, they can separate prized content into a distinctive space—possibly both temporal and physical—in which they can watch undisturbed, perhaps on the best set available. A viewer might also distinguish prized television or live sports by gathering an audience of friends or family to see a program that has been recorded or is shown live, or use the phone, Internet, or social media to chat about a show while it airs.

For the most part, though, theatrical technologies have evolved from established devices and have not dramatically affected how the majority uses television, particularly in comparison with convenience and mobile technologies. Enabling viewers to further enhance their experience with home-based theatrical technologies has perpetuated the trend toward cocooning that was already well established before the arrival of HD television in the U.S. market. Indeed, the most significant implications of theatrical technologies so far may be shifts in commerce, as those able to afford high-end home entertainment decreasingly spend money in traditional public venues, although the advent of the video rental market in the 1980s had already given rise to this shift in behavior and commerce early in the multi-channel transition. The opportunity to buy seasons of television on DVD (known as DVD sell-through), streaming services such as Netflix, and VOD capability continued this trend, although shifts from public to home viewing were much more a shift in film viewing than television. Households able to pay upward of two hundred dollars a month in service fees for new theatrical technologies and fast Internet are a limited population that then narrows access to these technologies. So far, it seems unlikely that the theatrical, convenient, and mobile television experiences, separately or together, will become as universal as the conventional mode of viewing of the network era.

Many have presumed that the theatricality of HD and the small screens of portable and mobile devices create contradictory aesthetic experiences. Although there are important distinctions, in truth, the small screens are also compelling—viewers just need to hold them much closer to their faces, which I don't mean flippantly; use of these technologies is a much more personal experience. Those who wanted live television on the go did initially have to sacrifice visual quality, but this resulted from download speeds more than screen size. By 2013, viewers streaming the NCAA men's basketball tournament on HD-capable smartphones enjoyed a visual experience every bit as good as could be achieved by a living room set. In any event, as the post-network era takes shape, convenience, mobility, and theatricality will not develop as mutually exclusive qualities. Rather, the growing availability of each has not only come to redefine the experience of television, but has also forced more varied and differentiated understandings of the fundamental characteristics of the medium.

Conclusion

Throughout the network era, viewers primarily watched television in the home and were acculturated to passively accept the limited programming choices and schedule mandates offered by a few networks. Where "watching television" meant selecting among the limited range of programs currently streaming through the set, a certain vernacular accompanied this mode of use; when people queried, "What is on television?" they expected an answer of a finite set of selections already in progress. New technologies slightly disrupted this "conventional" mode of viewing throughout the multi-channel transition, and dominant viewer behavior adjusted accordingly. Channel surfing, for example, became a common behavior as the array of channels became broader and remote control devices made shifting from one channel to another much easier.

In the network era, the conventional use of television was so uniform and unexceptional that it was not widely contemplated. Because such conditions as choosing among predetermined options came first, they appeared natural, and many theories of television unreflectively assumed them to be inherently characteristic of the medium. Developments during the multi-channel transition diminished the value of such theories. Adjustments in the technological attributes of television freed the experience of viewing from its confinement to a linear schedule and irrepressible flow of externally scheduled programming. New and varied ways of using the medium resulted. These dynamic changes required similarly dynamic changes in identifying and framing the issues at hand, as well as theorizing about them. As the multifaceted technologies and uses of television continue to burgeon, and television itself acquires disparate and unfamiliar attributes, we need to think of the medium not as "television" but as televisions.

Returning to practical matters, we might note that uses and ways of viewing identifiable by the early years of the twenty-first century may well persist, while others may be added as the medium converges with other digital broadband technologies. But here too these developments can be related to broader cultural concerns, even though industrial and technological formations may be in transition.[77] The selective adoption of new television technologies and ways to use them not only contributes to adjustments in other production components, but also affects the entire process of production, as well as the role of television in society.

New ways to use and view television provide bountiful opportunities for audience research. Existing studies of audience use provide little information about how viewers might use convenient, mobile, or theatrical television,

although some research in the uses and gratifications tradition suggests preliminary parameters. Is mobile or portable viewing dominantly a solitary
activity or is it shared, and what types of content or locations of viewing
encourage variant behaviors? Do people tend to guard their viewing when
they use personal screens in public spaces—wary of the cultural capital it
might expose—or do they openly flaunt that they can view in nondomestic
spaces and expose their viewing selection as a valued marker of their tastes
and preferences? Is there etiquette for both users and bystanders of portable television? And what content is most often watched outside the home?
Industry workers hoping to profit from new technologies seek answers to
these questions while scholars try to understand the cultural implications of
such changes in media use. Emergent technologies require research if we are
to understand the broader media field as well as future demarcations of the
boundaries of "television."

And yet the old, conventional set and its uses linger. We must remember how entrenched related viewing behaviors may be and not lose sight of
them. A network-era "default mode" of watching television will remain part
of experiencing the medium, yet even this conventional mode will not be
static. Just as the introduction of the remote control and the VCR altered the
conventional use of television during the multi-channel transition, so too is
the gradual penetration of DVRs having an effect. Even DVR users do not
behave consistently; they record far more prime-time programming and are
more likely to view morning shows, news, and sports live.[78] Likewise, data
revealed that the multiple sets that became more common in the home after
the network era mainly provided families with more convenience by allowing them to watch in varied rooms and that in 2006 roughly 80 percent of
homes have only one set turned on during prime time, suggesting far more
co-viewing than many assumed.[79] By 2014, assessing how many sets are on is
unlikely to be a good gauge of co-viewing given the proliferation of mobile
and portable devices for viewing, unless that count includes computers, tablets, and smartphone screens.

Various anecdotes inform my understanding of technologically facilitated changes in use explored throughout the chapter and illustrate changing behaviors associated with technologies and their consequences. I do
not suggest that these anecdotes represent larger behavior patterns or can
replace detailed and rigorous empirical study of these phenomena. At this
preliminary point in the distribution of many of these technologies—when
only the initial uses of an unrepresentative group have emerged—comprehensive analysis and understanding cannot yet be achieved. Though we have
a better sense of what technologies viewers are using in 2014 than we did in

2007, we still know very little about why or what audiences are actually doing when watching television while using a "second screen." Most simply use the devices to occupy themselves during commercials or moments that don't capture their interest, but we remain in early days. Existing studies haven't focused on particular types of content, and behavior is likely to vary significantly based on whether it is prized content, live sports or contests, or linear viewing. I treasure my prized content viewing and remove all manner of distraction to embrace it fully using the best audiovisual technologies I can access. Sports viewing, on the other hand, almost always requires a secondary device—to pass the downtime between plays in my case, or, for my husband, to trash talk with fantasy league members. And linear television—well, I haven't willingly watched like that since 1999; but when visiting family or in a hotel, there too I'm simultaneously on a device, but never to do anything related to the show.

At the same time that new technologies have enabled vastly augmented uses of television, the adoption of devices that enhance theatricality, mobility, and convenience have also made conventional behaviors strange, disorienting, and unpleasant. Once adopted, their use can become so encompassing and natural that it is challenging for even the most critical mind to step outside his or her own habitual practices and meaningfully evaluate the role the technologies have come to play in daily life. Some have adopted new technologies so quickly that it is difficult to "make them strange" or to realize that others may use the devices differently. In the midst of writing the first edition in 2006, I spent two weeks in a hotel room that was not equipped with a DVR, had very limited channel selection, and lacked an interactive program guide. I learned a great deal about how much my television behavior had changed since I had adopted the newer technologies, and during those two weeks I was surprised by what I adapted to, what most frustrated me, and how much less television I watched as a result of the comparative inconvenience of the experience. By 2013, I found that having fast Internet was a far more important hotel feature than even having a television. As one of the 38 percent of mobile phone owners without a smartphone, I miss access to e-mail and information more than entertainment, and when I'm ready to be entertained, I far more prefer to demand something from Xfinity or Netflix than settle for the linear experience offered by the hotel television.

It is difficult to consistently name developments and "a medium" in the midst of such substantial redefinition. Even as the author, I can identify tensions and contradictions in the way I write about television as a medium when one of the central arguments of this chapter and the book is that we can no longer conceive of the technology with such singularity. As the transition in use continues, new words and terms will emerge or be reallocated in order to make sense of television and its multiplicity that might ease the tensions still in evidence here. The distinctions of prized content, live sports and contests, and linear viewing begin this categorization. As I've explored the growing convenience, mobility, and theatricality of post-network television, I've attended to the relative relevance of these technological gains for these different viewing experiences.

Devices that allow viewers to enjoy a theater-quality experience in their homes or take their television on the go should be considered as a part of a portfolio of products that complement rather than compete with each other in a multifaceted technological televisual field. In the late 1990s, Nicholas Negroponte argued that the technological distinction of real significance was that between analog and digital, and the technological connections enabled by the digital transition have indeed proven to be profound.[80] The convergence among technologies uncertainly connected other than by their digital language raises ambiguity about whether something like YouTube is best categorized as "television," "video," "computer," or perhaps even just a "screen" technology. Certainly, the ability to deliver video unites television, computers, and mobile devices, but our residual acculturation may lead us to approach screens that feature familiar programs as television for some time, regardless of the technology we use to receive and view it. In the same way that "broadcasting," the "airing" of shows, and "tuning in" have remained part of the industrial and cultural vernacular—even though they precisely describe only a small part of television use—"television" continues to function as a meaningful term. As is the case of production components examined in subsequent chapters, the adjustments of the multi-channel transition and post-network era created multiple and competing uses rather than replacing one monolithic norm with another. We must now think about television as a highly diversified medium; even as "watching television" has continued to signify a set of widely recognizable behaviors, the singularity and coherence of this experience has come to be fleeting.

3

Making Television

Changes in the Practices of Creating Television

The business has changed so massively. . . . You will never have the market forces again that, how do I put this, that allow people to get rich. . . . The reality is you will never have the licensing fees negotiated again that resulted in *ER* getting [millions of dollars] an episode, and that's where a lot of people made what many would probably insist is an unconscionable amount of money. . . . The upside home runs for shows [profits from big hits] have been sort of flattened out by the new economic models of how shows are produced.
—Dick Wolf, producer, 2006[1]

The barrage of new technologies marketed to us in the early years of the twenty-first century has indicated much about the changing nature of television—so much so that even non-technophiles have realized that changes are at hand. Yet, though hours of television programming daily stream into our homes and now onto an array of new devices as well, the process of creating shows remains well hidden from most viewers. To be sure, by the late 1990s, the casual viewer could notice adjustments in types of shows and how networks organized them in their schedules. What most viewers may not have realized was why these changes occurred and how these shifts related to the broader structural adjustments that were revolutionizing the production of television.

This chapter's epigraph captures the perspective of Dick Wolf, arguably one of the most powerful (and richest) television producers of the last two decades due to the phenomenal success of the *Law & Order* brand he created. Here, Wolf replies to a question about whether he could be as successful if he were entering the business in 2006, and his response indicates the consequences of the changes in how television programs are made that this chapter explores. The "flattening out" of profits changes the type of programs the industry is likely to produce in significant ways and allows a much broader array of programming to exist and "succeed"—albeit by a variety of measures—than was the case in the network era.

One of the biggest changes in the making of television resulted from a regulation introduced in the early 1970s that was then eliminated in the 1990s. This set of rules—the financial interest and syndication (fin-syn) rules—altered the rules about who was allowed to make television programs, adjusted the relations of networks and studios, and affected who profited most substantially. The competitive environment that resulted from the elimination of these rules and many other deregulatory policies allowed expanded conglomeration and necessitated that those who create television devise new methods of funding. It also led to the erosion of the more monolithic norms of the network era, including the division of labor established between networks and studios, as well as the financial model according to which they operated. These adjustments and those of other production components not only affected the storytelling possibilities of the industry, but also led networks and studios to revise long-standard programming practices related to schedules, reruns, and program lengths and formats. These adjustments increased the scope of commercial storytelling and enabled the creation of what I distinguish as prized content. In treating these matters, this chapter also explores the increased need for innovative promotion techniques that networks have adopted in order to reach the splintering audience.

The Relationship between Financing Production and Production Norms

Though I examine the aspects of financing and economics involving audiences and advertisers in chapter 5, assessing the financing of television production—the making of television—requires understanding that the techniques used to fund productions routinely costing millions of dollars per hour have enormous implications for the shows that are made and many of the general dynamics of the television industry. Developing programs is one of the most difficult, uncertain, and therefore risky aspects of television production. In the early days of television, networks produced their own programming or received it from sponsors—as they had when they operated as radio networks—but television shows were far more costly than those of radio because of the added labor and complexity of visual recording. As the television era began, the networks sought to decrease the risks involved in creating programs by licensing them from film studios. This strategy was economically prudent: the film industry already had established facilities and structures for visual media. Consequently, a key practice of creating television involved dividing the process between two different entities: the studios that create the programs and the networks that organize and distribute them. Although these are distinct tasks, the networks still engage in "creative" activities such as selecting programs and often directly shape the creative direction of their shows.

Splitting the roles of studios and networks necessitated a means for financing television series appropriate to the varied risks and rewards inherent in the separation. A practice known as "deficit financing" consequently developed—an arrangement in which the network pays a license fee to the studio that makes a show in exchange for the right to air the show. The license fee typically allows the network to air an episode a few times (a first and rerun episode), but the studio retains ownership of the show. In effect, then, the license fee just allows the network to borrow the show, but this first licensor has considerable input on creative matters. Studios keep subsequent licensors' likely tastes and needs in mind, but these entities rarely imprint upon the television text to nearly the extent of the original licensor—at least in the system that has existed to date. Importantly, the license fee does not fully cover the costs of production, which creates the "deficit" of deficit financing. The studio absorbs the difference between the cost of production and the license fee, which can now amount to as much as millions of dollars for each season. If the network orders enough episodes, the studio can then resell the series in various other markets, though the studio typically has had to wait as many as five years before receiving the revenue of these sales.[2] As an illustration of the exceptional profits earned by a rare show, CBS's CEO, Les Moonves, noted in 2012 that *I Love Lucy*, a show last produced in 1957, still delivered $20 million in annual revenue.[3]

Deficit financing minimized the substantial risks and costs of developing programs for the networks while initially affording the studios considerable benefits as well. In the case of successful series, the studio receives a large return on its investment when it sells the show in a combination of later markets, called syndication windows, because the sales provide nearly pure profit: no additional work typically goes into the show and the network receives none of the payment.[4] However, if the show is unsuccessful and does not produce enough episodes to be syndicated, or if no buyers want the show, the production company must absorb the difference between the cost of production and the original license fee. For example, in the late 1990s, an hour-long broadcast-network drama typically cost approximately $1.2

million per episode to produce, with broadcast networks paying $800,000 to $1 million per episode in licensing fees.[5] Assuming the standard twenty-two-episode season, a production studio might lose anywhere from $4.4 to $8.8 million on a season of episodes. If the ratio of license fees to production costs remained constant—which is unlikely because producers usually renegotiate license fees after a few years and talent costs commonly escalate as well—the production company would assume $22 to $44 million of debt by the fifth season. At that point the series would reach the one hundred episodes then commonly necessary for "syndication," a process by which the studio resells the series in multiple markets, such as domestic broadcast affiliates, cable channels, and networks in any number of other countries to recoup its costs.

According to this scenario, a typical late-1990s drama would likely be sold both to a cable channel and international buyers, in addition, perhaps, to local stations for once-a-week airing, typically on weekends. For example, *CSI* was the last series CBS added to its schedule in 2000, and although the network expected little from it, *CSI* quickly became the season's breakout hit and regularly ranked among the most-watched shows each season.[6] The series was coproduced by Alliance Atlantis and CBS Productions. CBS Productions (through the commonly owned distribution company King World) sold the first domestic syndication run of the series to the cable network Spike for $1.6 million per episode and then sold the series to individual stations throughout the United States, while Alliance Atlantis sold the series in 177 different international markets for at least $1 million per episode in each major market.[7] The series also developed into a franchise—adding *CSI: Miami* and *CSI: NY*—and although the original *CSI* remained the most popular of the three in the United States, by 2006, *CSI: Miami* was the top U.S. show around the globe and had earned $6.4 million from DVD sales.[8] As late as 2013, the investment fund Content Partners purchased a 50 percent interest in the franchise from Goldman Sachs for $400 million.[9] This stake entitled Content Partners to a share in the future rights to the series, of which only the original—the Las Vegas–based *CSI*—remained in production. The price the group was willing to pay suggests the perceived value of the franchise, which included over 720 episodes when purchased.

The deficit financing system established early in U.S. television production continues to dominate norms, although important dynamics in the relationship between studios and networks have changed in important ways. In the 1960s, the networks had the upper hand in dealings because of their monopsony power as the only three buyers of content, and the networks attained greater control and less risk by forcing production companies to deficit finance their programs while also demanding a percentage of the syndication revenues. This "profit participation" by the networks caused many production companies to struggle financially, especially independent producers—those not aligned with a major studio—because they needed all of the revenue from successes to offset both the cost of failures and the substantial overhead expenses of production. Networks obtained profit participation in as much as 91 percent of programming by the mid-1960s, which led the government to intervene with regulations called the fin-syn rules at the beginning of the 1970s.[10]

The fin-syn rules prohibited networks from holding a stake in program ownership and having a financial stake in the syndicated programming they aired, as well as limiting the number of hours of programming per week that they could produce. Much of the power that the networks developed before the rules did not result from any formal collusion, but from their status as the only three potential buyers for series. The control of distribution by the three networks defined the relationship between studios and networks and significantly disadvantaged production companies that had little recourse against network strong-arming. The realignment of power between producers and distributors—the networks—in September 1971 by the FCC-mandated fin-syn rules and the consent decrees put forth by the U.S. district court might be considered the first disruption of dominant network-era production processes. Many regard the years subsequent to the rules' enactment as a golden era of independent production, marked by the heyday of MTM Enterprises (producers of *The Mary Tyler Moore Show*, *Lou Grant*, *Bob Newhart*) and Norman Lear's Tandem Productions (*All in the Family*, *Good Times*, *The Jeffersons*), among others.

But as the television marketplace began to change with the introduction of cable and the emergence of new broadcast networks such as FOX, the networks agitated for the elimination or reduction in the rules on the grounds that there were now more places to sell programming. The introduction of cable channels initially wasn't all that persuasive an argument, as these channels had particular foci (CNN: news; MTV: videos) and none had programming budgets comparable to the broadcast networks. Because FOX did not provide the full day of programming common to other broadcasters and only two-thirds of the prime-time programming, it received exemption from the fin-syn rules, which arguably marked the significant beginning of their demise. While a threat to eliminate the rules surfaced as early as 1983, it finally materialized in 1991, when the FCC began eroding the rules, which were completely removed by 1995 just as the WB and UPN entered the broadcast competition as non–full-service networks and superficially doubled broadcast competition from the network-era norm.[11]

The lifting of the rules immediately changed the dynamic between studios and networks. The mid- to late 1990s was a period of pronounced conglomeration in the media industries, and as soon as regulators eliminated the rules, each network began populating its schedule with new shows purchased nearly exclusively from studios owned by the network or from within the conglomerate owning the network—what I will refer to as "common ownership." This preponderance of common ownership, or "vertical integration" in the vernacular of economics, radically redefined relationships between studios and networks and adjusted financing norms. Networks prioritized content generated by commonly owned studios and again often demanded a share of syndication revenues in order for a show not produced by a commonly owned studio to receive a place on the schedule. These were typically called "coproduction" deals—a misleading term because the commonly owned studio that was added to "coproduce" often supplied minimal, if any, support, but still earned the rights to syndication profits. Such profit participation was a nuisance for major studios, while it substantially disadvantaged independent producers because they depended upon all possible syndication revenue.

The key reason the networks contracted with commonly owned studios was a desire to accrue syndication profits, but another reason was fear that the top shows would demand exorbitant license fee increases after their original three- or five-year agreements expired. High-profile cases emerged at NBC in the late 1990s and early 2000s with its license negotiations for the Warner Bros.–produced shows *Friends* and *ER*. As Wolf notes in this chapter's epigraph, the license fee for *ER* increased to $12 million per episode at one point, ten times the original fee; Warren Littlefield, then president of entertainment at NBC, succinctly notes, "We didn't own *ER*. We just rented it. After the first four years, we rented it for a lot more."[12] These shows could have been purchased by another network if Warner Bros.' license fee demands were not met, as happened in the case of *Buffy the Vampire Slayer*, which moved from the WB to UPN in 2001.[13] Although few cases of shows moving to a different network actually occurred, the threat—particularly for top audience-drawing shows—was significant enough that networks sought a share in as much programming as possible in order to maintain greater control.

Common ownership among studios and networks created a mutual interest in success, but the performance of the network and that of the studio were evaluated separately within the conglomerate, and each had to meet unit budget goals. Consequently, even intra-conglomerate deals remained competitive. For instance, Warner Bros. studio could not absorb a substantial loss on a show just to help the WB decrease license fee expenses. Some

advantage might be offered, as in the case of 20th Century Fox studio selling a syndication run of *The X-Files* to commonly owned FX for less than
market value. Such uncompetitive practices were occasionally revealed, as
occurred in this case, when the actor David Duchovny successfully settled
a suit against the studio because the cheaper sale of the program decreased
his residual earnings.[14] Common ownership of a studio and a network did
not provide either entity with carte blanche. Various aspects of the industrial and organizational structures of television production—whether the
independent evaluation of different divisions within the conglomerate or the
separate stakes in production profits often held by actors and producers—
curtailed the unrestrained provision of advantage that could develop.

The advantage of deals among commonly owned entities results from
the increased likelihood that they would negotiate with a sense of equity
although not necessarily discount. Paying $10 million per episode of *Friends*
was detrimental to NBC, and Warner Bros. may not have pushed for such
unprecedented payment had it been negotiating with a commonly owned
network.[15] Abuses certainly could and did occur, but many counterexam-
ples in which commonly owned networks lost out in bidding or commonly
owned networks beat competitive offers also emerged. Such varied evidence
made it difficult to sustain claims about the uniform behavior of conglomerates.[16] Still, the alignment of common ownership represented a considerable
shift in practice and created a competitive advantage for commonly owned
studios that made the financial model of scripted series creation untenable
for independent producers.[17]

The practice of deficit financing continued with few exceptions as a dominant practice for financing series despite the development of common ownership, though this adjustment in the balance of industrial power between
networks and studios produced unanticipated costs, quite literally. A panel of
studio and network executives at the 2007 National Association of Television
Program Executives (NATPE) conference agreed that common ownership
of studios and networks had contributed to rapidly escalating programming
costs—the typical cost for an hour-long broadcast show jumped from $1.2
million in the late 1990s to $3 million by the mid-2000s—as they noted it
was easier to approve incremental cost increases for productions when one
entity had a stake in production and distribution.[18] The common-ownership
model allows the conglomerate the opportunity for immediate revenue from
production expenditures—in the form of advertising revenue—as well as the
later revenue available from syndication, increasing the incentive for incremental spending that is perceived as the difference between success and
failure—though this can never be known with any certainty. And once one

studio makes a successful show with a $3 million budget, so must others to compete. Some on the panel noted that escalating production costs are also likely to have contributed to the troubled status of independent producers, who are less able to afford such incremental spending.[19] Common ownership and coproduction particularly disadvantaged independent producers—so much so that by the end of 2005 the production of prime-time scripted series was no longer a viable possibility for independent producers.[20]

The economic norm of deficit financing and then seeking syndication revenues remains relatively constant in television production, despite other adjustments in production norms introduced in the early post-network era. At the 2004 NATPE conference, Caryn Mandabach notably proclaimed that "deficit financing is dead." Mandabach was a partner of the once legendary (but by then defunct) independent production company Carsey-Werner-Mandabach, which created *The Cosby Show*, *Roseanne*, and *That '70s Show* among others, so her opinion on the matter earned consideration, and it was echoed by a number of executives speaking at the conference that year. Nonetheless, deficit financing has lived on and still remains dominant for funding scripted series in 2014. Rather than portending the death of deficit financing, what the executives were really speaking to was the inability of the system to persist in its existing form; demand for higher-budget content emerged at the same time networks were less able to provide the mass audiences desired by advertisers.

A variety of unconventional financing experiments has emerged during the preliminary years of the post-network era in an effort to reinvent production-side financial norms in relation to the distribution possibilities now characteristic of the industry. Though none of these experiments has become particularly common, their emergence provides evidence of the limitations of the status quo. As in the case of the textual differences of cable original series discussed subsequently, these different financing practices also introduce changes in production practices and the content the industry might produce.

Despite the significant shifts in the broader industrial relationships between studios and networks introduced by the creation and then elimination of the fin-syn rules, the particular allocation of risk and reward of deficit financing allowed a fairly consistent set of production norms for broadcast television. Though budgets expanded significantly over time, many of the other features remained constant, such as the norm of seven or eight days of shooting per episode, most of which is done on a set with fixed camera setups that are carefully lit to shoot action from one angle, then reset to shoot from the other and edited together in post-production. A full season included twenty-one or twenty-two episodes that would air from late

September through May, with rerun episodes interspersed. Many of the new costs were related to increased shooting on-location and the growing use of sophisticated visual effects, such as those featured in mid-2000s series such as *Lost*, *Heroes*, or *24*. Series such as these raised the stakes and expectations of television, though networks and studios quickly learned that audiences also required carefully crafted stories and characters, which proved more difficult to replicate than the visual effects.

Many of these norms of production and airing schedules established decades ago still undergird most hour-long, scripted broadcast television production; however, both scripted series produced originally for cable channels and unscripted television series introduced alternatives in the preliminary post-network period that coexisted with legacy practices and even suggested some evidence of eroding broadcast norms. There is not a simple cause-and-effect story for the flurry of simultaneous adjustments in the broader programming landscape, though the most significant force for change likely came from cable channels' moves into original scripted series production. As already established, throughout the 1980s and 1990s, cable channels gradually drew away many members of the broadcast audience, but in these decades, mainly did so by offering the same programming or type of programming offered by broadcasters. By the beginning of the 2000s, though, cable began to establish itself as a site for original programming and introduced a variety of new production norms.

A confluence of industrial differences encouraged cable channels to develop content that was a clear formal alternative to broadcast offerings. For the most part, basic cable did not keep up with the budgetary arms race of broadcasters and instead experimented with various techniques to contain costs, including handheld shooting, producing outside Los Angeles, and changing the series order and release schedule. A norm among cable series quickly developed of much shorter, thirteen-episode seasons, with episodes aired in consecutive weeks instead of the broadcast strategy of spreading content out by including several weeks of reruns. These shorter seasons were also important for luring writing and acting talent that had avoided television because of the intense demands of the twenty-two-episode season. A thirteen-episode commitment was also less demanding of viewers, and cable channels also often provided many reairings of a new episode each week, especially before VOD offerings became plentiful. The combination of fewer episodes and greater access to each episode defied the conditions that led broadcasters away from incorporating much seriality in their storytelling. Fewer episodes also allowed writers more time for plotting and story crafting, which in some cases yielded more "complex" television on cable.[21]

Further, cable channels' dual income streams from advertising and subscription fees allowed them to target more niche audiences and to find success by providing advertisers particular types of viewers rather than a mass audience. Many early cable originals consequently featured far less conventional characters, uncommon settings, and more serialized storytelling than typical of broadcast networks.

Adjustments in television style were also apparent, though they are not all related to the shifting industrial norms focused on here. During the early 2000s many series broke from network-era stylistic conventions. In the case of the half-hour comedy, this was evident in the erosion of the long common, three-camera, proscenium stage style that was often shot in front of a studio audience. After broad comedy hits recorded in this style, such as *Seinfeld*, *Friends*, and *Everybody Loves Raymond*, completed their runs, the networks struggled to create mass audience replacements. There was considerable discussion of the "death" of the sitcom as single-camera comedies shot "film"-style became more prevalent, though failed to achieve comparable success. These comedies, shows such as *Scrubs* and *Sex and the City*, used the opportunities of this style to incorporate storytelling techniques such as voiceovers, flashbacks, and fantasy sequences unavailable to comedies shot live.

Paradoxically, as single-camera comedies provided an alternative to the multi-camera norm in that genre, some dramas began to replace the single-camera, film-style norm of drama with multiple cameras. Traditionally, a dyadic scene in a drama was shot from one direction, then reset (a process of repositioning cameras and relighting that easily takes more than an hour) and shot from the other perspective. Though editing makes this process seamless, it can produce more staid acting performances because actors have to perform multiple takes and know when they are actually being filmed. In an interview with the journalist Alan Sepinwall, *Friday Night Lights'* producer Jason Katims explained the show's use of multi-camera shooting techniques:

> The single essential ingredient about the show was that we had three cameras shooting in this particular style. Through [the] majority of the filming, we would be shooting both sides of the scene at once. If it was a Tyra and Tami scene, one camera would be behind Tyra shooting Tami, the other side would be behind Tami shooting Tyra, but doing it in such a way that they weren't filming each other. It allowed the actors to have a very naturalistic way of performing. They were able to respond to each other in the moment. They didn't have to worry about matching what was done in the previous take, they didn't have to worry about overlapping with each other, because the artifice of filmmaking was taken out of the process.[22]

William H. Macy, who stars in *Shameless*, has also spoken of the way multiple handheld cameras can change an actor's performance. In an interview with Terri Gross, he reaffirms Katims's suggestion of greater naturalism because the actor has much less sense of what is being shot when scenes are filmed this way.[23] He explained that actors can be overly self-conscious about the close-up shots that punctuate a major scene, and these are specifically set up in single-camera shooting so that the actor knows when the big take occurs. In three-camera production, one camera might catch this version at the same time another has been shooting a wide shot, freeing an actor from forcing this key take in a way different from the rest of the scene.

In this case, the stylistic innovation of shooting with multiple cameras developed as a result of digital camera technology that allowed for handheld cameras that could capture images without the shaky style common of *The Shield* or *Southland*, shows that use the unstable image to reinforce a sense of spontaneous action. With stability controls, the cameras could move between the interactions in a scene, often without the actors knowing what was being shot. This technique can be employed as a way to shoot scripts more quickly; the time saved by not having to set multiple versions of the same scene can save enough time to reduce an episode's shooting time by a day, which was the motivation for using this style on *The Shield* (discussed more in chapter 7). The relevant point here is more of innovation and experimentation with established production techniques that developed for technological, stylistic, and economic reasons. In many cases, productions for cable channels tried these experiments as part of their efforts to craft distinctive content, and often with the limitation of lower budgets.

Unscripted Series

Notably, even before experimenting with scripted originals, many cable channels developed unscripted original programming, whether MTV's *Real World* (1992–), Lifetime's *Project Runway* (2004–), Bravo's *Real Housewives* (2006–), which launched Bravo's pop culture presence, or the exploits of *Jon and Kate Plus 8* (2007–2011), which established TLC. These shows, and the unscripted endeavors of their broadcast competition (*American Idol, Survivor, Dancing with the Stars*), introduced different production economics and an alternative to deficit financing.[24] Unlike scripted series, these shows typically have little syndicated value, so their costs must be recouped in their license fee or through the license fee and integrated commercial placement fees.

A wide range of unscripted programming has emerged since 2000, and that range makes it impossible to speak of uniform norms in the form.

Unscripted programming *generally* costs less to produce—for example, an episode of the docusoap *Ice Road Truckers* was estimated to cost "well under $500,000" an hour in 2007.[25] The same report notes that "conventional network reality shows" cost "$1.5 million to $2 million" per hour, indicating that unscripted programming costs could quickly escalate toward scripted costs, but without the subsequent revenue windows.[26]

Unscripted series introduced a variability in production costs that was important for networks balancing the increasing costs of scripted series. In some cases the unscripted series provided networks with programs with a live appeal similar to that of sports, which might better guarantee that advertisements would be viewed, and viewed soon after the live airing. Though there may be little in the production financing of unscripted series that might be applied to the scripted world, the emergence of this form in the early days of the post-network era has been important for expanding the array of television narratives in circulation.

Independent and Outside Financing

Part of the reason for the persistence of deficit financing has been that it allowed the studios to become veritable banks once the system was established. Where the manufacturer of another type of good would go to a bank for a loan to develop a new product, the unpredictability of cultural industries made bank lending or investment rare. There was little collateral to offset the bank's risk, and bankers could hardly assume the risk of predicting the likely success of cultural products. The studio, on the other hand, is expert in the creation of content and therefore better positioned to risk money earned from past successes in the pursuit of additional properties with long syndication lives and profit streams. The alternative to financing through a studio is so-called independent financing, which has a long history in the film industry, but this has only recently emerged as a way to finance television production. In the film industry, independent financing has been a key source for the creation of films deemed too risky or lacking in the properties of likely commercial success; directors who are not beholden to a studio for funding also have greater creative autonomy. The film industry also began using financing from hedge funds and other large investors in the 1980s and 1990s. Hedge funds invest in film "slates" that offer a mix of films. The outside funding supporting television to date has been project-specific.

The challenges of the evolving television industry led to experiments with independent financing of television. One of the most widely discussed cases was the financing of the third season of the TNT caper drama *Leverage* by Winchester Capital Management. This case was unusual for a variety of reasons: the series creator, Dean Devlin, had a preexisting relationship with TNT after producing a successful series of films and was able to self-finance the first two seasons; and the financing wasn't sought until the third season, by which point the series had established itself and its popularity. Most producers would not be able to finance the initial, most risky season needed to create a track record, so while this case is notable, it doesn't provide a widely applicable model of funding.

But this type of funding structure is important because it allows, in the words of Ted Sarandos, the chief content officer at Netflix, a "shift" in "the development burden to the producer," a risk borne more heavily in the past by the distributor (network).[27] Though placing such costs on a producer isn't likely desirable, it also creates a mechanism by which the passionate producer has a way to circumvent studios unwilling to take a risk on an idea—which has been one contribution of the independent film sector to that industry. Sarandos explains how Netflix utilized this strategy in developing *House of Cards*, noting that it gave the director, David Fincher, a two-season, twenty-six-hour commitment. The funding for the series, though, was secured through the independent studio Media Rights Capital, which is backed by Goldman Sachs, AT&T, WPP Group, and ABRY Partners. As in the case of *Leverage*, the expansion of Media Rights Capital—traditionally an independent film financier—into television suggests the transition of independent film norms into television.

The emergence of companies specializing in providing independent funding for television could have significant implications for the organization of the industry. Though taking on more financial risk may be undesirable for producers, the relationships typically crafted with this type of financier allow greater creative freedom and may provide increased latitude in the variety of deals made with distributors. For *House of Cards*, this meant that MRC paid the costs of developing the show and outlining the first two seasons before sharing it with—and receiving input from—any network. One account notes that Netflix's deal for *House of Cards* allows it exclusivity for four years, after which MRC can license the rights through conventional routes, though Netflix is also reported to have paid a steep license fee (above $4 million per episode) for the first two seasons.[28]

New Distribution Models

Related experimentation can be seen in the distributor Debmar-Mercury's disruption of the deeply entrenched model for comedy syndication. In 2006 the company established a $200 million distribution agreement for a hundred episodes of the first-run syndicated comedy *Tyler Perry's House of Payne* on the cable channel TBS, followed by a run on FOX-owned stations a year later; more recently Lionsgate TV (Lionsgate now owns Debmar-Mercury) used this same model to produce the Charlie Sheen vehicle *Anger Management* and George Lopez's *Saint George* for FX.[29] Such deals entirely circumvent the broadcast networks, which are normally the original licensors for comedies, and the arrangement, now called 10-90 sitcoms, guarantees an episode order substantial enough that the studio can be certain it will be able to seek revenue from subsequent windows and do so more quickly than the usual production pace, which requires five years before a hundred episodes can be amassed. These deals require a channel to license and schedule a test run of ten episodes, and if the episodes achieve a predetermined ratings target, the cable channel is obligated to license ninety more episodes over two years. Debmar-Mercury can then sell secondary rights within two years.

While innovative in many ways, this model also has significant implications for production. As the industry journalist Cynthia Littleton notes, it "leaves showrunners no margin for screwups, reshoots or unplanned hiatuses to rethink story arcs."[30] Bruce Helford, the producer of *Anger Management*, describes the process as "backwards," but favorably so. For the first batch of ten episodes, Helford prepared and shot each episode over two days, in comparison with the five-day schedule common for a broadcast network sitcom—though Tyler Perry's shows have even produced three episodes a week. Helford explains that a strength of the initial ten-episode order derives from

> not trying to stuff everything into a pilot to get it to test well. And you really have to go on your gut instincts. You don't have the time to over-think things, and the actors don't have the time to get tired of the material. That gives it an honest spontaneity that you don't always see in sitcoms.[31]

Helford's final point is a preference for the adjustment in the dynamic between the producer and studio executives that the 10-90 format requires. The fast and relentless pace leaves little opportunity for network involvement, which allows the creator more autonomy.

Many of the changes in production norms introduced at the transition to the post-network era resulted from financial crisis and the need to find ways to produce less expensive content. The 10-90 comedies produced to date have not been hailed as exemplars of creative innovation or advancements in television comedy, but they suggest an adjustment in the television business. In the case of Tyler Perry's shows, the 10-90 format has surely contributed to Perry's uncommon success in bringing majority-black–cast entertainment to television.

Online Originals and Other Still Experimental Forms

As Internet sites such as YouTube, FunnyorDie.com, Hulu, and Yahoo have endeavored beyond distribution into original "series" creation, yet other models for production are emerging. In some cases, the established studios produce the content in a scaled-down model of what is common for broadcast and cable production. In other cases, a corollary to the independent financier can be seen supporting the work of producers who have built an audience based on videos made as amateurs; in this case, producers have acquired Kickstarter and other fan-funded backing. For example, Freddie Wong, known better as FreddieW, produced the *Video Game High School* web series, raised over a million dollars for multiple seasons through a Kickstarter campaign, and secured a sponsorship deal with Dodge.[32] Wong was also able to leverage the Kickstarter success and substantial YouTube subscriber base to gain production support from YouTube, including use of its YouTube Space Los Angeles studio.

At least as of 2014, it could be argued that YouTube was creating a space for content different from what might be found on conventionally distributed television. Where Hulu's experiment with the original series *Battleground* and Yahoo's *Burning Love* looked much like low-budget knockoffs, YouTube provided a forum for voices not likely to be found on broadcast networks or cable channels. Freddie Wong is a good example, but so too are the YouTube channels YOMYOMF and Awkward Black Girl, which have developed large subscriber bases for content—or at least acting and writing talent—still missing from the post-network abundance. Though YouTube's "channelization" efforts have not been particularly successful, the funded channels do help amateurs transition into professional video makers, and shifts toward subscription funding may offer further support for this content. YOMYOMF, which receives funding from YouTube, well illustrates what Alex Cross, YouTube's head of original programming, describes as the goal of the funded channels: "YouTube encourages people who are smart, engaged, and curious, as well as underserved in other media."[33]

But by 2013, a much less optimistic vision of these online portals also emerged. Two of YouTube's most high-profile channels, Machinima and Maker Studios, were openly critiqued by some of their most popular talent for a range of offenses including inequitable ad revenue sharing and forcing contracts that required creators to give up rights in perpetuity, often on young talent lacking access to representation.[34] Many of these disputes garnered a lot of attention within the subcommunities of YouTube channels, and social media have provided opportunities for aggrieved parties to publicize the conditions of their contracts, though several such videos have also quickly disappeared or gone private. It is clear that digital distribution hasn't led to a democratic utopia that only the most naïve would have supposed. Only time will tell whether the conditions of producing in the conventional industry and the emerging industry will remain distinct and how they will differ. Disney was in talks to buy Maker Studios for $500 million as I finished the book, while Machinima had recently received an $18 million investment from Warner Bros., which suggested a coming need to rethink the place of both entities relative to new and legacy media industries. These developments mirrored the model of many other industries that outsource research and development activities to entrepreneurs and then purchase needed technology or expertise once it proves successful.

The model of web distribution and Kickstarter funding suggests the most direct funding model possible, with creators developing content funded directly by fans. This strategy remained most preliminary, though widely discussed, as the book was completed. In recent months, the industry establishment turned directly to these funding mechanisms that developed to aid those without access to studio support. The success of Rob Thomas's Kickstarter campaign to fund a *Veronica Mars* movie in early 2013 drew significant attention to the ability of social media outlets such as Twitter to connect fans of shows outside the most successful and created a heady environment that imagined a mechanism for funding television production that allowed audiences greater feedback and even a stake in the process, or at a minimum, could "prove" fan interest in concepts of which studios were unconvinced.

Though the array of online original series continues to expand, the bounty of content and the lack of certain economic models have left the future role of this content unclear. As of 2014, online originals seemed to operate similarly to baseball's "farm league" system. The online distribution provided a way for new voices to be heard, and those deemed "successful" typically used that success to secure subsequent projects on broadcast and cable. Though it remains possible that circumstances will be different in the future, advertiser-supported online video was primarily a means to enter the industry as of 2014. To some degree, the shows developed for online original subscription services such as Amazon or Netflix function differently than those supported through advertising because the goal of this original content is to drive subscriptions, and perhaps technology sales in the case of Amazon. Though it remains early for online original video, it also remains to be seen whether a profitable business can be sustained in online distribution alone.

A final experiment bridges established and emerging digital production and distribution and suggests the possibilities of this environment for established, though niche, performers. The comedian Louis C.K. drew a lot of attention in December 2011 when his direct-sale concert experiment netted $1 million in just three weeks. C.K. charged $5 per download and bypassed traditional distributors with a performance video filmed over two live performances and edited and directed by C.K. at a cost of $170,000. In a blog post recounting the details of the production, C.K. noted that he earned far more than expected. He donated a quarter of the profits, used another quarter as a bonus for his staff, and noted in a post made just four days after releasing the video and earning an impressive $200,000, that he'd made

> less than I would have been paid by a large company to simply perform the show and let them sell it to you, but they would have charged you about $20 for the video. They would have given you an encrypted and regionally restricted video of limited value, and they would have owned your private information for their own use. They would have withheld international availability indefinitely.[35]

The outcome of C.K.'s experiment was a valuable lesson for those who could afford to develop the digital distribution infrastructure—which C.K. priced at $32,000—illustrating that "conventional" pay rates can be achieved and surpassed for established artists, while also maintaining complete creative control.

As everything from Kickstarter campaigns to direct-to-consumer comedy shows suggest, the likely production norms of the post-network era remain far from certain. Most basically, the massive expansion in distribution explored in the next chapter has enabled a range of new practices to emerge. Many of the norms of the network era and multi-channel transition remain in place, though they are clearly being challenged, and within a few years, funding strategies that appear as mere experiments now will likely become more established. The multiplicity of types of "television" that now can be created and distributed suggests that it is unlikely that the industry will ever be so dominated by monolithic production norms again, so the question is not which of these or other funding strategies will replace deficit financing. The financial imperative of creating shows likely to succeed in syndication led studios to produce certain types of series—typically those with an established record, such as law, police, or hospital shows—and decreased the likelihood of producing less conventional fare. Multiple funding, production, and distribution models enable the creation of some content that might reach vast audiences and others that reach quite narrow segments, and both might be understood as successful by a variety of measures.

Labored Relations

Many activities are involved in the creation of television programming. Practices such as deficit financing and federal regulation such as the fin-syn rules operate at a macro level and exist as given norms of operation for those who work in the industry on a day-to-day basis. There are also many other important aspects of making television that are not externally imposed, one of which encompasses the working conditions and standard labor practices of the industry. By the end of the multi-channel transition, the Hollywood creative community at the center of U.S. television production featured many norms increasingly atypical of U.S. labor relations. Hollywood continued to operate with an unusual level of unionization; almost all work in the mainstream creative industries relies upon a collectivized entity to negotiate basic fee scales for work and residual payments on content. While the maintenance of union and guild centrality in Hollywood might be almost inexplicable relative to the union-busting and destabilization of workers' collectives throughout the United States, it does support the notion suggested by many who study creative industries and argue that this work involves features fundamentally distinct from most others.[36]

Despite the centrality of Hollywood's collectivized workforce, major labor disputes emerged throughout the multi-channel transition as studios and networks tried to save money by decreasing labor costs. The media scholar Chad Raphael notes that in the 1980s and early 1990s, five creative industry unions went on strike once (the National Association of Broadcast Employees and Technicians, the Directors Guild of America, the American Federation of Musicians, the Screen Extras Guild, and the American Federation of Television and Radio Artists); and the powerful Screen Actors Guild went on strike twice and the Writers Guild of America three times.[37] The 1988 Writers Guild strike lasted twenty-two weeks, delayed the premiere of the 1988 season, and cost the industry an estimated $500 million.[38] Indeed, the costs of this strike for both sides continued to weigh particularly heavily as new contracts created patchwork agreements for much of the multi-channel transition. Difficulty keeping labor agreements up-to-date with adjustments throughout the television industry resulted in a three-and-a-half-month strike by the WGA that concluded in February 2008.

New technologies and distribution windows have been particularly challenging for existing labor agreements. One issue developed from a deal agreed upon in the mid-1980s that established the residuals creative talent would earn on VHS sales that remained in effect through 2006—by which point DVD distribution had become a $4 billion industry, and various online distribution formats were exploding. The key contention for creative talent resulted from the categorization of new technologies as "home video" rather than as "pay-TV," which earned four times as much.[39] Likewise, deals for cable production crafted in the 1990s remained in effect despite the substantial change in the type of programming and budgets of these networks, and creative talent such as writers began losing valuable income from rerun airings when networks decreased this practice as audiences embraced the new control and choice available. As new digital distribution platforms enabled studios and networks to experiment with allowing viewers previously unimagined access to programs, the industry lacked policies regarding how this distribution should be considered and what rights different creative workers possessed. Networks and studios also began creating supplementary content distributed on show websites, and it was unclear how the labor involved with these efforts should be remunerated. The guilds wanted compensation appropriate to their contribution to these new ventures, which threatened studio and network shares of the new bounty.

The new technologies and distribution formats that made previous contracts outdated indicate just one point of tension for the television industry's uncertain labor market. Some television producers had evaded the cost of union production by fleeing Hollywood in a practice known as "runaway production" and led many to worry about the future of work in this "industry town," although it was the already diminished film industry production that was primarily responsible for reducing the available work in the immediate Los Angeles area. Television work began climbing slowly in the early 1990s in response to new broadcast and cable needs and reached a peak in 2002, the year it bested feature films as the area's primary production activity.[40] During this time, much television was also produced outside the city; while 75 percent of prime-time series were shot in Los Angeles in 2005, only 44 percent of surveyed cable programs were filmed there, and 2012 featured a record number of new drama series shot outside California.[41] One way cable has been able to afford original programming has been by moving production to Canada, where producers avoid union rates and, in the 1990s and early 2000s, benefited from a weaker Canadian dollar. Some dramas that aired on the WB and UPN (later the CW) were also shot outside Los Angeles in response to the efforts of various cities and states throughout the United States that offer tax incentives to encourage the financial boost of production spending; most of the shift has been to New York.

The labor conditions at the heart of the union and guild strikes and runaway production affected the creation of programming in various ways. Although some people may associate fame and stardom with working in television, the percentage of workers who achieve household name recognition is infinitesimally small relative to the number of people required throughout the production process. The guilds and unions have functioned primarily to secure basic rights and suitable working conditions for those paid at base level, as the irregularity of production and the high demand for jobs have created ample opportunities to exploit this particularly unstable labor force. As the economic conditions of the industry changed in relation to the adjustments examined here, networks and studios sought cost savings wherever possible, and, with wages forming a large component of production budgets, these savings often came at workers' expense.

Raphael argues that early in the multi-channel transition the economic conditions and practices of the newly acquired broadcast networks contributed greatly to labor unrest, and that skyrocketing star salaries drove production costs so high that the networks had to begin including some low-cost programming. Such programming was first evident in shows like *America's Funniest Home Videos*, then in the surplus of newsmagazines through much of the 1990s, and finally in the unscripted shows that did away with the need for many unionized employees, particularly actors and writers.[42] Cost saving was an important factor in the surge in reality programming in the early 2000s, but the novelty of reality shows, as well as the fact that they drew larger young audiences than other lower-budget programs, was also part of their appeal. Nonetheless, by the end of the multi-channel transition, the industry had begun running out of the stopgap solutions that had allowed it to continue operating by subtly modifying the network-era model.

One of the consequences of adjustments in program financing models and outlets for video entertainment has been a significant disruption in the norms of industry compensation. Certainly many eagerly eyed the new revenue streams available first from DVD sales and broadband distribution possibilities (which at that time seemed endless), but determining how those funds would be shared has proven to be difficult for the industry.

Transitions in Programming

Many of the features of television programming that we have long taken for granted—that shows should last thirty or sixty minutes and have commercials embedded throughout, that they should air at the same time every week, that sometimes a network will air an episode that we've already seen—result from network-era norms of program creation established by the broadcast networks. Many of these practices resulted from a negotiation of economic considerations in a manner that again underscores the intricate connections among artistic and commercial components of cultural production. For example, once it became technologically possible, the "rerun" was a key strategy of broadcast economics as it decreased the number of weeks that the networks needed to pay for new programming. Likewise, conventional program lengths developed to facilitate the constant flow of programming and included the use of commercial messages embedded at anticipatable intervals that became characteristic of U.S. commercial television. These norms differ from those of other countries in which networks might allow periods of blank screens to air because of irregular program lengths.

The television season is a prototypical network-era concept, fundamental to a linear viewing environment, and emerged from factors of competition, audience research, and program acquisition and financing. The twenty-one- or twenty-two-episode seasons the networks purchased throughout much of the network era and multi-channel transition allowed the networks an initial airing and at least one rerun airing to fill a time slot for roughly forty-three weeks of the year. Remaining weeks were left for specials, films, sports, and holiday programming. The television "season" provides a quintessential example of an industrial ritual with commercial and artistic ramifications. No external force mandated this practice, but once developed, it proved difficult to suspend. In the early 1960s the Big Three networks established the concept of the "television season" that mirrored the U.S. school year (more on this in chapter 5). Spanning September through May, the season remained dominant for four decades and then began declining in the early 2000s as a result of a number of competitive pressures.[43]

The three networks did not establish this practice through formalized collusion; rather, they developed and maintained the practice as an unofficial industrial norm of mutual benefit that freed them from the need and expense of purchasing new programming for the fifty-two weeks of the year. When the upstart network FOX sought entry to what seemed a zero-sum industry in the late 1980s, it achieved some success by launching new series during the summer, thereby counterprogramming the reruns of the Big Three with original shows. Competition from FOX initially was not significant enough for the Big Three to adjust their conventional practices, but in the late 1990s, cable networks launched original narrative series such as *Any Day Now*, *Sex and the City*, and *Witchblade* during the summer. The increasing loss of audience members to cable during summer months began to jeopardize the network-era model of the television season, especially once the cable model of the year-round, rerun-free season emerged. Although broadcasters' abdication of summer competition had been supported by industry beliefs in sizable drops in audiences during these months, such drops had become insubstantial by the early 2000s. In the 1950s, the HUT (homes using television) level dropped 28 percent during summer months, but the average summer use in 2003 measured just 5 percent lower than during the regular season.[44]

Apart from the tendency of decreased television use in the summer, another rationale for the "television season" was maintaining optimal audiences during the key "sweeps" months of November, February, May, and July, in which Nielsen collected national audience data. The Big Three networks consequently organized their schedules to debut programs in mid-September in order to acquaint viewers with new programs before the November measurement. They then scheduled new and rerun episodes throughout the year so that the season concluded with widely viewed finale episodes during the May measurement. By 1987, however, Nielsen had refined its technology to allow it to measure viewing practices on a nightly basis; the People Meter could sample enough homes nationwide to produce nightly ratings that were accurate nationally, but not in individual markets.[45] This theoretically made sweeps periods unimportant to the networks' national advertisers, but the networks continued sweeps-schedule "stunting" both because the period remained crucial for their local affiliates and because the networks earned most of their revenue from the affiliates that they owned and operated. In 2004 Nielsen began implementing Local People Meter (LPM) technology in the largest markets, which enabled it to produce accurate local data nightly as well. Sweeps became functionally irrelevant to LPM markets, which included most of the networks' owned and operated stations, and further diminished the need to maintain the network-era television season, though it has persisted out of convention nevertheless.

Adjustments to the television season affected other aspects of the process of programming, including the corresponding cycle of program development. In the network era, the broadcast networks all began to develop new series in the late summer months, known as "pilot season" to some, during which program executives scheduled countless meetings with hopeful producers who "pitched" new ideas. Based on their interest in the ideas and their needs, the networks committed to pilot scripts and even pilot productions during the winter, so that by early spring they would have a variety of pilot episodes or presentations to consider for the fall schedule they announced at the upfront presentation in May. (The upfront presentation immediately precedes the upfront advertising sales process, during which broadcast networks have historically sold 75 to 90 percent of the advertising time in their schedules for the upcoming season.)

The broadcast networks' shared schedule for program development affected the creative process itself, as well as power relations within the industry. On the one hand, the shared calendar afforded creative talent a level of power, as networks wary of the scarcity of certain ideas sought to lock talent and ideas in place so as to not risk losing them to another network. On the other hand, operating on an industry-wide schedule constricted talent availability, placing actors and other workers in the dubious position of committing to certain projects and "passing" on others, while having little assurance that the project they committed to would be chosen for the network schedule. In fact, talent working on a series in production were often eliminated from consideration for new series because they believed their series would continue when they faced the real likelihood of show cancellation and minimal job security. This system created difficulties for networks, too, as when they signed "holding" deals with actors and creative staff to ensure that they would be available to the network—and to prevent them from working for others. This practice created inefficiencies when networks paid talent they did not use or if ideas sat on a shelf because a network would rather pay for a concept it might develop than risk losing an idea to a competitor. The use of holding deals also led networks to prioritize series that made use of the actors they were "holding," creating a dynamic in some ways reminiscent of the Hollywood studio era.

The decreased observation of the television season forced adjustments in the norms of the yearly development cycle. At least one network, FOX, claimed to utilize a year-round development process by 2005, though program debuts remained focused on fall and January, and the irregular schedules of cable channels also led to the emergence of alternative development cycles. Other networks claimed to program year-round, but primarily achieved this by maintaining the September through May norm and airing short-run unscripted series during the summer months. Maintaining the network-era convention of the television season benefited the networks because it eliminated the financial burden of year-round original programming. The financial losses networks faced as audiences not only switched to cable programming during the summer but also decreasingly returned in the fall provided the impetus to revise strategies.[46]

Adjustments in these cycles have modified power within the television industry. Freeing specific parts of the series development process from certain calendar periods creates more opportunities for creative workers. Writers and producers might be willing to present more unconventional ideas to networks if they do not need to fear that pursuing the project might lead them to be locked out of the job market until the following development season. Yet networks have continued to create countermeasures—such as holding deals—to reassert their control of this process. The interrelations among the convention of the television schedule, the upfront advertising buying process,

and the annual cycle of development illustrate how a web of industrial practices continues to perpetuate undesirable consequences, and how the contin-
gency of interrelated practices tends to thwart change. While new norms can
(re)establish power relations among the various entities involved in the program creation process, change as significant as the erosion of the dominance
of the television season is rare and will have widespread, substantive effects.

Cable originals challenged the yearly network time frame because their
economic model did not require attracting audiences to each time slot each
night of the week. The cable channels gradually developed particular time
slots, and then particular nights in which new programming was consistently
available, even if a program might appear only thirteen times a year. Following a strategy originally established by HBO, which built Sunday as its beach-
head for original series, FX developed Tuesday then Wednesday at 10:00,
while AMC featured Sunday debuts. This tactic certainly wasn't a complete
success, especially if the program bench wasn't deep enough to maintain
new episodes year-round or if substantial variation existed in the quality of
the shows or audiences reached. Certainly these matters of schedule became
decreasingly important as more viewers adopted DVRs that allowed automatic series recording, and diminished yet further once robust VOD offerings enabled easier nonlinear engagement. In the intervening period, many
viewers were simply perpetually confused about when new cable series' episodes could be expected and whether a particular show had been cancelled.

Also important in adjusting production norms was cable's willingness to
provide creators with more latitude on episode length. This was first noted
as one of the storytelling strengths of subscription cable; writing for HBO
freed writers from having to write commercial breaks and from the "shorter
hour"—the forty-two to forty-five minutes of programming available per
hour once commercial time is included—which could be a significant amount
of additional narrative time. By the fifth season of *Sons of Anarchy*, FX frequently allowed for episodes with sixty program minutes, which it scheduled
in ninety minutes, rather than forcing producers to trim the episodes.[47]

Another adjustment in form emerged from the different ways viewers can
and do encounter video storytelling. The veteran director and producer Paris
Barclay notes that "a new adrenalized storytelling" has emerged "as a result
of Internet clips, and the ability for viewers to multitask on their iPads or
phones while watching."[48] Here he notes awareness among storytellers of the
value of small coherent bits that might be shared online and used to promote
the show. His comment also acknowledges the "second screen" increasingly
common in viewers' laps and the need to construct stories that hold viewers' attention to the main screen. Though the pressure to keep viewers off

Facebook and e-mail and engaged with narrative is still quite new (at least the electronic version of the attention problem), significant adjustments in story pacing, editing, and storytelling may emerge as a result.

Viewers' experience with cable channels' unconventional scheduling and season organization contributed to their changing expectations of broadcast programming and to broadcasters' willingness to deviate from network-era practices. Although shorter and more irregular seasons meant there would be less new programming than during the twenty-two-episode seasons common throughout the multi-channel transition—and twenty-two was a reduction from earlier norms—the variation in scheduling and season lengths expanded the types of stories that could be profitably produced for U.S. television. Before this, the production conditions advantaged narratives that could be organized into bits with episodic resolution, and thus limited the types of stories that could be and have been told. The demand for successful series to endlessly perpetuate themselves also resulted in many stale hours of U.S. television and made it difficult for the medium to explore stories that have a more finite narrative range, as is common in other national television contexts. The most recent development of Netflix releasing a full "season" of episodes at once has encouraged yet further innovation. The *Arrested Development* executive producer Mitch Hurwitz acknowledged that producing the last season for Netflix led him to think about the storytelling in completely different ways. He considered viewers' ability to watch and rewatch episodes at will as he constructed the series' stories and their organization.

The changing competitive environment has reinvigorated interest in the short-run or limited series that had been quite successful in the 1970s (*Roots*, *Winds of War*) and standard in many other countries. In many cases, networks—both broadcast and cable—produced limited-run series to test program ideas that seemed to defy conventional boundaries (NBC's *Kingpin*, *Revelations*, *Book of Daniel*; Showtime's *Sleeper Cell*; USA's *The 4400*; FX's *Thief*). Of these examples, only *The 4400* and *Sleeper Cell*, notably both cable series, proved to have the necessary viewer interest to warrant subsequent seasons of production, but even the others expanded the storytelling world for a few weeks. If not for the limited-run option, networks are less likely to commit programming budgets and schedules to such risky programming endeavors and consequently might avoid them altogether.

Another advantage of the closed-ended nature of these series is that they can attract creative talent unlikely to work in television otherwise. The director Steven Spielberg, who served as executive producer of the SciFi (now Syfy) miniseries *Taken*, is a case in point, as is the actor Kevin Bacon, who acknowledged that the thirteen-episode season was crucial to his willingness to accept the lead role in FOX's *The Following*. To be sure, the limited series has not replaced ongoing series, and successful limited series can be redeveloped to have multiple seasons; rather, the fracturing of the competitive environment allowed the return of programming forms that had become infeasible and indicated an important expansion in the storytelling that U.S. commercial television could encompass.

This erosion of norms has also been due to the distinctive economics and programming organization of cable channels, which forced them to defy the dominant programming and scheduling practices of broadcasters as they began to produce original scripted series. As the competitive environment adjusted, however, broadcasters increasingly borrowed from cable channels' experiments, adopting practices such as uninterrupted new episodes—especially for series with more serial organization—and intentionally ordering only thirteen episodes (which is different from cancelling a show after thirteen). The increase in production of more serialized series has also affected norms of syndication scheduling. The common scheduling practice for shows in syndication has been "stripping" them, or airing a new episode every day, Monday through Friday, at a certain time. When Spike purchased a syndicated run of *The Shield*, it "stacked" episodes instead, airing three episodes in a row on one night of the week to better enable the viewer to follow the ongoing storylines.

Although I argue that the multiplicity of practices emerging by the end of the multi-channel transition provided important new opportunities for television, it is also true that the variation in scheduling practices and season organization confused many viewers accustomed to network-era norms. The cable channels used no apparent logic in determining when to present new episodes of shows, which led to difficulty for casual viewers who did not regularly watch those channels in discerning when new seasons would begin. Maintaining network-era practices also became confusing amidst so many other varied scheduling and season organization strategies. This inconsistency in scheduling norms contributed to networks' enhanced efforts in promoting their shows.

New Challenges in Promotional Practices

They're [Netflix] also not spending $40 million a show on a marketing campaign. They have a guy in a room who writes an algorithm.
—Peter Micelli, TV packaging and literary agent, CAA, 2013[49]

Program promotion has tended to exceed the regular activities of making television, but the central role of the network in this process warrants assessment. For much of television history, few observers have considered networks' self-promotional activities. Throughout the network era and much of the multi-channel transition, networks commonly included clips from upcoming programs within their commercial blocks and, for the most part, limited their promotional activities to using network airtime. There were a few exceptions, especially with respect to particularly important markets: in this case, you could determine the value of your home television market by noting the number of billboards and other out-of-home advertisements on which networks considered it worthwhile to spend portions of their promotional budgets. Otherwise, the few viewing options of the network era made on-network promotion particularly efficient.

Adjustments throughout television production processes required new promotional techniques and increased the importance of this already essential practice. In the course of the multi-channel transition, broadcasters responded to expanding competition by increasing their on-network promotions; for example, a study of NBC and ABC found that an hour of each network's programming contained five more minutes of promotional content in 1999 than 1989, and another study estimated that the U.S. broadcast networks collectively aired 30,000 promos per year.[50] If the networks had sold that time to advertisers, they could have earned an estimated $4 billion—lost revenue that further suggests the economic significance of promotion.[51] Though annoying to audiences, this increase in commercial load—known to advertisers as clutter—was also known to decrease the effectiveness of their messages. An analysis by Needham and Company estimates that "marketing and advertising costs often add 40 percent to the costs incurred to produce [television] content," but also that "marketing spending only buys viewer trial, not repeat viewing."[52]

Broadcasters' reliance on their own network as their primary promotional venue meant that the emergence of cable competition produced twice the consequences. Cable programming lured broadcast audiences away from broadcast series and also removed them from the audience for promotions; the latter effect became particularly problematic as audiences missed promotion for the fall season during their summer cable viewing. As a result, broadcasters suffered decreased ratings for new shows and had fewer opportunities to pitch upcoming content to their target audiences. The diffusion of audiences into niche venues and to new technologies that also diminished the utility of on-network promotion required more varied and precise practices.

Networks began experimenting with new promotional strategies to find the audience members who were eluding their traditional techniques as the post-network era began and most programming no longer attracted a large and heterogeneous audience. First, they made use of "sister" networks joined through common conglomerate ownership to reach a broader audience with conventional strategies. These endeavors often illustrated the "synergy" the vast media mergers were intended to create, as in the case of MTV airing a special about the new season of *Survivor* just before its launch on CBS, when both networks were part of the Viacom conglomerate. A telling indication of the extent of sibling promotion emerged in 2002 with the news that the largest advertiser on AOL Time Warner media was AOL Time Warner. The conglomerate contributed 5.5 percent of the $8.5 billion AOL Time Warner reported in advertising and commercial revenue that year.[53]

In addition to leveraging cross-ownership, the networks also maintained conventional promotional strategies or enhanced efforts in established venues such as through television critics. The networks staged elaborate press tour events for critics in hopes that they would draw attention to new shows, as critics' columns provided a way to reach viewers who may not be watching the network. Critics became increasingly important as their reviews and "tonight on" recommendations provided promotional venues to alert viewers of programming on networks and cable channels they did not regularly view and as legitimate, unbiased sources within the cluttered programming field.

Irregular and infrequent viewing, which was an acute difficulty for cable channels from their launch, complicated their promotional efforts. Like broadcasters, the cable channels were their own primary venue for promotion of their content, but few cable channels could rely on regular and consistent viewing in the manner that broadcasters maintained. (There were exceptions such as MTV and ESPN, which cultivated regular viewing in their niche audiences.) Thus, cable channels would have to commit substantial budgets to off-channel promotion if they hoped to reach an audience broader than their few million regular viewers—a significant expense not incurred by most broadcast programs. Under the circumstances, common ownership proved particularly valuable for cross-promotion, and this provided one of the few places where conglomerates achieved their goals of synergy.

Although broadcast networks had the advantage of regularly attracting more viewers than cable channels, as audience segmentation expanded, they found it increasingly difficult to maintain their audience status in a promotional environment valuing niche appeal. Clear channel branding became a trademark of successful cable channels, but their very mission as "broadcasters" made such branding impossible for broadcast networks. For example, in some weeks of the 2005–2006 season, over half of the CBS schedule featured episodic crime dramas. Although these series greatly contributed to CBS's status as the most-viewed network at the time, the consistent success with a specific genre gradually decreased the diversity of the audience likely to "stop by" CBS, where they could be reached with promotions for other shows. In that same season, CBS launched an innovative series about a young music executive called *Love Monkey*, yet poor ratings for the series in its first three airings led the network to pull the remaining episodes and cancel the show. *Love Monkey* was very different from most CBS programming at the time and was therefore likely to reach an audience distinct from the one that viewed the CBS criminal dramas—viewers who liked crime drama were more likely to switch to NBC to watch *Law & Order: Special Victims Unit* during the hour when *Love Monkey* aired. Consequently, not only was much of the promotion for this show, which appeared in crime dramas, wasted, but also the lack of similar programs on CBS's schedule made it difficult to marshal an audience for a series different from those already airing on the network. Moreover, though crime and detection procedurals have been crucial to CBS's broad-based success, this would not be a productive way to brand the channel, since in total, these prime-time series are but one part of a network that also features news, morning talk, and sports programming, as well as comedies and other types of prime-time programs. Apart from illustrating the challenges of promotion in a more fragmented media environment, this example also indicates the importance to networks of relatively mass events more likely to draw heterogeneous audiences, such as *American Idol*, the Olympics, and live sports. In addition to garnering high ratings, networks can recoup the value of costly league license fees and exploit the value of such events by maximizing their broader reach in promoting coming network fare.

These possibilities notwithstanding, by 2004, the networks had begun experimenting with less conventional promotional strategies off the air. For a few years, networks tried a broad array of efforts to reach viewers, though quickly moved away from these once "social media" arrived and created entirely new ways to market to and interact with viewers. By 2010, social media and new forms of distribution had created many opportunities for connecting with audiences and drawing them in to programming. Refinement of these strategies continues at a fast pace, as trends in technological uses of both television and social media such as Facebook and Twitter evolve. A key strength of the new tools available to marketers is that they provide alternatives to the mass messaging once common. Many of the new promotion strategies can be as tailored to audience tastes as the shows they promote.

Of the pre–social media strategies, ABC instigated some of the most creative off-air promotion through campaigns such as its marketing of *Desperate Housewives* using dry cleaner bags printed with "Everyone has a little dirty laundry" in 2004; likewise, it promoted *Lost* by distributing messages in bottles with details about the show. Of course the success of these creatively exceptional shows might have been entirely unrelated to these unconventional promotional campaigns, but many networks followed the strategy regardless. Significantly, these uncommon strategies also yielded substantial public relations buzz, enhancing the effectiveness of the campaign without additional cost. Experiments grew more varied in the subsequent season: NBC strapped portable television screens showing previews of *My Name Is Earl* to young women in bars; the WB installed special mirrors with a paranormal effect in two hundred nightclubs in three cities to promote *Supernatural*.[54]

Though these strategies garnered public relations attention, some networks also took advantage of new technologies and distribution possibilities that better enabled achievement of their primary goal: getting viewers to just watch the show. Here, a key strategy involved expanding opportunities for audiences to sample content, which, in turn, involved experimenting with alternative distribution methods. The first experiment emerged over a year before the explosion in online distribution platforms and possibilities that began in October 2005. In September 2004, the WB made available the pilot of *Jack & Bobby* for free to AOL's 3.5 million broadband subscribers. Audience members viewed the episode more than 700,000 times in the eight days before the series' launch.[55] Although *Jack & Bobby* did not survive the season, alternative distribution that expanded viewers' ability to sample series proved a valuable promotional technique in helping the new series break out of the cluttered environment at the beginning of the season.

Promotional strategies utilizing digital distribution began in earnest as the networks introduced new shows in 2006. Many pilots were "leaked" to popular sites such as YouTube or peer-to-peer sharing networks such as TVtorrents as networks realized that the artificial scarcity they created for their products contradicted their goal to build audiences.[56] Not all efforts relied on digital distribution: a captive audience numbering over four million had the opportunity to view the pilot of UPN's *Everybody Hates Chris* aboard American Airlines flights; networks included DVDs of pilots in copies of *Entertainment Weekly* or gave them away in other promotional venues, while the studios' practice of releasing the previous season on DVD just before the launch of the new season also made use of new distribution possibilities to aid series promotion. The networks' promotional efforts were estimated to cost them as much as $200 million for the 2005 season.[57] Such efforts weren't isolated to program launch; as audiences and networks adjusted to the post-network era's nonlinear possibilities, networks enabled VOD and other viewing opportunities that may have appeared a scheduling strategy, but were more precisely a matter of promotion.

The range of off-air strategies that developed throughout the 2000s may have been new to broadcasters, but many of the efforts were pages from the book HBO had written on successful promotion in the late 1990s. Basic cable channels first replicated HBO's strategies in their effort to break into domains of attention controlled by broadcasters. In addition to the challenge of airing on a subscription network—and therefore being unavailable in the majority of homes—the unconventional and irregular seasons of HBO series required that the network engage in a major promotional blitz to remind existing subscribers of new episodes and lure new ones to subscribe. HBO used out-of-home and DVD previews years before broadcasters seemingly invented these strategies. The differentiation of the HBO product from that of other networks also enhanced its marketing options. The style of HBO promotions tends to replicate the network's value proposition of offering something of exceptional quality and clearly distinct from the rest of the televisual field. Despite the fact that HBO reaches only a fraction of television households, buzz about HBO programming has frequently dominated the popular culture space, including front-page coverage of the long-anticipated sixth season of *The Sopranos* in the mass audience–targeted *USA Today Weekend* supplement.[58]

Following HBO's effectiveness at achieving word of mouth about its programming, network marketers have also used "viral" marketing strategies enabled by Internet and social media communication as competition among broadcast networks has grown more intense. The networks have thus designed campaigns to reach "super fans," those peer-influencing viewers who might talk up a series in offices and chat rooms. Where the common viewing of the network era once led viewers to discuss the previous night's viewing around the apocryphal watercooler, the conditions of the waning years of the multi-channel transition and the opening years of the post-network era have required networks to utilize pop-culture opinion leaders to lead viewers back to their sets.

The next flashpoint in evolving promotion techniques drew lessons from these experiments and began to use social media to ignite further interest. FOX took the unconventional step of broadcasting its pilot of the musical series *Glee* immediately following *American Idol*'s May finale in 2009. The concept of *Glee*—a series about a high school choir group and largely structured as a musical—seemed most unlikely to excite audiences, but the pilot illustrated the series' vast entertainment value. Betting that the key to success was to get people to sample the series, FOX teed it up on its best platform and then made the series available for free streaming across a range of broadband distributors until the series' second episode was aired in the fall.

In addition to making the pilot widely available and allowing plenty of time for word of mouth to build, FOX's promotion team used emerging social media tools to stoke further excitement and leveraged many older strategies as well. Each character had a Twitter account that the FOX public relations team used for posting amusing tweets, as did many of the largely unknown cast members; multiple versions of the pilot were released, including a director's cut and a version featuring the cast tweeting about the episode; and FOX conducted a contest for the biggest GLEEk—the person who did the most to spread the series over social media. All the social media helped feed and expand some of the more traditional—though still unusual—summer promotion such as the cast's appearance at Comic-Con, a series of mall concerts, and an appearance in Los Angeles.[59]

Certainly *Glee* had particular features—such as the musical performances—that offered entertainment value to what might otherwise have been simple star-sighting promotion events. Further, its high school setting matched well with the younger generation's technology adoption. Since *Glee*'s success, program marketers have tried a variety of strategies to drive viewers to shows, to keep them engaged between episodes and seasons, and to encourage them to do promotional work for them. It seems unlikely that any magic formula will emerge, since different types of shows have different features that lend themselves—or don't—to social media campaigns. And it certainly seems the case that no amount of creative marketing can overcome a show that isn't similarly compelling. But in the post-network era, simply making the most likely audience aware of a show is a sizeable challenge in itself.

Promotion has also become more integrated into the basic processes of series creation. Many shows have developed additional content that networks make available on their websites to better serve viewers' desire for "more" of their favorite shows. Some series also utilize blogs written by a member of the series' writing staff as a way to communicate and engage their fans. In some cases the blogs present "extratextual" content—storylines and information related to, but independent of, the actual series narrative, while in other cases, series' staff use the blogs in the same manner as many fan forums that predated the blogs, treating the space as a means for talking about the show and joining in fan discussion.[60]

Networks have seen immediate results from their online promotions. Viewership of the CBS comedy *How I Met Your Mother* increased by one million viewers, an 11 percent increase, the week after showrunners posted a music video supposedly made by one of the characters on My Space, while the *Late Show with David Letterman* increased its viewers by 5 percent in the month after a promotional deal between CBS and YouTube began.[61] Even the public auditions for unscripted series can provide promotional value; although the series cast few "characters" in these venues, local press about them, as well as publicized casting calls, encourage existing fans to increase their stake in such shows.[62] Initially, show marketers had to rely on viewer interest to search out the content, but the development of social media created opportunities for many additional ways to reach viewers and build their interest in shows.

In stressing such innovations in promotion at the beginning of the post-network era—as well as the challenging conditions that gave rise to them—I do not mean to suggest that establishing successful shows in the network era was easy. Fred Silverman, the renowned programmer of that era, was once quoted as saying, "Fifty percent of success is the program and fifty percent is how the program is promoted."[63] But the new conditions of the multi-channel transition and emerging post-network era have certainly required adjustments in how programs are made, scheduled, and promoted, and here it is important to note that promotion does more than draw audiences to programming; it also prepares them to have certain expectations of the show and thus contributes to how they understand it.[64] For example, in its promotion of its 2002–2005 series *American Dreams*, NBC often emphasized nostalgia and conventional characteristics of family drama, despite the series' regular engagement with deeper conflicts and darker aspects of its 1960s setting. Not only did this promotion repel audiences uninterested in the saccharine stories that are common to family dramas promoted in this way, but not characteristic of *American Dreams*, it also contributed to how viewers who did watch the series approached and defined the show.

Even though programmers' promotional efforts illustrate new levels of creativity, these techniques do little to solve the core marketing problematic of matching interests or needs with particular "products." The linear television business would never be a distribution form that could maximize the potential of digital tools to develop effective recommendations in the manner offered by a retailer such as Amazon or a service such as Netflix, whose use of "because you bought/rented x, you might like z" formulations have proven to be particularly effective in cultivating sales and loyal customers. Data from Netflix show that viewers who selected rentals based on the recommendations that matched both their rental histories and their ratings of those earlier rentals had far higher satisfaction rates than viewers who selected rentals based on blockbuster promotion.[65] Post-network-era promotion could better take advantage of nonlinear viewing to emphasize marketing that is more targeted and that leverages the increasing volume of data obtainable through social networks to craft recommendation engines and tools for sorting content that apply viewers' specific tastes and preferences—which is the adjustment, still mostly specific to Netflix, that Micelli references in this section's epigraph. As Netflix experiments with content creation, this bank of data about viewer behavior and preferences also becomes a tool for selecting themes and genres of new shows.

Conclusion

It used to be that the hits paid for the failures. But now, as the margins get smaller and your upside is cut in half, the economics of doing business become much more challenging. We're extremely sober about being an independent in this climate but being independent may have also enabled us to weather this downturn better than some of the competition.
—David Kissinger, president, Studios USA, 2001[66]

In the epigraph above, David Kissinger reflects on how being an independent studio helped Studios USA survive the changing industrial environment for program creation in the late 1990s. Importantly, Studios USA's independence was short-lived; the studio became part of Vivendi-Universal less than two months after these comments. The transition into the post-network era has indicated that production norms can change quite significantly when adjustment is forced by new conditions, but also that change merely in anticipation of adjustment is nearly impossible. As we analyze the relationship among production, financing, and distribution, production seems the component that redevelops most in reaction to more deliberate evolution by other components. So while it seemed that the industry had changed significantly enough in response to the fin-syn rules in the early 2000s as to write the obituary of the independent scripted production sector, by 2013, new competitors relying on a different financial model (Netflix with *House of Cards*) and even those forcing evolution on some of the most staid areas of the industry (the 10-90 format) revealed again the cyclical tendencies of industries in which it is not only characters, but studios and industry sectors, that may rise from the grave.

Considerable experimentation with production practices that respond to new distribution opportunities and new funding mechanisms persists to a degree that makes it difficult to predict what ultimately might emerge as post-network-era norms. The evidence at this point suggests considerable innovation; low-budget production practices that can stretch a Kickstarter budget or speed norms of comedy production—as through the 10-90 format—indicate some adjustments, while both studios and independent financiers simultaneously produce television that sets budgetary records. Production is one place where the innovation encouraged by the climate of uncertainty is evident. "Unconscionable" profits may no longer be possible, but it is difficult to claim that the changes have diminished the quality of the products that are now possible.

The displacement of linear viewing led to substantial consequences in the content of program promotion. Rather than letting viewers know when a program would air and giving hints about "this week's" story, networks created ads more akin to film trailers, designed to rouse viewer interest in core aspects of the story.[67] The way certain types of television have come to be discussed in culture—mostly original cable series right now—has inculcated new viewer expectations. After reading endlessly about *Downton Abbey* in newspaper articles and Twitter feeds, the uninitiated viewer does not seek to watch the latest episode, but to begin the story from its pilot. Even a transition in behavior this small is important for imagining a much different television future.

Marketers' post-network-era strategies consequently continue to evolve from a task focused on driving viewers to watch a particular show at a particular time to helping viewers find content of interest amidst the vast programming abundance. Though all sorts of arguments can be made for the excellence and achievement of some programming in this era, those claims are tempered by the reality that simply "being on" television no longer assures the cultural relevance that was once the case. For all of the new potential the fragmented television space provides for the circulation of ideas and stories far beyond the limited mainstream of the network era, many series air as though they are trees falling in unoccupied forests.

The production and promotion strategies explored here intersect with the expanding distribution opportunities discussed in the next chapter. Given the changes in technology and distribution, it has become increasingly possible to imagine a future in which broadcast networks exist as advertising-supported venues for free initial program sampling that viewers could then subscribe to and view at a self-determined pace. Such a situation would disrupt many norms of program creation even more than they have been thus far. Programming decisions would no longer be subject to the need to find shows to fit an established schedule, and new financial models would develop. Indeed, the very place of networks—both broadcast and cable—could become uncertain in such an environment; studios and MVPDs could take on the distribution capabilities networks have long controlled.

And it is also the case that premiere ratings grow decreasingly relevant in gauging series. Though journalists continue to focus on audience size at premiere—a measure important for networks focused on creating mass audiences for high advertising compensation—the matrices needed to account for the complexity of revenue streams have yet to become part of the popular conversation about series "success." Studios now need ways of better evaluating the long life and many revenue streams series may have. For example, industry journalists devoted expansive attention to Netflix's *House of Cards* premiere—how many viewers watched how many episodes how quickly. Yet in the case of the niche exclusivity of Netflix as a service accessed by an audience measuring less than a quarter of television homes, this first release provided limited suggestion of the series' cultural relevance. Similarly, though many cultural discussions of success were focused on how many viewed the premieres of *Game of Thrones* or *Homeland*, a more sophisticated discussion including download revenue and DVD sales was certainly warranted.

Unlike the case of the production components considered in the chapters before and after this one (technology and distribution), where preliminary post-network-era practices were coming to be established by the time I completed this book, the practices of making and promoting television series lag somewhat behind these components—in some ways, in need of more permanent norms of distribution and economics in order to develop production norms that match these new realities. Adjustments in the distribution and financing of television programs will surely continue to alter the process of show creation in significant ways, and the continued erosion of the network season and schedule certainly suggests further steps toward the ultimate collapse of linear viewing norms for prized content. The freedom from the constraints of only telling stories that could be confined to thirty- or sixty-minute episodes, in twenty-two episodes per season, and in an ongoing but mostly episodic narrative began to illustrate the expanded programming possibilities of the medium, and the diversifying financial models explored in subsequent chapters will disrupt norms for program creation even further.

4

Revolutionizing Distribution

Breaking Open the Network Bottleneck

The future is about whatever I want, wherever and whenever I want
it. . . . and the more ways you do that, the more revenue there is for
everybody in the business.
—Josh Bernoff, Forrester Research, 2005[1]

An age-old debate within the television industry concerns whether content
or distribution is "king." Your position on this question depends greatly on
what sector of the business you work in, with favor going to your own role
as either a creator of content or a controller of the means by which content
reaches viewers—that is, a distributor. This debate was somewhat less complicated in the network era, when ways to distribute television were scarce.
Producers sold series either to networks or to local stations—a situation
that created a significant bottleneck that allowed only a limited amount of
programming to get through to viewers. After programs had an "original
run" on a network, producers typically resold the episodes in international
markets, to independent stations, and to broadcast affiliates to recoup the
costs of deficit financing. These opportunities to sell content after and even
during the original network run are called "distribution windows."[2] The
limited number of distribution windows in the network era greatly contributed to the ephemeral nature of television programming at the time, for
without personal recording capabilities and few alternative ways to receive

programming, viewers had hardly any opportunities to re-screen content, and never on their own terms.

The limited ability to reach viewers was such a fundamental aspect of the network era that few realized how considerably it defined the basic functioning of the medium. The post-network era, however, eliminates the distribution bottleneck. Previously unimagined possibilities for television have developed as new ways have emerged for video storytelling to reach audiences, including easy distribution of amateur and noncommercial content. New distribution methods ranging from the DVD to the myriad Internet video sites that exploded throughout the last decade have changed the nature of television. Television is no longer a linear trickle of programming dictated by network executives, but has swelled into a wide ocean of content that viewers can dip into at will—provided they are willing and able to pay directly for content or for services that enable such flexible, nonlinear use. New forms of distribution have also created new revenue streams for studios and adjusted the types of programming they develop. The growing variety of ways to reach viewers has decreased some of the risk of unconventional programs because new distribution routes provide opportunities to make money on shows that fail to achieve high ratings during network runs. Internet distribution both complements and competes with traditional television entities by offering at-will access to many shows produced first for linear distribution as well as an array of content produced only for broadband distribution.

The expansion of standard network-era distribution windowing to include various forms of Internet and on-demand (VOD) distribution produces consequences for all the other components of production—from business models that have altered the type and range of content that can be profitable, to creative processes that have responded to new opportunities in the industry. These new distribution methods have contributed significantly to inaugurating a post-network era of U.S. television and changed the nature of television as a cultural institution.

To be sure, ways to distribute television were already expanding appreciably throughout the multi-channel transition, during which cable networks rapidly proliferated and hungered insatiably for programming. The number of broadcast stations nationwide also expanded from 1,011 to 1,442 between 1980 and 1990; some of these stations remained independent, though most established affiliation with newer, non–full-service networks (FOX, the WB).[3] Videotapes initiated an affordable way for viewers to purchase programs— called "sell-through" in the industry—though this form of television distribution didn't become particularly viable until a decade later with the creation of the DVD. Cable systems then began offering limited programming on-demand, and soon after, the possibility of distributing video on the Internet gave viewers even more ready and varied access to programming. New technologies such as portable and mobile devices then allowed viewers to use wifi and mobile data plans to vary the places they viewed content. All of these developments opened new distribution windows and created new markets for producers to sell programs.

Importantly, "distribution" describes a broad range of activities, some of which are interrelated, others of which are fairly independent. Changes in distribution after the network era can be distinguished as either new *distribution windows* or new forms of *distribution to the home*. Distribution windows include the different locations producers sell programs after their original run on a network. In the network era, the only options were international markets and local stations. During the multi-channel transition, these windows expanded to include cable networks, direct sale on VCR tapes, and then DVD and, limitedly, VOD; more recently they have also come to encompass Internet sites from which episodes can be downloaded or streamed, such as Amazon or iTunes, subscription services that offer an array of programming, such as Netflix, or advertiser-supported sites such as Hulu and Crackle. These distribution windows are also easily confused with the expanded opportunities to view programming online, but while still in its original license window. Studios and networks now enable viewers who miss a linear airing to view for a brief period through MVPD VOD, on network websites, or Hulu.

Distribution to the home, though related, instead identifies how television content reaches the home and the convergence or competition among communication and entertainment technologies once it arrives there. While television once came into the home only through signals broadcast over the air, an increasing range of possibilities developed during the multi-channel transition. Cable and satellite became common mechanisms of delivery, and companies traditionally limited to telephony such as AT&T and Verizon joined the competition in particular markets by the mid-2000s. Even more significantly, broadband Internet distribution of video began in 2006—though it really exploded by 2011—and diminished the domination of cable and satellite as the only pipeline for most channels into the home, although cable services continue to provide the broadband connection for many viewers. Refinements in technology by cable providers—such as the transition to DOCSIS delivery technologies (data over cable service interface specifications)—reduced perceived scarcity by allowing an exponential expansion in content capacity, while wifi and mobile data networks have eliminated many of the place-based viewing limitations that long tethered

television audiences. Expansions in broadband speed, capacity, and afford-
ability created anxiety for traditional television distributors as fears of consumers "cutting the cord" of cable and going "over the top" (OTT) to watch
all their television through broadband-delivered programmers became
an obsession of the trade press beginning in 2010. By 2013, some MVPDs
began bringing a more revolutionary model to market. TV Everywhere initiatives developed that allowed the flexibility of viewing across devices and
at a self-appointed schedule, which had been a primary value proposition of
competitors such as Netflix.

Conventions of Distribution during the Network
Era and the Multi-Channel Transition

Like the film industry, which releases films in theaters, then to pay-per-view,
VHS/DVD, premium cable, and broadcast or basic cable, the television industry has also utilized similar standardized, time-delayed distribution windows
with tiered pricing that forces consumers to pay a premium for earlier access
to content.[4] This practice allowed the network paying the license fee to enjoy
a period of exclusivity in which viewers could find the program only on the
network supporting the original run. Throughout the network era and multi-
channel transition, viewers typically could not watch syndicated episodes
until a show had been on the air for about five years, and the buyer of that
first syndication run enjoyed exclusivity in the syndication market.

   To understand how the process worked, consider the example of the comedy *Friends*, which Warner Bros. studio produced for NBC beginning in
1994. The series was sold for its first syndication run in 1995, even though
syndicated episodes could not begin airing until September 1998 in order
for the show to have produced enough episodes for a buyer to air it Monday through Friday without too frequent replay. This first run of syndication went to individual stations in each market—the major metropolitan
areas with television stations.[5] Often the stations in a market bid against each
other for multiyear rights to a series, because being the exclusive provider
of a popular show was important for stations' schedules, especially in an era
of limited choice. Consequently, syndicated episodes of *Friends* might air in
the early evening on the FOX station in your area, while new episodes would
continue to be found nationwide during prime time on NBC. After selling
the series exclusively to local stations two or three times, Warner Bros. then
sold syndication runs of the hit show to cable, with the result that there could
be old *Friends* episodes on your local FOX station and on the cable channel
TBS and still new episodes weekly in prime time on NBC.

Such practices were highly standardized in the network era and much of the multi-channel transition. Series did not begin airing in syndication until a hundred episodes had been produced; stations commonly paid cash for the episodes and had exclusive rights to the show in their market.[6] Subsequently, however, what windows a show was distributed through, in what order, and how much money a show might make in each have come to vary greatly. Studios can sell shows and begin syndication runs before a series reaches a hundred episodes; stations can purchase the shows with various cash and barter agreements—meaning that they trade advertising time to the distributor in exchange for a lower cash payment—and they rarely maintain exclusivity, as many series become available on DVD and through online or VOD distribution before beginning syndication.

Unquestionably, the emergence of cable channels provided the first significant shift in distribution and marked the beginning of practices characteristic of the multi-channel transition. Budgets for original cable programs were diminutive relative to those in broadcast television, but the rapid proliferation of cable networks meant the creation of new buyers to which studios could sell syndicated programming, as well as cheaply produced original content. The average one hundred channels received in homes by 2003 created many opportunities for studios to sell both old and new programming.

Although the advent of cable seemed to revolutionize distribution, these developments appear quite subtle now that digital technologies have radically expanded viewers' opportunities to access video nonlinearly. First, direct sale of full seasons of shows on DVD began to change distribution practices in a way that eroded the exclusivity and ephemerality of programming; and then, in a few chaotic months in late 2005, industry workers threw out many of the old rules. New technologies ranging from VOD and broadband delivery to devices such as the newly video-enabled iPod eliminated the need for viewers to rely solely on networks and channels.

Outlets for streaming full-length commercially produced video and network and studio strategies for these distribution opportunities developed haphazardly throughout the late 2000s. Distributors such as iTunes allowed viewers to pay directly for episodes, which introduced a fairly unprecedented "transactional" model for paying for television. In the process, decades-old practices that derived their value from exclusivity and delay among windows were tossed aside overnight—with regret and uncertainty—but new technologies and viewers' embrace of them forced studios and networks to begin letting go of network-era practices. To be sure, many feared the consequence new ways to access programs would have on later windows, but with the very real possibility that programs could illegally circulate online worldwide

Table 4.1. The New Distributors

Type of Distributor	Functions	Examples

MVPD (multi-channel video programming distributor)	Provide wire that carries video programming into the home
	
	Cable and telco companies also provide Internet service (in many cases)
		Cable
		  Comcast
		  Charter
		  Cox
		Satellite
		  DirecTV
		  Dish
		Telco
		  AT&T U-verse
		  Verizon FiOS

BDVS (broadband-delivered video service)	Deliver "television" or video programming over a broadband Internet connection, but do not own or control the wire into the home	Advertiser-supported
		  Hulu
		  YouTube
		Subscription
		  Amazon Prime
		  Hulu Plus
		  Netflix
		  some YouTube channels
		Transaction
		  Amazon
		  iTunes

within hours of airing, studios and networks eventually realized they had no choice but to experiment—recalling the consequences of the recording industry's unwillingness to alter distribution practices a few years earlier. Banking on the notion that the best defense against piracy has been to make well-organized and high-quality content legally available for purchase, networks made shows available within hours of their original airing, though often curtailed the length of time they are available until they are released in traditional secondary windows. The television industry thus anxiously began its free fall into new norms of distribution that allow viewers their desired access to content anytime and anywhere, doing its best to limit content availability and whenever possible exact a fee (see table 4.1).

Close a Window, Open a Door: Shifting Norms of Television Distribution Windows

The proliferation of networks throughout the multi-channel transition created many new buyers for original and syndicated programming, but until the whirlwind of new viewing devices and platforms arrived, networks and studios did not experiment extensively with varying distribution practices. As they have come to do so, many in the studios feared that new distribution methods would destroy old models of revenue that the industry relies upon, but in the course of developing and adopting distribution experiments, they have also found unanticipated benefits. The opening of myriad distribution windows has provided networks with new promotional tools to reach audience members, as well as created revenue-producing opportunities for both studios and networks to amortize failures. Many of the new distribution experiments began emerging in 2006 as the networks tested nonlinear access and the studios looked for new opportunities to profit from their libraries. But nearly a decade after Apple rocked the norms of television distribution and economics with the $1.99 episode download, the lack of clarity and consistency characteristic of previous eras left many television creators and distributors anxious. And for good reason.

Changes in distribution shifted production economics enough to allow audiences that were too small or specific to be commercially viable for broadcast or cable to be able to support niche content through some of the new distribution methods, particularly those featuring transactional financial models. Just as cable had radically expanded the array of content that could be found on television, the new distribution windows promise to again rewrite the possibilities for what can be found on television. Fearing added competition, networks initially attempted to quash some of the new distribution opportunities. The true push to change came from other industries—such as consumer electronics—and from viewer uptake of the technologies the consumer electronics industry made available. As the tech-savvy began to redistribute content on various online peer-to-peer services without network clearance, the networks realized they could no longer slow the evolution and began openly experimenting with new models of distribution and financing.

By 2010, a decade of varied strategies that seemed revolutionary at the time were revealed as mere transitional practices in the broader process of disrupting distribution norms. Few in the mid-2000s anticipated how quickly broadband speeds would improve and enable high-quality images without extended buffering times over Internet connections, let alone wifi or mobile phone networks. As shown in the time line of digital distribution in table 4.2, a game-changing development has appeared nearly every year since 2005, which has left many in the industry in a persistently reactionary mode as the ground continues to shift too rapidly to develop long strategic horizons. This section explores three major shifts in television distribution norms that have occurred so far in the post-network era. Though the first strategies of repurposing and reallocation now seem minimally significant, given that they still demanded linear viewing and were not conceived as a break in long-dominant distribution windowing, the revolution in distribution that came later built upon these preliminary steps.

Repurposing and Reallocation

The practice of original run repurposing that began in 1999 marked the first significant adjustment to distribution practices after the industry adapted to cable.[7] "Repurposing" refers to a practice in which content providers crafted deals that allowed a series to earn additional revenue during its original run either by airing multiple times on the broadcast network licensing the series (more than a typical rerun) or by airing concurrently on a cable network. Repurposing consequently shortened the previous window between original run and syndication to as little as a matter of days.[8] In the case of one of the earliest examples, NBC and the USA cable channel arranged a financing deal giving USA rights to air *Law & Order: Special Victims Unit (SVU)* within two weeks of its broadcast airing (Studios USA produced *SVU* for NBC).[9] In the same season, Lifetime aired the new series *Once and Again* the same week it appeared on ABC. The industry journalists Deborah McAdams and Joe Schlosser credited the vertical integration of the entities involved—Disney's Touchstone produced *Once and Again*, the series first aired on Disney's ABC, and was repurposed on Lifetime, half-owned by Disney—with the deal's rapid resolution.[10] But the role of conglomerated ownership caused concern as networks announced repurposing arrangements that appeared to favor products from commonly owned studios. Not only did commonly owned broadcast and cable networks frequently develop these pacts, but repurposed series were often produced by studios also owned by the broadcast network or its conglomerate. To many in the industry, it seemed repurposing developed from the deregulatory policies of allowing conglomeration and eliminating the fin-syn rules. Nonetheless, the practice expanded in 2001–2002 as networks and production studios established repurposing deals for the WB series *Charmed* on TNT, NBC's *Law & Order: Criminal Intent* on USA, and FOX's *24* on FX.[11]

The transitional strategy of reallocating content between broadcast and cable networks also related to repurposing and the conglomerated ownership of broadcast and cable networks. "Reallocation" involves shifting shows developed for broadcast networks that failed to find commercially sufficient audiences to cable networks, where they might reach niche audiences and be considered more successful. The opportunity to shift programming to a commonly owned cable network allowed the conglomerate to amortize development and production costs that it would otherwise have to absorb and provided an important distribution option. Reallocation consequently decreased the risk inherent to program development; it also encouraged programmers to pursue shows that would otherwise be deemed too uncertain by offering an additional opportunity to recoup production expenditures. In rare occasions, programming developed for cable has aired as a special event on broadcast networks or during otherwise rerun-heavy summer schedules, as in the case of special episodes of Bravo's *Queer Eye for the Straight Guy* aired by NBC and the reallocation of USA's *Psych* on NBC in the summer of 2006.

Common ownership among production studios, broadcast networks, and cable channels facilitates the reallocation of programming, but was not a requirement for the practice. In the early 2000s, cable channels owned by conglomerates also owning broadcast networks often operated in a manner similar to baseball's farm team system, as a space to test boundary-pushing or niche-focused content or as a venue for demoting struggling programs to amortize their costs. In other cases, networks used reallocation to provide additional promotion for underperforming content or to recover the cost of particularly expensive productions.

By 2005, countless examples existed of series cancelled by broadcast networks whose unaired episodes were shipped over to cable channels and aired with little promotion. Common ownership was often a factor in this practice, as in the case of the critically lauded *Boomtown*, which was produced by NBC Studios and originally licensed by the NBC network, but was then reallocated to the commonly owned cable network Bravo. But common ownership was not always essential, as the example of *Pasadena* demonstrates. In 2001, FOX cancelled this show after just four episodes; four years later, SoapNet, a channel owned by Disney, resurrected the Columbia TriStar production for an airing of all thirteen existing episodes, presumably basing its decision to do so on the quality of the show, its fit with the network's brand, and the intervening success of similar shows (*The OC, Veronica Mars, Desperate Housewives*).

Repurposing and reallocation provided an important initial disruption in norms, but quickly became insignificant as outlets for programming expanded beyond linear broadcast networks and cable channels. For the most part, repurposing simply allowed viewers a second chance to watch a program, a practice that became unnecessary as DVRs reached broader penetration and VOD offered the ability to screen multiple episodes at will. As such, these strategies may be seen as more characteristic of the multi-channel transition than the post-network era. As nonlinear programming norms emerged as characteristic of the post-network era, cable channels' reliance on broadcast network content became less beneficial and even deleterious, as airing more generally targeted, broadcast-originated content prevented more brand-specific content development.

A somewhat different form of reallocation emerged in cases where networks and channels cancelled shows and outlets considered a better fit for the content and audience scope emerged to take over the license fee: *Southland* (NBC to TNT, 2010); *Damages* (FX to DirecTV, 2011); *Cougar Town* (ABC to TBS, 2013).[12] Rather than the scheduling and amortization ploy of earlier reallocation efforts, these are better understood as a variation on the development process in which the second network decided that the creative property was a good fit for its brand and its initial run likely would bring greater audience sampling than would an entirely new series.

Repurposing and reallocation were linear solutions that marked an initial disruption, but prove to be but footnotes in the broader story of post-network distribution. Notably, though, the practices of original run repurposing and reallocating programming among networks revealed the importance of conglomeration to many emergent distribution practices, especially since it was mainly, if not exclusively, the commonly owned entities that were most able to experiment with new methods of delivery. Both repurposing and reallocation developed in response to shifting industry economics introduced by audience fragmentation among cable channels during the multi-channel transition, a development that decreased broadcasters' dominance and profit margins. As audience size diminished, the networks sought multiple revenue streams to maintain budgets and recognized that viewers needed more opportunities to view in a context of such programming abundance. Adjustments in distribution practices such as repurposing and reallocation provided initial post-network strategies for cost savings, generating new revenue, and taking greater risks.

DVD: Own All the Episodes of Your Favorite Series

The next industrial development that adjusted conventional distribution windowing and revolutionized the possibilities for profiting from content resulted from the DVD sell-through market; however, as in the case of repurposing and reallocation, DVD sales provided an initial post-network distribution form that was soon overshadowed by VOD and broadband-delivered streaming services. In the early 2000s, the success of full seasons of shows packaged on DVD surprised many in the industry. By 2005, DVD sales of television shows reached $2.6 billion and accounted for nearly 20 percent of the overall DVD sales market.[13] For a popular television-on-DVD series such as *24*, the nearly three million DVDs of the series purchased by 2006 were equivalent to the DVD sales of a movie that earned $50 million from ticket sales; 24 generated $72.1 million from its first three seasons of DVD sales by 2005.[14] High sales were even possible for less popular series, such as *Buffy the Vampire Slayer*, which— though never a top Nielsen performer during its network run—earned $123.3 million on sales of its six seasons of episodes by the end of 2004.[15]

For audiences, DVDs conveniently aggregate multiple episodes—unlike VCR tapes, which can include only two or three episodes. Likewise, DVDs are commonly sold in complete seasons that require limited shelf space, which makes them attractive to fans who want to create libraries, to new viewers who seek to catch up on previous episodes, and to anyone who wishes to avoid television conventions such as commercials and one-week gaps between episodes.[16] But once services such as Netflix offered the alternative of streaming episodes instead of renting DVDs and as studios and MVPDs made VOD offerings more robust, these viewing opportunities free from the hassle of any physical media form quickly replaced DVDs for many. Certainly, owning a physical form was important for collectors and those wishing to view content multiple times; but for the many who just wanted to watch a show once, DVDs were comparatively costly and cumbersome.

From an industry standpoint, DVD sales provided a new revenue window for successful shows as well as new economic support for boundary-defying ones that did not succeed in their original airing. DVD distribution—and later the ability to sell rights to subscription services—also enabled studios to recoup production costs on shows unlikely to be distributed to conventional subsequent windows such as cable channels and broadcast stations. This was the case with the creative and innovative 2003 FOX series *Wonderfalls*, which failed to find an audience quickly; the network cancelled the show after four airings, leaving nine episodes unaired. The series' studio, 20th Century Fox, later released a DVD set of the thirteen episodes in February 2005 and sold 25,000 copies in two weeks, rewarding fans with some narrative closure and the studio with added revenue.[17] Similarly, the fan favorite *Firefly* sold 500,000 copies of the complete series of fourteen episodes less than two years after its release. FOX cancelled the series after just eleven episodes in 2002.[18]

The bigger DVD story for FOX/20th Century Fox was the case of *Family Guy*, a series the network aired from April 1999 through April 2002. FOX decided to take the series back into production and began airing new episodes in 2005 after the unexpected performance of the DVD, which ranked as the number-two single-season television DVD release as of May 2005, and the sizable audiences drawn by the show's syndication on Cartoon Network.[19] After receiving the request for new episodes, *Family Guy*'s creator,

Table 4.2 . Key Developments in Digital Distribution of Television

2005 June	YouTube launches.
2005 October	iTunes announces $1.99 downloads.
	Apple makes a deal with ABC to start streaming episodes of *Desperate Housewives* and *Lost*.
2005 December	"Lazy Sunday" video from *Saturday Night Live* gets 1.2 million views in ten days on YouTube.
2006 May	ABC starts streaming ad-supported episodes on its own website.
	CBS launches its InnerTube service (but the original incarnation does not include prime-time episodes, just auxiliary content).
2006 June	*Lonelygirl15* premieres on YouTube. The show gets half a million viewers within forty-eight hours, and although it was revealed to be a hoax in September 2006, the show aired through August 2008.
2006 August	CBS starts offering some prime-time episodes on InnerTube.
2006 September	NBC begins to offer select episodes for streaming.
2006 October	Fox on Demand service is launched through limited affiliates and on MySpace. It initially offers only a few shows, is separate from the main Fox.com site, and tries a transactional model.
	Google buys YouTube for $1.65 billion in stock.
2007 January	Netflix starts streaming television shows and movies (PCs only; Macs in 2008).
2007 May	"Charlie Bit My Finger" becomes the first amateur YouTube video to spread widely.
2007 July	ABC makes episodes available on a mobile platform and upgrades its player to allow for high definition.
2007 October	Hulu (originally a joint venture between News Corp and NBC) launches in beta mode.
2008 January	Netflix offers unlimited streaming plan.
2008 March	Hulu becomes available publicly in the United States.
2008 September	Tina Fey returns to *Saturday Night Live* to play Sarah Palin; the clip posted to NBC.com becomes the network's most-watched video.
2008 October	Netflix streaming becomes available on Macs, TiVos, and Samsung Blu-ray players.
2008 November	Netflix begins HD streaming.
2009	Hulu becomes culturally relevant. It doubles its content library, adds Disney as a partner, and by October 2009, has over 855 million video views.
2009 April	Susan Boyle video spreads internationally. Nine days after the *Britain's Got Talent* airing, her performance had over 91.6 million views, while overall views for all of her performances had surpassed 103 million hits.
2010 February	HBO GO launches.
2010 April	Apple releases first iPad.
2010 June	Launch of Hulu Plus for $9.99 a month. Price is reduced a few months later to $7.99 a month.
2010 September	Beginning of the "Netflix Surge." Netflix begins offering streaming-only service in Canada in September; announces that a similar plan will be available in the United States by the end of 2010.

2010 October	Showtime launches its Showtime Anywhere app.
2010 November	Comcast releases its Xfinity TV app for the iPad.
2011 March	Time Warner launches the iPad App with thirty-two channels that allows TWC subscribers to watch live streams of the channels. Immediately faces backlash from Viacom, News Corp, and Discovery. Removes eleven of the channels two weeks after launch, but then fights for the right to include them.
2011 April	Cablevision offers Optimum App for iPad. Subscribers download it 50,000 times in four days; various content holders demand to be dropped and file legal action.
	Time Warner and Viacom file lawsuits against each other.
2011 June	Viacom files lawsuit against Cablevision for the Optimum App.
2011 August	Viacom-Cablevision lawsuit settled. Viacom channels remain on Optimum App.
2011 December	Louis C.K. nets $1 million in three weeks on $5/download direct sale of Beacon Theater performance.
2012 May	Time Warner Cable–Viacom settle disputes. Time Warner is allowed to include the Viacom channels in dispute on its app.
2013 February	Netflix releases *House of Cards*.
2013 May	Netflix releases fourth season of *Arrested Development*.

Seth MacFarlane, noted, "The DVD market barely existed when we were cancelled. But now, fans can protest the cancellation of a show with their wallets, buying DVDs, rather than just writing letters to the network. . . . It's completely changed the economic model."[20] Jeff Zucker, then chief executive officer of the NBC Universal Television Group, acknowledged in 2004, "The numbers are already affecting how some shows are developed."[21] DVD revenues could be an enormous boost for shows; DVD earnings for the early seasons of HBO's *The Sopranos* were significant enough that the studio recouped the entire cost of producing those seasons through DVD sales alone.[22]

An important aspect of DVD sales was that top sales of DVDs often did not mirror the top performers in original airing. The CBS series *CSI* spent much of the early 2000s as the most-watched scripted show and drew large audiences and fees in syndication, but produced lackluster DVD sales; this was also the case for shows in the *Law & Order* brand.[23] DVD release provided economic support for program forms such as serials and cult hits that were often marginalized by standard syndication practices in which episodic programs earned premium rates because they drew more substantial audiences and offered scheduling flexibility to stations. This new way for audiences to view and for studios to monetize content was crucial in changing industry lore about the features of commercial content and contributed to enabling the more serialized and complicated narratives that began emerging. Long-touted industry lore claimed that most viewers saw only 25 percent of episodes for a series, and network decision makers used this lore—created in the network era, no doubt—to justify the hegemony of episodically structured television. The emergence of the distinction of prized content is unquestionably tied to the new ways of viewing that DVDs first allowed and to studio and network attempts to maximize the increased storytelling possibilities of this distribution form.

Studios also sought to make the DVD purchase an additionally attractive proposition for audience members by including special features that are available only on the disks. The series *24* filmed a brief sequence that occurred between seasons 3 and 4 for the season 3 DVD set, and the DVDs were the only place fans could see this bit of narrative. Lisa Silfen, senior vice president of program enterprises at MTV Networks, noted the value of special features and extra content: "It's a great opportunity to give the viewer that added value—what got left on the cutting-room floor, what we didn't have time to put on the air, extra photos, contests, games—all different things to create that package for them."[24] Even more conventionally structured series such as *Friends* provided different content versions in their DVD release. Though the "extended versions" weren't significantly different—often including scenes and footage trimmed due to time constraints or subtle network censorship—the release of content different from that widely available in syndication could be a selling feature for devoted fans.

Regardless of the added content, many viewers indicated that convenience was of a higher order of importance than the bonus footage as the television season DVD ceased to be the primary way to view out of synch with the linear airing, and became instead an archival or collector's mechanism. Though DVD sell-through hasn't provided the ongoing economic boost studios may have hoped, it played a crucial role in helping the industry realize viewers' pent-up demand for alternative viewing experiences. Experiments with DVD release also allowed the networks and studios an initial, comparatively controlled experiment in selling programming directly to the viewer before the possibility of the sale of individual episodes more significantly disrupted distribution practices.

Content on Demand: VOD, Downloading, and Broadband Streaming

In the mid-2000s, myriad opportunities to distribute content electronically on demand seemingly developed overnight, revolutionizing distribution practices and considerably advancing the development of the post-network era. How networks and studios deliver content to viewers—over the air or through a wire connected to a television or a computer—quickly became unimportant relative to viewers' ability to access what, when, and where they wanted. A key characteristic of advances in distribution such as VOD (on-demand offered by MVPDs), downloading (retail sites allowing permanent ownership such as iTunes or Amazon), and broadband streaming (Netflix, Hulu) resulted from their intangibility—the viewer never accessed a disk or hard copy of the content—and therefore didn't need to go to a physical store to purchase it or even await its arrival by mail.[25] These distribution windows consequently reduce production and distribution costs, which are, of course, critical concerns in the industry's economic models. Further, the elimination of limitations such as retail shelf space and sizable reduction in manufacturing and transportation costs makes a range of content that was previously outside the market, what Chris Anderson calls the "long tail," an increasingly valuable component of the industry and one that requires reconsideration of development strategies.[26] Despite the reduced production costs and expanded markets, issues of content rights, licensing, and concerns about the implications of superseding traditional windows have prevented the technological possibilities now available from being as significantly realized as will likely come to pass upon the full establishment of post-network-era norms.

The video-on-demand (VOD) capabilities introduced by MVPDs arguably marked the first step into the content-on-demand world for the established television industry. Cable systems identified VOD as a strategic enhancement that offered added value to their subscribers and provided the cable systems themselves with a competitive advantage over their satellite challengers, which were technologically limited from providing as robust on-demand offerings. Cable VOD services experienced a significant increase in 2005 in both use and frequency of use, as 88 percent of homes with the service used it in 2005, an increase from 65 percent in 2004, while 53 percent of those viewers used VOD at least once a week, compared with 24 percent in 2004.[27] But even by 2006, the content available for free VOD remained limited and largely isolated to films rather than series. Until 2011, MVPDs primarily offered just short-form versions of content available on existing networks, such as extra footage and cast interviews as well as some low-budget content including fitness, education, and niche interest fare. MVPDs and first-run rights holders initially tried to force viewers wishing to use on-demand capabilities to catch up on current series to pay per-episode fees in the limited cases such content was made available. Difficulty monetizing content through advertising slowed VOD development. The systems for loading content were technologically cumbersome and advertisements had to be embedded within the content, while no mechanism existed to "count" the VOD audience.

Subscription services led experimentation with full-length on-demand offerings. By 2002, HBO and Showtime realized the particular value of on-demand capability for fee-based services and offered the feature, although not all cable systems carried it. Creating on-demand opportunities reduced the frequency with which viewers had the experience of finding there was "nothing on" when turning to the service, which had been a perennial complaint and cause of "churn"—the industry's term for the canceling of subscriptions; and premium channels didn't have to worry about convincing advertisers to fund their experiments. On-demand portals allow access to a constantly rotating slate of films and original series that enable viewers to time-shift their viewing of these networks. Advertiser-supported networks remained wary of risking the commercial skipping likely to result from making their programs available on demand, and all were aware that such a system would introduce even greater complications to the already challenging task of Nielsen measurement of time-shifting homes. Nonetheless, as VOD adoption and use grew, networks approached licensing deals with attention to enabling some on-demand distribution, particularly as a promotional tool that might lead more viewers to use VOD to catch up on series and then join the linear viewing audience. Negotiations involving commonly owned content again proved most flexible in many cases.

As late as 2011, many still regarded MVPDs' VOD platforms an incredible failure, particularly as emerging broadband streaming services such as Netflix and Hulu drew far greater attention for their "on-demand" services. This began to change as MVPDs struck deals with original license fee holders (the networks and channels) who had secured allowances from content owners that made full episodes of recent series as abundant as the back catalog and catch-up offerings of broadband-delivered services. For example, in 2013 the "Watchathon Week" offered by Comcast—which aimed to move more subscribers into tiers offering access to its Xfinity on-demand service—provided viewers access to "more than 3,500 episodes of 100 TV series" on multiple screens and devices.

Two types of VOD emerged, and though many viewers might not perceive a distinction between them, it is important to understand these as two industrially different practices. The first strategy used VOD as a promotional tool to expand the linear audience for the show. By 2014, most networks enabled a limited number of episodes for free streaming through MVPD-provided on-demand, network websites, and Hulu's basic platform. In most cases, between three and five of the most recent episodes were available, though no episodes might be available if the series was not currently airing. Though a viewer certainly could view a new episode each week through this type of

VOD, the limited number of episodes generally discouraged streaming as a replacement for linear viewing or recording; in other words, viewers couldn't build up a cache of episodes and view the whole season consecutively as Netflix viewers had grown accustomed to doing. Alternatively, some services provided the full past season just as a new season launched. This too served a promotional function of enabling viewers who hadn't found the series the previous season to catch up and join the current season. Using on-demand distribution for promotion primarily benefitted the original license holder (network or channel), though the studio arguably benefitted indirectly since on-demand viewership could indicate broader interest for the show.

The second VOD strategy was that of library building. In this case, MVPD VOD as well as subscription broadband-delivered services created a new distribution window and licensed content from studios. This allowed viewers access to "back-catalog" or "library" content, perhaps all the episodes of a show no longer in production. Shows that were not in demand for linear cable syndication could even appear here while still in production. As an experience, VOD library content is similar to watching an "old" episode in syndication on a cable channel, though may be more valued by viewers because they can watch episodes on their own schedule and in the order they desire.

Though the average viewer may not distinguish between streaming an old episode of a show on Netflix and a newer episode of the same show on Hulu or the network's website, the different economics that allow the show to appear in these different VOD platforms distinguish these as different distribution strategies. The networks are able to offer recent episodes on VOD as part of their original license fee agreements. If there is any commercial revenue from advertisements in these windows, it is shared by the player (MVPD; Hulu) and network. In the case of library content, selling the broadband license provides the studio that created the series with a new syndication window to help amortize costs or derive profits from programs created at a deficit, while none of the fee is likely to go back to the network originally licensing the series.[28]

It is in many ways too soon to have a clear vision of the success of using video on demand for either promotion or monetization. Even in 2013, the commercial insertion capabilities in free streaming players such as Hulu went un- or underutilized. In the case of MVPD on demand, MVPDs could disable fast-forwarding through commercials in VOD playback, but in my experience, rarely did so until 2013. Moreover, most VOD and broadband distributors offered much lower commercial volume, often featuring pods with a single advertisement that merely promoted the channel that licensed the show being viewed. Evaluating the monetization of on-demand was still

Figure 4.1. Screen shot of current (2013) on-demand interface.

difficult because it remained unclear how the availability of programs in this earlier window would ultimately affect the viewing and rates of subsequent, more traditional syndication windows. Of course, it was also unclear that these traditional windows would persist as the transition to the post-network era continued to develop—for example, would viewers choose to watch a linear episode if they had not seen previous episodes when they also could select multiple previous episodes of that same program from a VOD interface? Also, the phenomenon of a so-called Netflix effect began to be noted, as *Variety*'s editor in chief, Andrew Wallenstein, posed the query, "Should Netflix be paying studios for content or the other way around?" in the wake of *Breaking Bad*'s 5.9 million viewers for its final season premiere.[29] The effect Wallenstein notes is the previously unprecedented season-to-season growth experienced by cable shows that had past seasons distributed on Netflix, through VOD, or even full marathons on cable channels' linear schedule that enabled a viewer who missed prior seasons to catch up with a show's past and then become a linear viewer. *Breaking Bad*'s final season premiere doubled the audience of the previous season, and given its highly serialized narrative, it seemed unlikely that final season viewers tuned in without viewing past seasons.

In some notable cases, networks have limited on-demand availability much more stringently. FOX's *American Idol* remains the best illustration of this strategy, in which the network clearly seeks to maintain this show's cachet as a live contest by not making it, or its easily disintermediated performances, available through any other outlet. If it weren't for the

Figure 4.2. Screen shot of next-generation Xfinity X2 interface.

live competition component as a primary reason for audiences to view live, FOX's ability to suppress viewer demand for alternative ways to access the hit show would likely be far more difficult. In other words, FOX can use this tactic due to the particular features of this show, but would unlikely be successful with such a rigid rights clampdown in the case of a show with an ongoing narrative and with less motivation for immediacy.

Though a range of efforts by MVPDs and new aggregators made it far easier by 2014 for viewers to create a nonlinear television experience without resorting to illegal file sharing, significant problems remained. First, on-demand practices were highly inconsistent. Viewers could not expect any reliable on-demand access, as there was significant variation in the number of episodes available, when they became available, and how long they would remain available. To some degree, common network strategies could be discerned, but variation even within a network's strategy existed to an extent that made on-demand an unreliable repository. This variability enabled the DVR to persist in importance by allowing greater viewer control over when and how to view.

A second major limitation of on-demand derived from the poor interface of many MVPD systems, which were so cumbersome as to deter viewers from attempting to use them. MVPDs were very slow to bring visually intuitive interfaces with sophisticated search capability to market—in this case, again being shamed by new competitors such as Netflix into innovating in this way. The weakness of the interfaces required viewers to deliberately search for specific content rather than pushing content consistent with what a viewer has liked—or even just general content—as a suggestion. Anecdotally, I had mostly given up on my Comcast-provided on-demand service after repeatedly finding a paucity of offerings. By the fall of 2013, after taking what I experienced as a lot of time to flip through the seemingly endless and poorly arranged interface, I found the offerings far more expansive, but still rarely used it (see figures 4.1 and 4.2). My behavior began to change once Comcast began "promoting" series by placing particular series among the offerings accessible on the first page of its interface. Gradually I began to think to check the on-demand offerings and later found that a much more robust selection with a much smarter interface could be used if I streamed shows through the Xfinity website. Here I had exceptional choice, but not the large screen of my living room set or a smoothly streamed experience.

Despite these problems, some suggestions of "best practices" also emerged. From its launch in 2010, HBO's HBO GO application received rave reviews from users; the only complaint—and one echoed widely—was that it wasn't available without a linear HBO subscription. HBO GO is an application that required authentication of HBO subscription and provided access to all episodes of all HBO series, original films, and many of its licensed theatrical films on laptops, tablets, and mobile phones. Moreover, it featured a visually rich and easy-to-navigate interface. In many ways, HBO GO could be seen as the model for a nonlinear, post-network future: content was predictably and perennially available, and viewers could watch on a variety of screens. The notable problem was that in terms of rights and economics, such a platform was very difficult to reproduce outside the very particular context of HBO as a subscription service and as an entity that almost always operates as producer and distributor of its original content. As in the case of spurring earlier on-demand development, HBO's status as a subscription service enabled it a freedom from advertiser involvement and security of economic model that allowed it to innovate with much less risk. Further, uncommon for most networks and channels—either advertiser- or subscription-supported—HBO produced its original content, and thus was the rights owner, which freed it from complicated rights negotiations, even if only with another entity within the same conglomerate.

HBO GO was so beloved by users that it spawned a website entitled Take My Money, HBO!, a consumer campaign to encourage HBO to make HBO GO available as a stand-alone app that viewers could pay for without paying for a linear subscription. When the website launched, in 2012, 140,000 supporters signed up in forty-eight hours and responded to its creator Jake Caputo's request to tweet out what amount they would pay for a monthly HBO GO subscription. The issue for many of the petitioners was that many MVPDs required viewers to at least purchase a basic digital tier of programming in order to also subscribe to HBO, and though marketed as "basic," this tier typically cost a minimum of $20 per month. HBO earned considerable goodwill among MVPDs for creating a mechanism that didn't circumvent them—although this strategy was in HBO's interest as well. The company depends on MVPDs to manage subscriptions and market its services; also as a cog in the broader Time Warner enterprise, HBO is wary of the ramifications an action adversarial to MVPDs could have for carriage of Time Warner basic cable channels such as CNN, TNT, and TBS. Importantly, despite the popularity of HBO GO among some subscribers, the industry journalist Joe Flint reported in 2013 that only 4.5 million of HBO's subscribers registered for the HBO GO service and 98 percent still consume HBO on traditional televisions.[30]

Notably, this remains a highly dynamic area of television reconfiguration. By the time the first edition of this book went to press in January 2007, Netflix was still mostly seen as a competitor to the video rental industry; Hulu appeared as an interesting experiment, though one widely expected to fail; and YouTube, just a year past its late 2005 launch, streamed about thirty-five million videos a day and drew an audience of more than nine million people per month.[31] The buzz at that time was instead about various efforts by the established television industry to distribute content through websites, such as Motherload (Comedy Central), Overdrive (MTV), and Pipeline (CNN), and endeavors from outside the conventional confines of the industry, such as AOL's In2TV and iWatchNow.com. In just seven years, the landscape of broadband content delivery has shifted enormously, and it remains too soon to predict how much of the current configuration is likely to remain in place by 2020. Radical adjustment is not out of the question, though the disruption is more likely to come in the form of new distribution infrastructure, such as Google's Fiber initiative, or through cellular capacity currently only imaginable in the United States.

Consequences of Nonlinear Distribution

I view YouTube as a glimpse into the future of video distribution, completely untethered from media companies and linear distribution models based on schedules. I don't think it's a flash in the pan as a concept, but rather it opens the door to a landscape that allows consumers to be content providers, creating a new form of community particularly of common interest.
—John Lansing, president, Scripps Networks, 2006[32]

If the universe changes and they [viewers] want us to bring the content directly to them, then we can.
—Les Moonves, president and CEO, CBS, 2012[33]

What we'll see in the Internet is most cable networks will become Internet networks—we'll still call ESPN a cable network, but it'll be mostly delivered over the Internet in 10 or 20 years. The fundamental advantage of the Internet is individualization, control, being able to watch on any screen. It's just a much better technology substrate for video.
—Reed Hastings, CEO, Netflix, 2013[34]

Lansing's words were prescient when he made these remarks in 2006, and here Moonves and Hastings grapple with not only the present, but the future of video distribution. Indeed, part of what enabled YouTube to grow so exponentially as a video provider was the freedom from commercial interference that derived from its base of amateur creators. But as viewers began using applications such as HBO GO, it became clear that unshackling television from the tethers that had been central to its organization throughout the network era and multi-channel transition posed the greatest challenge to incumbent distributors. Changes to the structure of network-era distribution windows radically disrupted established norms of how studios and networks profited from their content and required similarly radical shifts in the economics of the industry. The industry hesitantly experimented with new distribution practices, particularly in the aftermath of iTunes' 2005 establishment of a transaction streaming business, largely out of the sense that they could not risk being left behind. There was little certainty about short- or long-term economic consequences of enabling transaction purchase, nor was it certain that the traditional windows would persist as nonlinear viewing expanded. While conventional wisdom forecast likely outcomes, such as a decrease in the value of later distribution windows from the use of these new early ones or a cannibalizing of the linear audience, such concerns weren't borne out. For example, one unanticipated result of DVR deployments was that the homes that used them actually watched more television, not only because of time saved through commercial skipping, but also because viewers could more easily and effectively access content of interest. The innovation posed by these new distribution windows was unprecedented enough that they too were likely to produce unexpected results, as early evidence suggested. Just as some industry discussion posed mobile, theatrical, and convenient television use as opponents in a zero-sum competition for viewers, an equally viable outcome was expanded use of television across all of these technologies.

The availability of on-demand distribution has only begun to affect the rules of financing and distribution that dominated the industry for more than fifty years. In testing new distribution methods, the networks and studios have been uncertain about the consequences for subsequent traditional windows such as syndication, but then syndication as it was known in the network era and the multi-channel transition seems likely irrelevant in a post-network era. Syndication has been used to fill out linear schedules, and this seems a strategy that may serve a residual linear audience, but one less likely to persist as more gain ready access to rich, well-organized libraries of content provided by subscription services like Netflix or by MVPDs. In an environment in which networks and channels do not need to "fill" a twenty-four-hour schedule, they are likely to have less need for programming associated with another entity, such as the network that first licensed it, for example, USA as the place to go to watch *Burn Notice* and other programming original to the channel instead of *Law & Order: SVU* and other shows that originated on a broadcast network.

It is already the case that cable channels derive greater competitive advantage by focusing funds on distinctive, original programming that matches their brand than by paying high license fees for shows already associated with a broadcast network; but the need to spread limited programming budgets to cover the artificial requirement of a twenty-four-hour schedule has necessitated purchase of such programming. Likewise, local stations might be more willing to use their budgets for producing original local fare instead of purchasing off-net sitcoms, and studios might distribute their content directly to viewers—as evident in channels built around U.S. studio content available outside the country. For example, the Warner Channel in Latin America features *Friends*, *Big Bang Theory*, and *The Vampire Diaries*—all series that were produced by Warner Bros. Studio, but that aired on various networks in the United States. Time Warner experimented with such a "channel"—albeit the post-network version of a channel—in 2013, when it launched the WB Instant Archive, as a subscription-based service accessing primarily its film archive, but also television series that were no longer in any form of syndication. Similarly, Sony has experimented with its advertising-supported site Crackle, which draws primarily from Sony-owned film and television properties. Such a move, in which the content rights holder distributes directly to the consumer—as technology now allows—provides a far more efficient economic model that could help consumers manage costs and allow creators greater revenue.

Just as distribution aggregators such as iTunes, Motherload, Overdrive, and In2TV initiated nonlinear viewing, YouTube, Netflix, and Hulu have changed the television experience. The nonlinear access to programming they provide plays an important role in changing viewers' experience with and expectations of content. Negotiating the chicken/egg conundrum of what comes first—content or viewer adoption—slows or thwarts the dissemination of many industrial innovations. For example, little HD programming was available for quite a long time because so few homes owned HD sets—perhaps much like the case of 3D and 4K/Ultra HD programming now—but viewers had little motivation to buy HD sets as long as there was so little content available. Early forays into broadband video, whether those already swept by the wayside or those fighting to dominate the post-network era, help viewers realize that new ways of experiencing video are possible and that there is nothing natural or inherent about the ways we've viewed until now. The fact that revising this chapter seven years after the original required extensive deletion and rewriting reminds me that the entities that are dominating broadband program delivery in 2014 may too pass from relevance before long. The details of particular services are consequently less important than understanding that, with the exception of live sports and contests, television viewing is steadily moving away from its linear origins.

All of these new distribution windows upset the long-existing norms that derived value from time delay and exclusivity. If the number of windows and the patterns content took through them became increasingly varied during the multi-channel transition, such variety has become even greater at the beginning of the post-network era. What windows a show will pass through and how much money it might earn depend upon the nature of the content (serial or episodic), the type of audience attracted (niche or mass), and the status of the producer relative to the multinational conglomerates dominating the television industry—with great diversity possible.[35]

Changes in Distribution to the Home

It's not lost on everybody that the market is shifting. It is inevitable that consumption of this form of entertainment television is going to be on the Internet. And there is a significant shift; . . . at some point this has to break, with or without Aereo.
—Chet Kanojia, CEO, Aereo, 2013[36]

At the same time that VOD and streaming increased viewers' ability to access an array of content on their own schedule, subtle changes in the technologies used to distribute that content to the home occurred that adjusted the dynamics of competition at the structural level. The arrival of cable inaugurated the multi-channel transition, and the establishment of a competitor to cable—namely, satellite—marked the maturing of this competitive environment. A decade after the Telecommunications Act of 1996 enabled telephone companies to compete in providing video and allowed cable to offer telephony, traditional telephone companies finally began to experiment with video service offerings. These additional competitors, known in the industry as "telcos," integrated phone, video, and data services by replacing the copper wire that the industry had used for voice transmission with fiber-optic lines that initially were far more robust than the coaxial cables used by the cable industry. As the first edition of the book went to press in early 2007, it remained unclear how quickly telcos such as Verizon and AT&T would make their product available or how significantly the added competition might affect the industry, but by 2014, it was clear that the telco competition was unlikely to provide significant adjustment. The telcos largely ceded major markets to the most established cable companies, and though their fiber optic DSL (digital subscriber line) was a superior technology when the telcos entered the market, it was "obsolete in comparison" with the wired speeds achieved by the cable companies using the DOCSIS 3.0 technology (data over cable interface specifications) released in 2006.[37] Just about 10 percent of homes with television received telco service in 2014, but both companies had functionally stopped trying to wire new markets, and the "competitors" Verizon and Comcast even established a joint marketing agreement in December 2011.[38]

The telco foray into the "cable" market proved too costly once it became clear that the DSL technological advantage was short-lived. In this same period, the explosion of mobile phone use and then smartphone technologies led these companies to realize that their established businesses in wireless provided a stronger asset going forward and warranted infrastructure development. The telcos had a stronger competitive position in the wireless world; as of 2012, AT&T and Verizon controlled 70 percent of the U.S. wireless market, and it is their infrastructure that many of those wanting their television "anywhere" rely upon once leaving home wired and wifi connections.[39]

Perhaps the most important aspect of the entry of Verizon and AT&T into video service was as a threat of competition in the early 2000s that may have helped encourage faster technical innovation than would have occurred otherwise, given the limited regulation and monopolistic structure of the cable industry. The chance that the fiber lines the telcos were building would provide notably faster speeds encouraged development of Internet protocol (IP) technologies that could be deployed over existing coaxial cable lines to allow the more robust VOD offerings of the 2010s. Without descending into too much techno-jargon, I'll note here that "IP" denotes a specific way of sending messages that involves breaking them down into packages that can then be conveyed separatcly—as is the case on the Internet. Where television is concerned, IP distribution is significant because it enables providers to transmit only the signal for the television channel that you want at that particular minute, in contrast to the method of sending all the channels to your home all the time, which had long been the technology cable used. If you have a provider using IP technology, and at 5 p.m. on May 3, you wanted to watch CNN, then CNN would be the only signal coming through the wire into that television. If you had conventional analog cable service, all the channels in your package would be coming through the wire, even though you could watch only one at a time. Although the difference may be imperceptible to the user, IP technology is more efficient and makes possible distribution beyond the bandwidth limitations that once restricted cable systems to roughly three hundred channels. The "Internet" part of the term IPTV has created confusion, since many casually describe Hulu and YouTube as Internet television as well. IPTV distinguishes the technical means for the distribution of a signal, which differs from "television" being distributed on the "Internet," or what is described here as broadband-delivered programming.

The more recent development in competition to the home is a small but potentially significant experiment that Google began in 2011 to rewire Kansas City, Kansas, with fiber to the home; the initiative was expanded to Austin, Texas, and Provo, Utah, in 2014, and the company was in talks with thirty-four other cities. The Google Fiber initiative promised to increase Internet speeds over a hundred times that being offered by MVPDs, and also developed a package of cable channels to more directly replicate the offerings of MVPDs. As the telcos had a decade earlier, Google threatened competition and a technologically superior offering, though, by initially seeming to mirror the MVPD model of channel packages, Google may have entered the market as a wolf in sheep's clothing. Competitive primacy in the post-network era would depend on controlling the "pipes," and the pipes Google provided suggested that the company anticipated a future of video distribution that more radically disrupted the existing linear model that MVPDs still prioritized and that it initially appeared to replicate. In an unbundled environment, Google's gigabyte service would be a valuable offering. The industry analyst Craig Moffett also speculated that Google's goal wasn't to take over the MVPDs' business, but to disrupt the slow pace of innovation in markets lacking competition to spur the development of high-speed networks in all markets that could be the backbone for many Google products and services.[40]

In addition to fearing new competitors such as Google Fiber, cable systems worry about "disintermediation," or separating the content and the delivery system in a manner that could allow programmers to bypass cable operators—and even broadcast networks—and go straight to the consumer through the Internet, a development more widely referenced as going "over the top" (OTT). The reality of this concern grew considerably by 2010, by which point compression and delivery technologies had solved many of the problems such as long buffering times and grainy, pixilated images that plagued early Internet video distribution. By 2010, the possibility of a new era of television distribution was being realized as providers and applications such as Hulu, HBO GO, and Netflix offered full-length, commercially produced content at a level of quality indistinguishable from conventional television—at least on the more personalized screens of laptops, the soon-to-market tablet, and even broadband-connected living room screens. It was at this point that concerns about viewers going over the top became so cacophonous in the trade press that one would have imagined the audience was deserting MVPDs in droves. The concern was enough for the longtime analyst Moffett to assert that cord cutting was "perhaps the most overhyped and over anticipated phenomenon in tech history."[41] Though a new breed of television household did begin early adoption of new distribution opportunities in this period—what Nielsen would dub the "zero-TV" home when it began measuring them in 2013—homes accessing television only through broadband were actually a very small segment. Some of the decrease in "television homes" that fueled the OTT anxiety could be explained by contraction in "homes" as the United States hit the bottom of the housing crisis that led many to lose homes and consolidate living with extended family and friends, as well as those who maintained their homes but eliminated cable due to its cost.

The anxiety about a potential over-the-top revolution that permeated industry discourse from 2010 to 2013 was not unfounded, but needed to be contemplated with far more nuance than was common. First, though viewers might cut or reduce their cable subscription, they typically maintained the same broadband service provider. Because 80 percent of cable subscribers received broadband from the same MVPD that provided their cable service, the consequence was that most of these viewers simply stopped paying the cable portion of their MVPD bill.[42] MVPDs' success in signing subscribers to "triple play" bundles of cable, broadband, and home phone services through the mid-2000s paid off as the providers could shift relative fees around so that they were making enough revenue from the broadband service to diminish the cost of losing a content subscriber. Alternatively, they constructed pricing schemes that made dropping the cable service financially disadvantageous—for example, raising the fee for Internet-only service above that of a combined Internet and cable bundle. Susan Crawford reported in 2013 that Comcast's fees for stand-alone data—in other words, a subscription to Internet access but not cable—was twice as expensive than if packaged with cable.[43] Though MVPDs could protect themselves from a loss of content subscribers by increasing fees on broadband, the loss of sub-scribers would have much greater ramifications for the cable networks that received carriage fees based on the number of subscribing homes and sold their time to advertisers based on the number of homes they reached.

The aspect of the OTT fear that was well-founded was concern about who the zero-TV household tended to be. In their first detailed evaluation of this group in 2013, Nielsen reported that zero-TV households had grown from about 2 million homes in 2007 to 5 million homes in 2013—out of a televi-sion universe of 115 million.[44] Of these 5 million homes, 75 percent still had at least one television set and 48 percent watched television content through a subscription service (Netflix; Hulu Plus). Almost half the zero-TV homes were under the age of thirty-five, which was the demographic feature that inspired particular concern. What also drew attention was that over the four quarters of 2012, Nielsen saw a break from the tendency of heavy stream-ers to also be heavy television viewers. A new behavior of heavy streamers who watched little television began emerging and began to suggest stream-ing as a replacement behavior, rather than just supplementation.[45] Without a precedent for this type of behavior, it was difficult to forecast whether this phenomenon largely spearheaded by young adults was a function of age, earnings, and life stage or whether this approach to content would persist as they proceeded to higher earnings and through subsequent life stages. (Of Nielsen's sample, 36 percent of zero-TV respondents noted cost and 31 per-cent noted lack of interest as the reason they didn't subscribe to cable.)

As this book goes to print, it is difficult to diagnose the long-term ramifica-tions of changes in the behavior of a particular audience group in the short term. In the end, it may not matter whether the young households of zero TV would have been "cord nevers" (those who would never tether themselves with a cable cord), because the threat of this possibility—along with techno-logical innovation and the daring of a few industry leaders able to see a future of television distribution that differs from its past—may be enough to por-tend industry-wide change. Experiments in new models of distribution have begun to emerge to an extent that it seems clear that the days of the tiered bun-dles of cable packages are numbered. Though this form of packaging content will likely persist for some for a long time to come, evidence suggests we've reached a tipping point from which this type of use will grow steadily residual.

But what is to replace it? Though MVPDs have been impugned for only selling content in bundles, the blame is shared with—and maybe even more deserved by—the owners of cable channels that have forced chan-nels unwanted by MVPDs and subscribers as the cost of access to desired content. For example, Viacom knows that every MVPD wants to offer MTV and Nickelodeon, but in order to get these channels, Viacom requires that the MVPDs also carry the ultra-niche channels Palladia and NickToons—and maybe even requires that they be bundled on the same tier. Despite the hundreds of channels, nine companies controlled 90 percent of professionally produced television in the United States in 2013.[46] This consolidated control of programming left even the MVPDs with little room for negotiating as these companies demanded higher retransmission fees. The bloated bundles have left viewers feeling that their subscriptions lacked value because it seemed they were paying for things they didn't want and had no recourse against. By many measures of leisure expenditure, however, the cable packages, especially those including robust VOD, TV Everywhere functionality, offer considerable worth. As content creators seek ways to maintain access to zero-TV homes by making content available outside the bundled cable tier, viewers will be able to assert their displeasure with bundling more directly.

The arrival of new distribution technologies and the merging of once distinct pathways of cable and Internet into the home have introduced many new questions for telecommunication policy. A yet unresolved debate about the potential for and ramifications of "net neutrality" persists, and questions about how broadband will be priced and whether and how MVPDs will impose usage caps remain pressing concerns that could significantly alter the future use of television. The dawn of the post-network era has largely featured uncapped home broadband plans so that users can upload and download as much as they desire at a particular speed through home Internet connections. The quick expansion of Netflix streaming in late 2010, however, quickly drew the attention of MVPDs, who, perhaps seeing an even more lucrative economic model, began wringing their hands and suggesting an inevitable coming broadband shortage that would be wrought by these "broadband hogs."

As the only pipe into most homes, MVPDs had secured a deep arsenal to maintain their profits regardless of industry reconfiguration: if viewers cut cable subscriptions and shifted to broadband-delivered programming, they could make up the lost revenue by requiring a broadband package that would allow the expanded usage. Or, if they wanted to maintain some cut of programming, they could partner with a broadband service for a percentage of the viewer subscription in exchange for exempting content from that service from caps. For example, if a new subscription video service such as Warner Instant Archive launched, it might partner with the MVPD and give the MVPD a percentage of its subscription fee in exchange for use of Warner Instant Archive not counting against a subscriber's broadband usage cap. In 2014 Netflix joined other video providers in paying MVPDs to connect directly to their servers to prevent network congestion resulting in slow buffering and other poor experiences for viewers. This arsenal largely had been built through the advantages provided by their monopoly status as the only ample pipe into most homes at the dawn of the digital era. Indeed, the introduction of meaningful regulation or competition had the potential to significantly alter the industry at any time; but with the exception of Google's Fiber initiative, neither seemed remotely forthcoming.

Consequences of Changes in Distribution to the Home

One of the primary consequences of the new forms of video distribution now possible is the utter fracturing of the mass audience. In the network era, the limited spectrum and the regulatory choices made in its division created a marketplace of few providers. The expansion of cable and additional broadcast networks throughout the multi-channel transition enabled the emergence of niche television audiences, though it is less the technological capability of additional channels gained in the mid-1980s and more the financial stability of these channels by the early 2000s that allows wide-scale development of original series that begin to really adjust the norms of television content. The question that remains at this preliminary point in the establishment of the post-network era is whether niche, loosely synchronous audiences will persist as a norm, or whether the audience will fragment further yet to a state that it is more accurate to speak of television as a personalized medium— more like the "audiences" of books, though with continued rates of viewing that far surpass reading?

The challenge faced by various entities of the television industry is that the economics of industry practices remain built on mass-audience, network-era norms. The fundamental challenge to the future evolution of television is the disjuncture between an economic model built in the network era and distribution practices characteristic of post-network technological possibilities. As of 2014, a certain answer to this challenge remains unclear, though the existence of this dilemma has become inescapable. Many possible ways forward exist, but each bears consequences for some established entity that is unlikely to easily sacrifice the status it has enjoyed in the previous regime.

One early possibility emerged on Halloween 2006, when Fearnet, a "channel" featuring horror films from the Sony/MGM library, debuted on Comcast cable systems. Though Fearnet transitioned to delivery as a linear channel in 2010, its initial iteration as a VOD channel that made available horror-themed programming for viewers to watch on their own schedule suggested a notable innovation in distribution and the continued importance of a network brand. Even though the "channel" did not operate an outlet that streamed predetermined content at certain times, it did function as a branded folder in which viewers could look for programs with particular characteristics.

A truly post-network environment is precisely that, television without networks—or at least without networks in their current configuration. Program aggregators—those entities or locations in which viewers can find programming of a certain sensibility or about a certain topic—will remain crucial, but the future of networks as aggregators that schedule the delivery of programming at certain times appears dubious. Post-network practices in which the viewer's pursuit of content dominates the process of selecting what to view increases the value of studio or producer reputation and diminishes the centrality of networks (depending on the mechanism for distribution). It also affords a competitive advantage to the types of programming that viewers particularly want to watch, instead of what they've watched simply because it's been "on." These new conditions, which can enhance the status and reach of what have previously been "cult" hits, should encourage studios to shift support from broad slates built on a strategy of intentional overproduction to smaller lineups with programs that all offer some distinction. Whether this will mean an increase in programs that are creative and innovative or those that attempt to tap into broad-based tastes, or perhaps both, is not clear. What is clear is that post-network television programs will not succeed simply because a network makes them available to viewers at particular times.

The next chapter explores the emerging economic possibilities enabled by the distribution strategies considered here. It is very clear that the remaining hurdles have little to do with technological solutions; instead the challenge lies in developing funding strategies that match emerging post-network uses of television. As Evan Shapiro, president of Pivot (the television channel launched by Participant Media in 2013), explained, "The industry is so sick and so not recognizing that they are sick, it is kind of spectacular."[47] And yet, despite efforts by incumbents to deny coming change, some established entities—HBO GO, for example—extend tentative propositions that can nevertheless become major disruptions; while others that have never been part of the game, such as Netflix, exploit the dissatisfaction and inefficiency of the existing system to an extent that forces broad-scale response.

Indeed, the notion of TV Everywhere remains hyperbolic, but it is certainly the case that TV is in a lot of places it wasn't before, and even more so, that viewers have far more choice over what plays on their living room screen.[48] The changes in distribution have allowed new players entry to the established industry, while others—Google, Aereo, Apple—knock resolutely at the door. Although entrenched commercial interests have the greatest assets through which to reassert themselves, the scope of change may well create new relationships among cable and satellite systems, telecommunication providers, and technology manufacturers.

Although I want to avoid the "blue skies" rhetoric that early on forecast cable as a democratizing force, it is nonetheless important to consider some of the implications of broadband video distribution, which has dismantled the bottleneck that afforded substantial cultural power to the gatekeepers and agenda setters of the network era. By 2007, broadband distribution had already offered previously unimagined opportunities to distribute video— including those beyond the confines of commercial profitability. To be sure, the gross surplus of content has made finding messages, videos, or stories of interest increasingly difficult. In response, broadband distribution companies have busily refined search and distribution applications that aggregate content and decrease the difficulty of sorting. In the process, though, they may offer advantages to certain content and providers and thereby contribute to reestablishing commercial media control.

Though this book focuses on the commercial television industry, perhaps the biggest change that new ways of distributing television have introduced to television as a cultural institution is the creation of new means for independent or amateur productions to find audiences. Of all the art and storytelling forms, television was under the tightest commercial grip due to the stranglehold of the networks on distribution. For decades, public television provided the most viable outlet for unconventional or noncommercial content, but its lack of independence from state funding manipulation and general underfunding ultimately afforded it limited additional breadth. Local cable access provided another alternative to commercial television and the networks' dominance, but was geographically, technologically, and financially limited. By contrast, broadband distribution enables a radical disruption in television's norm as a medium limited to commercially created content, even if audiences haven't migrated to primarily viewing non–industrially produced content.

Another substantial adjustment that has resulted from the dismantling of network-era distribution practices is the displacement of the "network schedule" as the dominant means of content organization. This schedule, and the conventions of other production components related to it, restricted variation in program length, with 30, 60, and 120 minutes (minus commercial allowances) being the only options in most venues. But as commercial content providers tested new distribution platforms such as VOD and broadband channels, many offered shorter forms and other programming outside the bounds of conventional program length and series structure. The growing variety of distribution methods and the nonlinear structure of VOD have now also enabled the further diversification in content forms considered in the previous chapter. First, DVD distribution created an additional market for limited-run series that video on demand has subsequently expanded. Some content created for Internet distribution has even come to be redistributed on cable networks (Bravo's *Outrageous and Contagious: Viral Video*, Comedy Central's *Tosh.0*); in other cases, a YouTube sensation received a network contract for a series of films and a television series (Lucas Cruikshank on Nickelodeon), and a television series based on a Twitter feed emerged with *$#*! My Dad Says*, although it had a TV life about the equivalent of 140 characters.

As suggested in the previous chapter as well as the next, the increasing multiplicity of ways of paying for and circulating programs has substantially expanded the range of programming that can be produced within the dictates of a commercial media system. Multiple opportunities for producers to recoup production costs allow a much greater variety of forms than the standard windowing process that demanded consistent program lengths and at least one hundred episodes. In the network era, international syndication provided the only way to recover production costs on miniseries—a circumstance that contributed to the telling of only certain types of stories in this format.[49] Moving away from a television experience organized by linear daily schedules frees storytelling from norms that were arbitrary and characteristic of network-era scarcity. Morgan Hertzan, chief creative officer of LXTV.com, an online network that features short video clips on New York shopping, nightlife, food, arts, and other categories, explains the advantage of nonlinear video distribution: "You can service a niche audience and only deliver a couple hours, well-focused and well-produced."[50] The expansion of his company to additional cities and its purchase by NBC indicates the potential for local commercial and informational content that was largely absent throughout the network era, the multi-channel transition, and the start of the post-network era as well.

As the post-network era continues to become established, most of the distinctions such as VOD versus broadband-delivered programming or streaming versus download that I make here with painstaking deliberateness will erode for audiences. Already, some early adopters move fluidly among Netflix streams, MVPD-provided VOD, DVR recordings, and YouTube videos and might refer to any or all as the "television" they watched last night, though the struggle among content, distributors, and an array of middlemen will continue. Who will wear the crown in the post-network era: content or distribution? Speaking at the Annual Livery Lecture at the Worshipful Company of Stationers and Newspaper Makers in March 2006, the News Corp magnate Rupert Murdoch proclaimed,

> Power is moving away from those who own and manage the media to a new and demanding generation of consumers—consumers who are better educated, unwilling to be led, and who know that in a competitive world they can get what they want, when they want it. The challenge for us in the traditional media is how to engage with this new audience. There is only one way. That is by using our skills to create and distribute dynamic, exciting content. King Content, *The Economist* called it recently.[51]

But the multiplicity of post-network technologies and distribution windows that has enabled an expanded diversity of content has also suggested a new competitor for status as king. Indeed, much of the more revelatory rhetoric about changes in the television industry asserts that viewers will be sovereign in the post-network era as industries compete to provide them with the content they desire on their own preferred terms. The Comcast CEO Brian Roberts declared as much at the 2006 National Cable Show, while a few weeks earlier the Disney CEO Robert Iger announced, "We've concluded the consumer is king. Remaining a slave to fixed consumption would be a huge mistake and at Disney we're refusing to do that."[52] In the more recent wake of Netflix's release of *House of Cards*, a journalist working out the economics of the endeavor shifted the adage to argue that "exclusive content is king."[53] Viewers do indeed appear likely to benefit to some degree as the cultural institution of television evolves and the emergence of new distributors reduces the control of the few and limited gatekeepers of the network era. New distribution methods allow more viewer choice, so that they can watch commercials or not, pay directly for programming or not, view content at self-determined times and locations, and have more ready access to content outside that created by commercial conglomerates.

Yet Roberts's fellow panelist, the Time Warner CEO Richard Parsons, tempered the "consumer is king" assertion, recalling the words of Gerald Levin, who had orchestrated the once heralded but by then negatively regarded merger of AOL and Time Warner. Levin posited that content might be king, but distribution was the power behind the throne, a maxim that again leaves the viewer out of the equation. Although expanded viewer sovereignty still seems possible in this nascent stage of the post-network era, the history of distribution tells a different story. All too frequently, emergent technologies provide multiplicity and diversity in their infancy, only to be subsumed by dominant and controlling commercial interests as they became more established. The contradictory interests of various industries may create room for viewers to win some victories, but so long as most of the country has only one choice for broadband service, the spoils will be minimal. Stay tuned: a battle royal has just begun.

5

The New Economics of Television

Madison Avenue is stuck in a 1950s time warp. While the era of mass media has long since departed—just glance at the hundreds of cable channels and thousands of special-interest magazines if you require proof—most ad agencies still operate the same way they did during the Eisenhower administration: Toss a single TV spot at millions of random viewers in the hope that a small fraction might be interested in that new Chevrolet or life insurance from Prudential.
—Paul Keegan, "The Man Who Can Save Advertising," 2004[1]

The advertising business has not matured in the past thirty or forty years. I wouldn't blame the current need for change on TiVo. It's an evolutionary process that has stagnated because advertisers and networks have been slow to recognize and adapt to changes in the consumer marketplace.
—Lee Gabler, Creative Artists Agency[2]

The commercial model supporting U.S. television has remained fairly stable since its establishment in the mid-1960s. As many have criticized, the lack of innovation and change in the relationship among television networks and their Madison Avenue supporters indicated a stunning lack of dynamism. Certainly, shifts occurred as audience measurement systems grew increasingly sophisticated and cable networks introduced new options throughout the multi-channel transition. For the most part, however, dominant practices remained in place until the late 1990s, when it became apparent that changes of prodigious proportions were approaching. Most tried to ignore them. Others attempted to halt them or hoped for some sort of intervention that would offer reprieve. A few boldly looked forward.

Advertising has always been central to the economics of U.S. television, but an unusual confluence of immediate economic crisis, programming innovation, and cultural uncertainty combined with the established consequences of expanded viewer choice and control to elicit the variety of responses attempted by advertisers in the early years of the twenty-first century. Historically, U.S. commercial television was dominated by certain advertising norms such as the thirty-second commercial. But this convention resulted from particular industrial organizations and competitive strategies of the network era and multi-channel transition, and was no more inevitable than the emerging post-network norm in which multiple advertising strategies, including product placement and sponsorship, began to coexist with the thirty-second ad. Such a multiplicity of strategies corresponded with the increasingly diverse practices, diffuse industrial organization, and distinctive programming experiences characteristic of the post-network era. The multiple television advertising strategies explored here—product placement, integration, branded entertainment, and sponsorship—did not "kill" the thirty-second ad, as so many trade articles suggested. Rather, they reflected the increasing variety of practices and types of television common throughout the production process, although, again, the transformation was not instant.

The scope of coming changes was clear to all by the late 1990s, and one might expect advertisers to have the greatest interest in identifying new models and norms because they paid for the system. Certainly all of the relevant players observed the data that trickled in during the multi-channel transition. Broadcasters, with their diminishing audiences and successful demands for higher rates, were not going to suggest a change in the status quo. Cable channels had much to gain and regularly agitated for more support from advertisers. Despite cable's multiyear existence, the channels did not develop compelling, word-of-mouth-generating narrative series programming until the late 1990s—particularly in the key prime-time period—which helped perpetuate broadcasters' dominance. The cable networks offered advertisers a new multiplicity of advertising sites, but the expanded choice of the multi-channel transition alone was not significant enough to cause a reevaluation of the commercial funding practices of U.S. television.

Indeed, it was one of those boxes viewers connected to their sets that brought about notable hand wringing and initiated some experimentation from advertisers and buyers. As the epigraphs to this chapter suggest, it was less the DVR box itself than the fear of the DVR box and the empowered consumers who owned them that shifted Madison Avenue out of fifty years of complacency. Over a decade before the first DVRs entered the home, the technology's analog predecessor, the VCR, sparked the industry to a similar panic.[3] Yet the end of the world of commercial advertising predicted in the early 1980s never transpired, which made it all the more curious that advertising agencies and networks so quickly forgot their unfounded fears when the DVR debuted. Despite the fact that the DVR, like the VCR, enables viewers to record and later play back programming, DVR early adopters—many of whom worked in the industry—knew something was different. The technology was too easy to use, its digital capabilities involved too substantial a leap, and its ready program guide was far more likely to entice viewers to actually view the shows that they recorded and thus become a default mode of viewing. Industry experts also knew that video-on-demand technology was maturing. Viewers were no longer going to be satisfied with a mere range of options; once allowed to sample the new technologies, they would demand control over when and how they would watch, and they were no longer going to be captive for commercial breaks. Instead of the 300-channel universe, the control technologies and distribution adjustments provided a 10,000-hour universe of instantly available programming.

Blaming the DVR for the experimentation in advertising techniques and program financing norms that emerged in the early 2000s makes for an elegant argument, but it is a grand overstatement of the impact of the device. Certainly the reassessment of dominant advertising models was overdue long before DVRs enabled advertisement skipping. The future uncertainty fueled by the DVR only helped the industry toward the "tipping point" at which the risk of trying something new appeared less dangerous than blindly maintaining the status quo.[4] Broader economic factors, including the 2000–2002 dot-com crash, the 2008 recession and Lehman Brothers bankruptcy, and the lingering economic malaise, also had substantial consequences for the advertising market. Consider that local television advertising was down 14.7 percent from the first half of 2000 heading into the fall of 2001 when the attacks of September 11 produced further uncertainty in the market.[5] After the attacks, analysts revised forecasts to predict even greater declines in advertising spending, and advertisers feared for the future of their industry.[6] Just as the equilibrium returned following these disruptions, the housing crisis, massive unemployment, and the so-called Great Recession again created economic uncertainty and an environment of contained advertising spending.

Alone, the disruption of new technologies and forms of distribution that bombarded traditional television industry norms would have been a lot to weather, but the fact that new, potentially industry-rupturing developments such as Netflix, Facebook, and Google's AdWords emerged concurrent with broader economic crisis made the questions of how to react even more stupefying. The growing echo chamber of industry "analysts" tended to lack a long view about the scale of change portended by "new media," a myopia in some cases encouraged by the simultaneous emergence of self-anointed industry bloggers who rarely provided the reasoned investigation of more grounded industry journalism, which was experiencing a crisis of its own. Though it was clear that the future would look different, maybe even *very* different from the past, it was entirely unclear what that future would encompass, which led to the overvaluation of potential disruptors and the undervaluing of established industries—such as MVPDs in the mid-2000s, for example. Yet still, in 2013, PricewaterhouseCoopers estimated that the U.S. television ecosystem would report a total revenue of approximately $142 billion.[7] Of this revenue, about 47 percent was derived from advertising and 53 percent from subscriptions to MVPDs. And despite the direness sometimes suggested by industry accounts, these remained incredibly profitable industries—just not *as* profitable in some cases. Analysis by SNL Kagan showed profit margins that averaged 40 percent for cable channels and 10 percent for broadcast channels. And while MVPD margins for video dropped into the mid-20s, most now earned profit margins of nearly 60 percent on Internet service.[8] So though some margins were diminished, they remained enviable relative to many other industries.

Well before the onslaught of new means of distribution and the broader economic crisis, the advertising industry was aware of the inadequacies of their measurement tools. All knew that network-era audience estimations offered only a suggestion of those who might or might not view a commercial; trips to the bathroom and the refrigerator had long stolen audiences before VCRs and remote controls exacerbated challenges that DVRs would further exploit. While introducing a panel discussion of new advertising practices in 2004, the industry commentator Jack Myers described the content of a trade advertisement hanging in his office. The ad reminds its audience—that is, advertisers—that only through "creative ingenuity" will they reach the "disappearing America," such as those fleeing to the kitchen or elsewhere at commercial breaks. Myers's punch line: the ad was created in 1953.

The problem of reaching the right viewers with the right advertising messages was thus by no means new, and by the early years of the twenty-first century, advertisers had more tools to aid them in this task than ever before. But the management structure and culture of the agencies had become far more "corporate" and reflected the structures of their clients, which had become lean, post-Fordist corporations that would not tolerate any economic inefficiency or uncertainty and sought guarantees that no advertising dollar would be wasted.[9] This concern about the efficiency of money spent on advertising drove an obsession with return on investment at the same time that the industry experienced unprecedented change in its advertising techniques and consumer research methods. The margins throughout the business were shrinking. The profits of the 1960s and 1970s were so great that the industry could maintain substantial revenue even as the multi-channel transition fragmented audiences and the middlemen between studios and audience expanded; but this could not continue forever. Audiences would not, could not, continue to pay significant hikes in fees for cable as well as broadband, and mobile services that quickly came to seem more essential than television programming. Likewise, financially secure consumers might have the leisure dollars to purchase one or two subscription services such as HBO or Netflix, but Hulu Plus, Amazon Prime, and YouTube channels as well? Piecemeal adjustments would last until the margins eroded, but at some point, these businesses would have to reinvent their models to correspond to an era of digital distribution.

The industry press often framed the redefinition of advertising practices as a question of the life or death of the thirty-second spot, but the relevant questions were far more substantial and nuanced. The advertising industry needed to respond to the challenges of an increasingly fragmented and polarized audience empowered with control devices that enabled them to avoid commercial messages in a variety of ways. An assortment of new and old strategies emerged or reemerged haphazardly during the waning years of the multi-channel transition. Although advertisers experimented with a distinct range of strategies, little consensus existed within the industry about what to call them: anything other than a thirty-second spot was often labeled "product placement" despite the significant variation in the strategies used.

Shifts in dominant advertising practices can substantially affect television programming and, consequently, the stories the medium provides. This chapter, then, distinguishes among different practices and notes their ramifications for the advertising industry and beyond. As in the other production components considered throughout the book, the adjustments in the operation of the advertising industry have significant implications for television as a cultural institution. Advertisers' desire to reach young, upscale demographic groups enabled the production of content that defied previous norms, while the multiplicity of financing strategies likewise diversified the range of programming commercial models could support.

Though advertisers have provided the dominant source of funding for U.S. television for the last sixty years, significant uncertainty about the future of television advertising persisted in 2014, consequently calling into question the future organization of the television industry. Already, the terrain of audience and advertiser financing looked different in 2014 than it did when I wrote the first edition, published in 2007. The early through mid-2000s indicated a lot of experimentation with advertising techniques, but while not failed experiments, no strategy has emerged as a replacement. Indeed, various models of product placement and integration have become crucial to and commonplace in unscripted television, and sponsorship has become core to live sports, but prized content resists network-era practices such as the thirty-second advertisement.

Rather than any of the television-based experiments, it has been the emergence, adoption, and integration of smartphones that may portend the most radical change for U.S. television's advertising-based norms. It seems unavoidable that some entrepreneur will discover a mechanism that harnesses the mobility and ubiquity of the smartphone, perhaps in combination with social media, to create an advertising platform that leaves the inefficiencies and uncertainties of most television advertising far behind. The thirty-second embedded advertisement that has been the norm of television advertising has always been highly imperfect, but it was a decent mechanism given the tools available in the network era. It remains only a thought experiment at this point, but we might imagine the post-network era as one that takes shape with only a small fraction of the previous era's advertising dollars spent on television, and that support is most likely to remain in programming that provides an environment in which advertising is most effective, such as live sports. It is the belief in a coming, superior advertising vehicle that contributes to my skepticism about the long-term viability of linear television.

Though in 2007 I imagined a diversity of advertising strategies to undergird the post-network era, the emergence of the television experience I describe in relation to prized content and evidence of more optimal advertising vehicles based on mobile and social media technologies suggest that the post-network era will feature a combination of advertising and direct-pay economic models. As noted in the last chapter, what is crucial to the future evolution of television is establishing an economic model appropriate to what are becoming post-network technological and distribution norms, and the progression since the network era has featured audiences paying more and more for television programming. As the expansion in original cable series has brought about an era of true programming abundance, carriage fees from MVPDs have become an increasingly substantive component in network/channel balance sheets. This is not to suggest there is much room in viewers' budgets for incremental direct pay on top of existing cable and broadband subscription, but to forecast that the inefficiencies in the bundling of content into channels and then channels into tiers cannot be sustained in the post-network era. Making content affordable for viewers and adequately profitable for creators will require contraction in the middlemen between these entities.

But the work of this book is not prognostication, and its arguments are based on available evidence. It is clear that long-established inefficiencies have begun to produce crisis: the exuberant initial response to Netflix and the doubling of zero-TV homes are a meaningful indication of discontent, though still small-scale phenomena. Yet it will likely require far more crisis—think the recording industry circa 2002—or some visionary industry leaders to force those still comfortable with the status quo to endeavor upon preemptive change. At this point, the strength of works such as this is their ability to take a long view on how television advertising developed and to seek insight about this coming era from its attempts to negotiate change up to this point.

U.S. Television's Varied Economic Models

Before examining the nuances of past and present national network television advertising, we need to tease apart the many different economic models that now support television. Though viewers may experience a wide array of content simply as television, the different economic models underpinning its creation have significant impact on the content that appears on the screen. Chapter 3 explores the economic relationships between the studios that make content and the distributors—networks and channels—that deliver it to audiences. Here, I examine the variation among the economic models that connect audiences and distributors and explain why those differences matter.

At the dawn of the post-network era, most U.S. television could be categorized as advertiser-supported, subscriber-supported, or a combination of these. Even broadcast networks began relying less on advertiser support, as they began receiving significant retransmission fees from cable systems beginning in the early 2000s. A 2012 financial analyst report estimated that CBS would collect $1 billion each year from distribution fees paid by MVPDs for retransmission by 2016. The entirety of the fees negotiated between owned-and-operated stations and their MVPDs would go to CBS, but the national network would also receive 50 percent of the fees non-owned affiliates negotiated.[10] Entities such as HBO and Showtime earned revenue exclusively through subscriptions. Importantly, though viewers typically paid $15-$20 per month for these services, the channel received only about half of these fees, with the other half going to the MVPDs who market the service and handle billing. Finally, so-called basic (nonsubscription) cable channels utilized a hybrid of these models, earning revenue both from advertising and from fees paid by MVPDs to include the channel in their bundle. The per household, per month fees paid by MVPDs varied widely: some new channels might receive no carriage fee; most received between $.20 and $.50, while those with a lot of original scripted programming such as TNT or USA earned around $1.00, and exclusive sports channels could charge upward of $7.00.[11] In total, MVPDs paid $30 billion in these fees in 2011, fees largely passed on to their subscribers.[12]

Though the post-network era remains preliminary, one economic model particular to this era is the direct-pay or transactional model. In this economic model, which modifies some of the features of the subscriber-supported model, the viewer pays for a specific piece of content. This might be a single episode—as first made possible through iTunes in 2005—or a full season of episodes downloaded or purchased on DVD. When I wrote the first edition of the book, direct pay seemed best understood as an expansion of the varied syndication windows that have long helped studios earn back production deficits. In other words, direct pay didn't initially create a new route for the creation of content, but a new way to earn revenue on content originally produced for a network or channel. The subsequent developments in original content creation by niche distributors such as Netflix and DirecTV and funding experiments such as Kickstarter, however, now suggest that this conception of direct pay as merely a secondary revenue stream imposes residual assumptions on television economics that place undue primacy on linear distribution. In theorizing the role of direct pay in this still-nascent moment of post-network norms, we must imagine the possibility of creating content primarily for a direct-pay market, though these efforts are still most preliminary.

Though video distributed online may seem fundamentally differently from the established television distribution norms of broadcast and cable television, leading online video aggregators such as Hulu and YouTube rely upon the same advertiser-supported economic model long used by broadcast networks, while the hybrid advertiser/subscriber model of Hulu Plus and some YouTube channels mirrors the dual-revenue model of basic cable. YouTube's model has evolved constantly; as of this writing, YouTube primarily relied on an advertiser-supported business model, but had begun experimenting with subscription fees for some of its "channels." Its per-user rates (CPM—cost per exposure to 1,000 viewers, often of a specified demographic group) were generally higher than television, because it was able to offer advertisers a way to reach light television viewers that was cheaper than the buys on broadcast or cable.[13] Advertising revenues were shared with the content creators, with YouTube retaining 45 percent—a slightly higher than typical distribution fee—about which some creators expressed discontent; though other analysis reported that YouTube retains as much as 70 percent of advertising on its premium channels.[14] It also featured a lighter commercial load, which was attractive to advertisers and viewers.

YouTube's share of video advertising remains small, estimated at $4 billion in 2012, compared to $60 billion spent on U.S. television advertising, and of that, approximately 12.5 percent was attributable to ads in premium content (its supported channels) and the rest from advertisements in user-generated content.[15] Financial analysis by Needham in 2013 estimated that YouTube had spent $350 million creating premium channels in the previous two years, money spent on "grants" of $1 million to $5 million per channel that functioned much like an advance against royalties. In the midst of writing, YouTube announced a subscriber component and plans for a pilot program of fifty subscription channels that will charge on average $2.99 per month and would feature a slightly higher share of revenue to the content producers.[16] Needham reports that YouTube "did not renew approximately 60 percent of the channels they funded in 2012, and renewed no channel which targeted audiences over 25 years old," which, paired with the subscription experiment, suggests that though YouTube is the dominant global online video platform, its business strategy and model are far from permanent.[17] Hulu had been experimenting with the option of subscription-level access since November 2010 and amassed four million subscribers.[18] Yet, though subscriptions and advertising revenue earned $700 million in 2012, Needham notes that Hulu had not earned a profit.[19]

These matters of how distributors earn revenues are important to understanding the programming decisions they make and how some strategies are viable for some models, but not others. The next section discusses the nuances of advertiser-supported models in great extent. But though advertiser support and subscriber support are both methods for funding television, they really are fundamentally different businesses. In many cases, the differing financial models of subscription and basic cable have enabled the profitable production of series with new ideas or a capacity to speak to particular demographic groups. Whether one focuses on the edgy content of FX's dramas or on character-driven shows such as HBO's *Boardwalk Empire* or Showtime's *Dexter*, it is clear that though passion for these shows was high among a subset of the mass audience, this audience was not large enough for broadcasters to profitably produce such series. By 2005, successful subscription cable, basic cable, and prime-time broadcast series exhibited clear distinctions that marked them as characteristic of their distribution outlet. Though broadcasters incorporated some of cable's norms—such as the thirteen-episode season—this was still uncommon. Broadcasters tried to replicate *Mad Men*'s period drama (*Pan Am, Playboy Club, Vegas*), but most efforts at trying to make a mass hit of a niche cable phenomenon have failed dismally enough to keep broadcasters replicating the broad-based legal, medical, and detective franchises that have long proven successful. Even more remarkably, online distribution enabled an "indie" television sector to emerge that told stories more different yet from the network-era broadcast fare.

The business model of subscription services mandates that they provide programming of such distinction—whether by measures of quality or value of niche address—that viewers are willing to pay directly for the content, thereby negating the need for advertiser support. Cable networks have consequently sought to develop programming that establishes their narrowly focused brands and allows them to deliver high indexes of particular demographic and psychographic groups of consumers. By contrast, broadcasters' businesses rely predominantly on advertising revenue, so their ability to earn more money depends upon delivering larger audiences. This aspect of their economic model prevented broadcasters from adopting a competitive strategy of addressing audiences that were too narrow.

Another economic dynamic plays out between the subscription fees that allow viewers access to a range of programming and the direct-pay model of buying particular series. The subscription service seeks to offer enough value to maintain subscribers. HBO does this by offering a variety of distinctive programming otherwise unavailable—it offers uninterrupted theatricals, a rich array of documentaries, original series that target diverse audience tastes (*Girls* versus *Boardwalk Empire*), sports competitions (boxing), and commentary unavailable elsewhere, and current events shows with a distinctive sensibility (*Real Time with Bill Maher*). Just as HBO began as a distributor of theatricals, Netflix offers a more recent illustration of the subscription strategy as it has evolved to offer back catalog television content that makes up much of cable channels' schedules, a library of films, and now original content as well. Before MVPDs developed their VOD libraries and increased license fees required Netflix to increase its subscription price, it provided a strong value proposition. Once MVPDs made similar offerings available as part of top subscription tiers, the value of adding on this service was less clear—necessitating Netflix's move into original production, to again distinguish its value proposition.

Digital distribution has created new choices for viewers in terms of how they buy subscription content. Viewers can choose either the full subscription or pay directly and buy each show on DVD or through a digital retailer such as iTunes or Amazon. Viewers choose the per-show proposition when the subscription bundle provides inadequate value—typically because it bundles much unwanted material and the desired content is more affordable if purchased à la carte. Lack of immediacy is also a "cost" to viewers who opt for the transaction purchase, as they often have to wait a few months to access content in this way; but as television becomes more nonlinear and audiences ever more fragmented, synchronousness seems less a priority. Though the vast press attention to Netflix's *House of Cards* release may have piqued the interest of nonsubscribers about this show, the value of immediate access didn't prove much of a motivator as Netflix did not experience a notable subscription increase to access the show. Many likely expected that the content would be freed of Netflix exclusivity; indeed, the presale on Amazon noting DVD availability six months after the original release assured many that the wait would not be too long. It also should be noted that industry practices such as exclusivity and time-delayed windows also encouraged interested viewers to seek unauthorized access to content. As explored in chapter 4's discussion of the "Take My Money HBO" campaign, viewers were not solely motivated toward unauthorized access by the costs of content, but by being forced to pay for a broader package of undesired goods. By 2014, the cost of purchasing full series on DVD or through download has kept most HBO and Showtime subscribers paying for the service, but those who desire only one or two series from these outlets might forgo the subscription and wait for the series' transaction release. Also, HBO's strategy of not licensing its content to other non-transaction entities—such as Netflix—provides further motivation for viewers to subscribe, though its recent deal with Amazon may suggest a change to this strategy.

As we think about the economic models that support U.S. television, it is important not to afford advertising-supported television a superior status simply because it emerged first. There are several valuable features of the advertiser-supported system, but many of those correlate with the broader network-era norms of the medium. The creative advancement of the medium that has occurred outside this economic model informs debates about the artistic limitations and possibilities of the medium. It may be that what were long held to be the limits of television were simply the limits of advertiser-supported television; and it may also be that advertiser-supported television simply needed some competition. Though accounting for the economic model that supports the production and distribution of particular television content is crucial to assessing shows relative to others, it also seems foolhardy to uniformly advocate any one model of financing: each has strengths and limitations of varying significance relative to one's priorities.

In the network era, a common financial model made assessing television much easier than the case now. As argued throughout the book, the expanded variation in industrial norms across industrial practices necessitates nuanced discussions of the medium's output and greater attention to features such as those that allow distinctions among prized content, live sports and events, and linear viewing. Subscription and direct pay are particularly valuable for prized content and perhaps some live sports. Despite the notable importance of both these forms of television, linear viewing remains the residual norm, and this experience has long been adequate for advertiser support.

Advertising Practices during the Network Era and the Multi-Channel Transition

The norms of radio determined the commercial basis of the U.S. broadcasting system while television was still just an imagined technology in the hopes of inventors. As a result, many of the key debates about and experiments with possibilities for financing broadcasting were established before television functionally existed. In television's early years, however, the inherent differences between the two media required some adjustment of practices inherited from radio. Primarily, the cost of television production relative to radio introduced complications to the established system of commercial funding.

The dominant commercial model of radio utilized a single-sponsorship system in which a corporation paid all of the production costs of a show and was the only product or corporate entity associated with it. While initially this system carried over to television, it soon became apparent that a single sponsor could not feasibly pay for the many facets of visual production on a continuing basis. Some genres with lower production costs, such as game shows, still enabled single sponsorship, although that contributed to other difficulties, as became apparent in the quiz show scandals of the late 1950s, when it was revealed that advertisers rigged the shows to support popular contestants and used other disingenuous strategies to maintain viewers. Reaction to these scandals, as well as the networks' desire to control their schedules, further contributed to the development of a new advertising model using a "participation" or "magazine" format; the latter term refers to the way television shows came to be supported in the same way as magazines, with advertisements for many different products mixed in with the programming.

By contrast, the sponsorship system of the 1940s and 1950s afforded advertising agencies and their clients considerable command over program content and even networks' schedules.[20] From the beginning of television, the networks objected to this arrangement, but they could not institute an alternative quickly enough to prevent it from migrating to the new medium in the early 1950s.[21] As they discerned that a deliberate and strategic schedule was as important as the quality of the programming placed in that schedule, the networks became eager to displace advertising agencies' centrality in program development and to gain control over their schedule—which included ending the norm of "time franchises" that allowed agencies to control specific slots in the networks' schedules. William Boddy's research recounts clear evidence of network pressure to end single sponsorship by the mid-1950s, although multiple-sponsor shows did not become dominant until 1962–1963, when 55 percent of the ninety-four shows on the air used this commercial format.[22]

The shift away from single sponsorship increased a network's control of its programming content and schedule and diminished the sponsor's role in both. Advertisers became less invested in specific content issues once they became one of many companies with commercial messages in a program. As networks assumed authority over their schedules and show selection, the change also had advantages for advertisers, including spreading their risk across a number of shows each week, while still providing agencies with substantial revenue opportunities.

Participation provided a far more beneficial advertising system for all involved, except perhaps the viewers. Although it responded to the problem of the cost of producing television, which was substantially higher than that of radio, a number of other forces contributed to the transition. As Boddy notes, shifting corporate strategies regarding the nature of television advertising messages and the type of corporation likely to advertise occurred concurrently with the move away from sponsorship.[23] Whereas large manufacturing corporations had dominated sponsorship and used this as an opportunity to promote their corporate image, they themselves began to rethink this "corporate angle" or "company voice" strategy at the same time that networks started to want to have a broader blend of advertisers less likely to be uniformly affected by periods of recession.[24] Increasingly, television became a medium more desired by packaged-goods companies that used advertisements to explicitly sell the attributes of a product or to sell the lifestyle they wanted consumers to attribute to the product. Although a large packaged-goods advertiser such as Proctor & Gamble could easily afford sponsorship, such an arrangement, which privileged the P&G name, would not provide name recognition for the substantial variety of products it sought to promote, such as Tide, Crest, and Palmolive.

The networks' identification of the competitive importance of schedule control enabled the establishment of many network-era norms, including the creation of programming that rendered television more than a haphazard assortment of disconnected programs. The motivation of sponsoring programs to further a certain corporate image or angle led to the production of a different type of programming than characteristic of most of the network era. Single sponsorship encouraged distinctive programming that expressed prestige. However, this too began to change as the networks began to seek out much cheaper programming, such as the kind Christopher Anderson examines in his study of Warner Bros. film studios' efforts to produce for television.[25] The studios mainly wanted to monetize old background footage and more efficiently use their back lots, and hoped to do so by recycling old and promotional content as television programming. The shift from sponsorship to participation also contributed to the networks' pursuit of profit participation (discussed in chapter 3), which made the environment difficult for independent producers and resulted in the fin-syn regulations. A weak program aesthetic dominated much of the 1960s, as the FCC chairman Newton Minow noted at the time and television historians have since affirmed. This programming resulted from characteristics of production practices of the era that overemphasized cost saving and led the networks to pursue only modest programming achievement.[26]

Various norms of the "television season" and the timing for selling advertising time developed once magazine-format advertising established its dominance. The annual September debut of programs led to the related annual process of securing advertising commitments in the spring, in what came to be known as the "upfront" market. During this period, which once lasted eight weeks but may now span only a week or two, broadcast networks sell 75 to 90 percent of the advertising time in the upcoming season on tentative, but fairly reliable commitments.[27] Networks sell the remaining advertising inventory throughout the year in the "scatter" and "opportunistic" markets. The upfront market is advantageous for networks because it affords them committed advertising spending before they begin producing programming. In exchange for the reduction of risk, the networks offer "discounted" rates on advertising purchased upfront. The upfront functions as a speculative market, as later scatter prices may be significantly higher depending on advertising demand.[28] Most advertisers purchase time upfront because the limited supply of programming in certain programs and on particular networks makes some buys available only during the upfront and because the networks came to offer guarantees on upfront purchases. The later scarcity traditionally has led to scatter rates that average roughly 15 percent higher.[29]

The upfront process generally benefits the networks, although they have developed some practices in response to advertisers' more substantial concerns about the uncertainty and potential inequity of the process. The networks decreased the uncertainty of the upfront purchase—a key concern for advertisers—and received higher rates in return when they initiated "guarantees" beginning in 1967.[30] Under this arrangement networks began guaranteeing a certain audience size for the advertisers' purchase and providing "make-goods" or supplementary advertising slots if they failed to achieve the guaranteed audience reach with the initial purchase. According to the veteran media buyer Erwin Ephron, this led to a shift from advertisers buying specific shows to their purchasing CPMs, an acronym for cost per thousand, or cost for one exposure to one thousand viewers of a certain demographic type. The CPM became a standard industry currency in national sales, which operates as follows: if an advertiser wishes to reach fifty million viewers and had established a CPM of $15, or $15 per thousand viewers, a network would put together a package of advertisements that would reach the fifty million viewers for $750,000.[31] Networks usually guarantee audience delivery only if advertisers purchase CPMs in the upfront. The networks benefit from this method of purchase because they distribute advertiser support between both popular and less established programs.[32]

The dominance of the upfront as the means by which advertisers allocate their spending is important because this in turn affects network practices. If networks sold individual commercials instead of the exposure to a certain number of viewers spread across multiple programs, they would likely make different programming decisions. The upfront allows the networks to package their new or weaker shows with their established hits, a practice not entirely dissimilar from the film industry's practice of block booking, in which studios required theaters to show lower-budget films in order to get the high-profile films they most desired. Although advertisements in hit shows might command even higher prices if sold individually, this would weaken the networks' ability to nurture new shows and could further discourage the pursuit of unconventional stories or formats.

The process of upfront selling garnered significant critique and debate from its inception, but regardless of perennial declarations of its death, and despite the substantial adjustments in other production components throughout the multi-channel transition, the practice has proven remarkably resilient. Paradoxically, the hold of the upfront only grew stronger as competitors for broadcast television's advertising dollars emerged. First, cable channels began holding their own "upfront" presentations ahead of the broadcasters each year in the early spring. Then, in 2012, the Interactive Advertising Bureau began a "Digital Content NewFront" at the end of April to pitch advertising opportunities on Hulu, Yahoo!, Google, and the interactive divisions of conglomerates such as Disney and CBS.

Major shifts in advertising and the economic support of television more broadly began early in the multi-channel transition and then shifted again at the beginning of the post-network era. The arrival of subscription-financed television marked the first major rupture from the network-era model of monolithic advertiser support through thirty-second ads. It required two decades for the subscription network HBO to produce a successful "television series," but the subscription experience prepared viewers for the pay-per-transaction distribution opportunities that subsequently became available through outlets such as iTunes as well for nonlinear subscription experiences such as Netflix. Subscription television also established a content creation environment very different from the one that pervaded advertiser-supported content. The financial mandate of drawing and maintaining subscribers led subscription networks to create programming of such distinction that viewers were willing to pay for it—if not in subscription, then perhaps in the various subsequent markets such as DVD sales or streaming transactions.[33] Again, the question of who pays for programming and through what financial model has substantial effects throughout the production process.

Challenges for Advertising at the Beginning of the Post-Network Era

The first decade of the twenty-first century featured an uncommon variety of advertising experiments. Advertisers long had reason to explore alternative television advertising strategies, but risk aversion prevented the allocation of substantial funds to methods other than the legacy model of the thirty-second advertisement. DVR diffusion and audience dispersal helped push advertisers to the tipping point, but there were other important factors at work. Crises resulting from fragmenting audiences, rising production costs, and commercial-skipping behaviors enabled by control technologies compounded until advertisers could no longer rely on the presence of an audience during commercials. Consequently, they began experimenting with product placement and integration, branded entertainment, and sponsorship, while continuing to support the thirty-second commercial. Calls for the end of the upfront system continued, yet remained unheeded despite the sizable adjustments in nearly every other industrial practice. The only real threat to this buying practice emerged once advertisers began shifting money out of thirty-second advertisements because strategies such as placement and sponsorship could not be developed and sold in this way. Many worried about the consequence of advertisers moving spending to the Internet, but even by 2012, advertising in online video amounted to only $2.89 billion, in comparison with $64.54 billion in U.S. television advertising spending.[34] By

2014, online video advertisements continued to struggle because the industry lacked a stable reporting structure.

Internet advertising derived its early norms from print rather than television, and video ad buyers were less interested in data about monthly views than a structure more similar to television. Buyers also lacked the assurance they had in television that web publishers placed ads in the intended content and that reported views were real and not "bots." A 2014 report from the Interactive Advertising Bureau estimated that about 36 percent of web traffic that advertisers pay for as views results from computers with viruses programmed to visit sites to inflate traffic figures.[35] Certainly, none of these problems were insurmountable—they just explain the slow shift to online advertising despite the increasingly robust ability to deliver broadband video content by 2014.

Advertisers also made greater demands for accountability and return on investment information as they faced a variety of new platforms on which they could reach consumers. Agencies dealt with clients who made increasingly contradictory demands as advances in marketing research and new media venues allowed the more creative and precise messaging that clients' marketing divisions prized, while their procurement divisions—those who allocate the clients' advertising budget—demanded definitive information that remained elusive about how advertising spending correlated with sales. Advertisers' interest in alternative data—such as measures of viewer loyalty and engagement—indicated the rising disillusionment with continuing network-era advertising practices in a dynamic and cluttered media environment, though contributed little to the general norms of buying television advertisements.

As the post-network era developed, conglomeration affected industry operations in a variety of ways. Although few have explored the conglomeration that has occurred within the advertising industry, this process developed alongside the concentration of media ownership in the content and delivery businesses.[36] By the mid-2000s, four holding companies (Omnicom Group, WPP Group, Interpublic Group, and Publicis Groupe) controlled most of the industry's business, owned forty of the top fifty U.S. agencies, and, in 2005, earned over $31 billion in revenue from their various advertising and media, public relations, marketing communications, and specialty firms.[37] By 2012, that figure had grown to $45.02 billion, and the industry consolidated further in 2013 with a merger of Publicis and Omnicom.[38] Since the consolidation, some agencies have unbundled different components and created separate independent media departments. As part of the trend toward "communications planning," agencies have eliminated segmentation by media, or the use of separate teams for television, magazine, and point-of-purchase, and instead combined all media for more integrated planning.[39] More recently, agencies have created product placement and branded entertainment divisions particularly charged to develop ideas for clients and networks and to expand these growing practices.[40]

By the early 2000s, advertising agencies provided a range of services, including creative development, strategic planning, media buying and planning, and account management. Sometimes a single agency supplies a client with all four types of service, while in other cases the work might be spread throughout different arms of consolidated holding companies or among entirely different companies.[41] The creative staff develops advertisements and the content of point-of-purchase or other brand communication within the mandate of a carefully researched and tested brand strategy that is typically developed by the strategic planning staff, which researches consumer behaviors and attitudes about the product. Media planners develop strategies for reaching particular consumers through targeted media buys—often across multiple media—and media buyers negotiate the purchase with the networks. Television buyers develop and purchase the best plan in the upfront and scatter markets and then monitor those buys throughout the year, tracking make-goods and overseeing the buys as networks adjust their schedules. As ratings data have grown increasingly sophisticated and the number of networks has expanded, media buyers have collected and sorted through much more information in order to predict likely series performance prior to the upfront to determine the best purchase to reach the client's target consumer. The account management component of the agency deals directly with the advertiser and facilitates communication among the other units, particularly in the increasingly common case that a single agency does not house the media and creative development divisions.

As Jack Myers has noted, even before agencies began creating separate product placement specialists, industrial shifts caused by new media and challenges to the status quo operation of television resulted in a shift of power within agencies from creative divisions to media buying and planning.[42] Such developments repositioned this facet of the industry from what one executive described as an "assembly-line factory" business to one more craft-oriented. He added that the additional "creative" role now common for planning and buying divisions suggested a need for a compensation model based on outcome, which entailed a significant adjustment in the financial underpinnings of the existing, but eroding, norm in which the agency collected a fee based on a percentage of an advertiser's buy.[43]

The conglomeration of the advertising industry has created a complicated environment because a client typically will not allow its agency to represent another company in its competitive sector—for instance, Wendy's will not allow its agency to also house the account for McDonald's, Burger King, Subway, and so on. In addition, the conglomeration of the corporations that support the commercial television industry—such as the purchase of Gillette by Proctor & Gamble or the merger of Sears and Kmart—has also affected the industry, making opportunities for new business for agencies increasingly limited and decreasing the clients available to agencies as the merged companies integrate their advertising.

Increased cost efficiencies have been a key outcome of the consolidation of all disciplines of advertising into a handful of holding groups, and the agencies also benefit from the expanded information about the health of the market they gain from aggregating so many clients within one holding company. This bottom-line, tight fiscal management has become apparent in the broader U.S. economic environment in this period as well, as market maturity has resulted in few opportunities for expansion and required corporations to manage costs precisely. Accountability to budget and attention to performance measures have come to be tightly observed throughout the economy in general, but these developments have had specific consequences for the particular economies of advertising and a television industry in the midst of substantial redefinition.[44]

Adjustments in the industrial organization and norms of practice in advertising affect television production and network operations in many ways, especially since advertising dollars support much of the industry. These changes in advertising are among the most substantial explored in this book. Unsurprisingly, advertising agencies have been particularly vocal prognosticators of the significance of the changes occurring and have been among those most willing to accept and embrace changes in economic models and technological possibility. This became increasingly necessary as advertisers received decreasing returns on their commercial dollars throughout the multi-channel transition. As a result, they were in a position to adopt and to benefit more from new commercial practices than other areas of the industry—like broadcasters—for which a network-era status quo remained preferable.

Advertisers faced the loss of a mass audience as a result of the steady increase in choice of new networks and leisure devices throughout the multi-channel transition. Expanding viewer control technologies such as DVRs and VOD applications also diminished the viability of decades-old practices, including the thirty-second commercial. But even as these factors provided particular impetus for adjusting norms, they should not be viewed as the only causes of the paradigmatic shifts that began to occur early in the twenty-first century.

Since then, advertising possibilities unimaginable in the 1940s—graphically inserted ads on sports fields and stadium backgrounds, digitally created promotions of products placed in shows produced decades ago, tags added to the bottom of personal correspondence such as e-mails, sponsorship of every substantial cultural event and even component parts—have become routine and, although annoying and disconcerting to many, commonly accepted. The advent of social media has integrated advertising even further into daily life with opportunities to "like" and "friend" various products as well as request a running stream of marketing messages by following a company or product on Twitter or signing up for regular e-mail messages. The increasingly precise data about viewers and their behavior that these technologies provide have offered advertisers and media planners ever clearer pictures of whom they reach and how, while experiences with Internet advertising have provided new models for creating alternatives to legacy practices. Media buyers also choose among a much greater range of media than in the network era, as the Internet in particular introduced a new venue to reach potential consumers in a very targeted and often participatory way. Finally, a significant shift in cultural sentiment toward commercials and commercialism has occurred in the twenty-first century with the unprecedented expansion in the venues in which we tolerate commercial messages. The advertising researcher James Twitchell estimated that even by the mid-1990s, Americans observed three thousand commercial messages each day—a figure that also points to the cluttered nature of the advertising space.[45]

Distinguishing Post-Network Advertising Strategies: Old Methods, New Names

Following the quiz show scandals and the decline of sponsorship arrangements, networks and producers avoided including brand-name goods in television shows during much of the network era. This avoidance helped to prevent conflict among those buying commercials in the show—whether during the original run on broadcast or later in syndication—but it also arose from a sense of social unacceptability, and in some cases, government regulation.[46] Many shows of the network era consequently featured families with kitchens stocked with generic "cola" and "beer." In contrast, the high-profile appearances of the Aston Martin in 1960s James Bond films and, in what many note as the great success story of product placement, Reese's Pieces in *E.T.* illustrate the use of this strategy in film. In television, the practice of including named goods in both scripted and unscripted series reemerged at the end of the multi-channel transition, when it came to involve several distinct forms such as placement, integration, branded entertainment, and sponsorship.[47]

In much 1950s programming, advertisers made no distinction between the practices that we now differentiate as product placement and sponsorship. Reference to the sponsors' goods may or may not have been included in the content of a series, and the titles of series did not consistently name the sponsor. After magazine-style thirty-second advertisements became the norm, the differences between the two became more distinguishable, although trade publications and even the books that have explored the emergence and reemergence of alternative advertising strategies have not been consistent in defining them as I do so in the following sections.[48] I draw on actual practices that developed in the early 2000s and attempt to delimit distinctions through a more precise vocabulary. Importantly, though trade publications might refer to a number of practices as product placement, the financial underpinnings of these deals vary significantly.

Placement

Product or brand "placement" refers to situations in which television shows use name-brand products or present them on the screen within the context of the show; however, even in this simple advertising strategy there have been significant variations.[49] Placement can be either paid or unpaid, a distinction that highlights how two different aspects of business drove growth in this practice. In the case of unpaid placement, or what Twitchell refers to as "product subventions," companies donate products needed on the set for reasons of verisimilitude—if a scene takes place in the kitchen, that set needs to be dressed with products that make it recognizable as a kitchen.[50] Set dressers developed relationships with the prop companies that supplied them with their needs during the multi-channel transition, and many of these prop companies moved into the product placement business as they found manufacturers willing to donate the needed product in exchange for making the brand name apparent.[51] In contrast, paid product placement commonly originates in a deal created by an entity representing the advertiser and developed through negotiations with a network or studio. These deals could involve arrangements to feature a sponsor's product or name across the network or a night of programming, but most often they focused on a particular episode or show. In some cases, these deals evolved as part

of "added value" to a purchase of commercial time: instead of sponsors paying a fee specifically for it, the network supplied the placement in return for
another transaction. A network might offer "added value" opportunities in
exchange for an advertiser increasing its annual upfront spending or just in
recognition of a regular large spending commitment to the network.

In addition to paid and unpaid placement, there is another level of distinction that I term "basic" versus "advanced" placement. In the case of basic
placement, set dressings make the logo or brand of products clearly apparent, but the narrative or dialogue does not call attention to the product or
brand. In contrast, an advanced use of placement mentions the product or
good by name. In an episode of the NBC comedy *Scrubs*, for example, the
doctors played the game Operation, while in *The Office*, restaurants such as
Chili's and Benihana were used for staff lunches and parties, and not simply
as identifiable settings: significantly, the restaurants are explicitly named and
discussed on the show.[52] Subway also has used this strategy extensively with
shows such as *Chuck* and *Hawaii 5-0*.

To be sure, such distinctions are not ironclad, and there is not necessarily
a correlation between whether basic and advanced placements are paid or
unpaid. Thus, it may be difficult to determine the point at which a placement
might be considered advanced or precisely when placement becomes integration (see below). Furthermore, in some cases placement can result from
an advertiser's sponsorship of a show, which in turn can also create variations in the practice. For example, Coca-Cola and Ford used both placement
and sponsorship in their support of the FOX talent-competition show *American Idol*. Large cups of Coca-Cola sat in front of each of the three judges,
and the couch shared by the competitors featured the trademark swirl of the
company's logo, but it was this subtle placement in tandem with the video
shot in the "Coca-Cola Red Room" that particularly highlighted the brand
relative to the program. No contestants requested a sip of Coke before going
on stage in a manner more characteristic of advanced placement, but Coke's
presence on the show seemed more than that of basic placement because of
the blending of placement and sponsorship. Identifying these kinds of variations allows for more precise analysis of the increasing range of advertiser
participation in programming.

Though many examples of paid, unpaid, basic, and advanced placement
appeared across the networks throughout the early twenty-first century,
these techniques mostly supplemented rather than replaced thirty-second
advertisements. PQ Media reported that advertisers spent $2.83 billion
on U.S. television integrations in 2011, which was an 11 percent increase,
mostly due to growth in genres such as unscripted reality, how-to, scripted

comedy, and telenovelas.[53] Among them, basic placement has become an especially common practice, although little is known about effectiveness and advertisers continue to work at developing "best practices." By the end of the network era, advertisers and social scientists both had expansive research about how to maximize recall and effectiveness of thirty-second advertisements, but less data existed to explain the need for recall in placement situations or to otherwise evaluate the efficiency and outcome of placement strategies.[54]

Regardless of whether an advertiser seeks a basic or advanced placement, the key attribute repeated by advertisers, production executives, and networks regarding the viability of placement is that the product must be "organic," meaning the product must not seem out of place, be too obviously a commercial message, or call too much attention to itself. For example, placement of a cereal might "organically" fit in a breakfast scene of a domestic sitcom, though having the characters discuss their home insurance might not. Of course, all of the things that make placement organic also make it less noticeable to audiences accustomed to encountering thousands of brand messages every day. By contrast, an "inorganic" placement calls attention to itself in a negative way and seems forced and awkward. Inorganic placement exposes the constructed nature of placement but also breaks the viewer's submersion in the narrative, and different viewers likely have different thresholds and responses.[55]

New digital technologies also allow advertisers to rewrite the television past as companies such as Princeton Video place contemporary products and brands into existing series. Products can be placed in the kitchen settings of old situation comedies using the same technology that imposes the first-down line on television broadcasts of football games or brand logos onto the field of soccer matches.[56] The capability of adding and deleting products provides an important level of control, given the sometimes unanticipated subsequent markets for television series.[57] Likewise, this capability provides a response to concerns about placement deals for original-run programming. Many have wondered what happens to the good in subsequent markets and how its presence could produce additional revenue. Digital technologies enable studios to create new placement revenues in syndication by reselling the placement opportunity in these secondary markets.

The particularly noteworthy aspect of placement is the speed with which it became a common practice. Although articles in trade press and panels at industry meetings were exploring the strategy by the mid-2000s, the topic—as related to television—was largely absent from industry discourse until 2001.[58] Despite the seeming omnipresence of placement by the end of 2005, this advertising strategy is a new tool in advertisers' arsenal that is beneficial because viewers can't skip over it, yet it is best used only in particular genres and instances. Placement has nonetheless reemerged as a strategy that advertisers are willing to pursue, and there is little evidence that the trend will diminish.

Integration

Product or brand "integration" is an additional category of advertiser support in the post-network commercial economy. In cases of integration, the product or company name becomes part of the show in such a way that it contributes to the narrative and creates an environment of brand awareness beyond that produced by advanced placement. Because of their generic attributes, unscripted series have been more successful in integrating products than scripted series. For example, beginning in its second season *The Apprentice* challenged contestants to develop an advertising campaign or a similar activity for a known and real product. Thus, the series utilized the organic marketing potential that it effectively wasted during the first season's use of unbranded activities such as selling lemonade. In addition to selling commercial time within the series, *The Apprentice*'s producers added millions to their budget by selling advertisers the opportunity to be featured within the storyline of the show.[59]

The development of both placement and integration has thus provided unscripted series with important financing, especially in view of the escalation of production costs arising from competition in the form. According to conventional industry wisdom, most unscripted shows have little potential to recoup production deficits through syndication and consequently require producers to fully fund production through license fees or placement. Integration and placement revenues enable shows to afford impressive concepts or hire the limited skilled editing and production talent in this area of the industry, despite lower license fees and lack of deficit financing. Organic integration is unquestionably easier to achieve in unscripted formats, but notable examples of integration in soap operas and scripted series exist as well.

Industrial discussions of the growing practice of integration have focused on creative workers' fears that they could be forced to construct storylines to include brands. Such fears are justified, although to date, both the advertising and network sides of the business generally have shown restraint, aware that producing bad television diminishes the reputation of all involved. Still, some producers have accepted placement as a necessary compromise that can expand budgets in valuable ways. Peter Berg, executive producer of *Friday Night Lights*, acknowledged that "anything that gives a little financial relief, you can't ignore . . . it's all about giving them [the network] what they need in a way that doesn't violate the integrity or offend the audience."[60] His show frequently set meals at Applebee's, among other placements, and featured football players wearing Under Armour, and the extra revenue from placements allowed the series to renovate the dilapidated Texas stadium in which it filmed. Yet the challenge of successfully negotiating advertiser desires and the sensibility of savvy audiences makes it likely that integration will remain a tool of unscripted programming that coexists with conventionally supported scripted shows.

Branded Entertainment

"Branded entertainment" is a third advertising strategy that has grown increasingly commonplace from the beginning of the post-network era.[61] In this case, the advertiser creates the content of the show, which then itself serves as a promotional vehicle—somewhat akin to a long-form commercial or infomercial. Branded entertainment shifts more toward a sponsorship model and has taken a number of forms. *The Victoria's Secret Fashion Show*, aired by ABC in 2001 and in subsequent years by CBS, provides one example. The "entertainment" of the hour-long show served the promotional function of revealing the attributes of Victoria's Secret lingerie.[62] Here, the financial model featured ABC and Intimate Brands splitting the advertising time in the hour, and Intimate Brands paying for production fees—estimated at $9–10 million by the time the show aired on CBS—so that the event had no cost for the network.[63] Although this deal may make the show appear to be little more than a long advertisement, Andrea Wong, ABC vice president of alternative series and specials, defended the program by noting, "Clearly this is more than an infomercial. You will see Victoria Secret product but also entertainment in the show," which also included popular musical performers.[64]

Branded entertainment in the mid-2000s experimented first with original content and series in which brands created their own movies, series, or specials such as *The Restaurant* (NBC, 2003–2004), *Blow Out* (Bravo, 2004–2006), or the *Victoria's Secret Fashion Show*. But by 2013, little such content continued to be produced. Instead, branded entertainment—which Frances Croke Page, vice president and director of entertainment media at RJ Palmer, defines as "brand messages in the content, rather than being confined to the ad space"—had moved to smaller-scale "entertainment," including things such as slideshows on websites, but also had become widely used by companies with much smaller advertising budgets. The last decade and a half has led to multi-platform advertising and opportunities online and to branding events that also might be supplemented with a conventional advertising buy. Page notes that there are "all kinds of creative things you can do that support what's happening in the advertising." She thinks of the "content space as being able to augment and help complete the story of the brand. Sometimes you have to tell the story from scratch, sometimes you are just adding on to a larger story," with the ultimate goal of leading a consumer who hasn't used the product to think of it in a new way.

Branded entertainment marks a fundamental shift from intrusive advertisements pushed at audiences who are engaged in other content to advertising of such merit or interest that the audience actively seeks it out. Given that branded entertainment involves very different viewer behavior and perception of content than thirty-second magazine-format advertising, the genre may well require a wholly different understanding of the psychological processes involved, as well as new terms for assessing advertising effectiveness. As far as advertisers are concerned, branded entertainment also requires a massive shift in where they commit their money.[65] *Advertising Age*'s Scott Donaton explains that in traditional advertising, the advertiser allocates 90 percent of the budget to distribution—or buying time or space—and 10 percent to content production.[66] To develop content of such interest that audiences seek it out requires that advertisers spend 90 percent on production, leaving only 10 percent for distribution.[67] The emergence of social media as a tool that can push audiences of potential buyers to seek out content allowed advertisers to spend even less on distribution costs, as they could place branded entertainment on a company website and drive viewers to it through social media. In a somewhat paradoxical example, Kmart created a commercial and placed it on YouTube to test the response to its potentially offensive marketing of its online shipping though the attention-grabbing line "I shipped my pants." Social media enabled the quick spread of the advertisement, which led to twelve million viewings of the ad in just a few days. This example is paradoxical because the branded entertainment is really just an ad distributed on YouTube, and the viewer response encouraged the company to then purchase television time for the spot and expand the campaign.[68] The economics of television will likely need to shift more significantly for branded entertainment to become more common on the medium, particularly on broadcast networks and during prime time.

Single Sponsorship

Finally, the long-established practice of single sponsorship has also become more common than it had been since the establishment of the network era. Sponsorship involves a single corporation financing the costs normally recouped by selling advertising time. Sponsored shows often include no in-text commercials, but feature "a word from the sponsor" before or after the show. By the end of the multi-channel transition, sponsorship did not take a uniform pattern, and it may or may not have included product placement or integration in the narrative. In most cases, a company sponsored a single episode rather than an entire series. But even here, episode sponsorship was rare enough that it could create a distinct promotional attribute for the series, as well as providing the sponsor with a comparatively uncluttered environment for its message. Many of the sponsored episodes were for cable series, which involved a lower cost to the corporation than sponsoring a broadcast episode. For example, FX featured "commercial-free" first episodes for each of its original dramas beginning in the summer of 2004, and Ford sponsored the season opener of *24* in 2003 and episodes of *American Dreams* in 2005. Significantly, Ford blended branded entertainment in the *24* sponsorship by wrapping the episode with a six-minute film that began before and concluded after the episode and presented a *24*-like plot that utilized a Ford vehicle. Yet for scripted entertainment, sponsorship remained rare.

The premier example of sponsorship support on television is found in live sports programming. As the rights deals for sporting events have escalated, the broadcast and cable entities that win these license agreements have sought expanded income sources to recoup these costs. Thus it is the case that not only do audiences tune in to sponsored sporting events such as the Discover BCS National Championship (and importantly, here the sponsorship of the event goes not to the television provider but the sports league), but they also see a countless array of paid brand mentions, from the Tostitos Half-Time Report, to the Geico First Quarter Summary, to the awarding of the Allstate Player of the Game, that are arranged by the telecaster. And it is the exposure to the television audience—millions more than those in the stadium alone—that gives value to sporting goods manufacturers' sponsorship of teams and provision of all manner of uniforms and gear. The field of sponsorship messages in televised sports has proliferated so extensively because sporting events persist as a rare form of programming that draws broad audiences who watch in real time and is a type of content that can include an array of sponsored messages within the entertainment content.

Importantly, sponsorship situations in which the advertiser chooses not to disrupt the show for brand messages allow for the creation of narratives of some distinction—as possible with subscription-supported television. Magazine-format advertising requires story construction that can be interrupted, and the practicalities of maintaining audiences require writers to manufacture plots with points of climax before the prescribed commercial breaks. Sponsorship, by contrast, can eliminate the need for such interventions. As Tom Fontana, writer and executive producer of many series, including *Homicide, Oz,* and *The Borgias,* explains, "When you don't have to bring people back from a commercial, you don't have to manufacture an 'out.' You can make your episode at a length and with a rhythm that's true to the story you want to tell."[69] Many identified the freedom from content restriction and advertiser meddling as key causes of subscription cable networks' success in garnering a lion's share of accolades for the quality and aesthetic form of their series. Such institutional factors are significant, but the freedom from the tired narrative structures that require plot climaxes before commercial breaks also contributed fundamentally to their distinction. No scripted series has yet been produced with a regular sponsor, but such an opportunity could yield dynamic possibilities for storytelling unavailable to writers constrained by the narrative plotting required by commercial pods.

The renewed use of placement, integration, sponsorship, and branded entertainment has affected advertising companies and the networks in various ways. For one thing, agencies have redeveloped in-house divisions focused on creating integrated deals for branded entertainment beyond the standard thirty-second advertisement. The advantage of locating such divisions inside conventional advertising agencies is that it provides coherence with other ongoing strategies and also creates more continuity in the relationships between clients and their agencies. Another effect has more to do with technique and style. Many in the agencies quickly identified the need to integrate and place products in such a way as to not repel consumers, and worked to scale back the commercial messaging requested by overly eager clients. In 2003, when Frances Page was principal, strategy and business affairs for MAGNA Global Entertainment, she argued that the best uses of integration and placement include subtle penetration into the entertainment narrative, but also noted the need to utilize other public relations venues to maximize the exposure of the placement.[70] In 2013, she noted even further integration between the traditional task of public relations agencies, advertising, and media buying, which fits well with the

"360-degree communications planning" that the consolidated holding companies seek to provide.

For example, in 2004 Tylenol constructed a placement deal with CBS's *Survivor* in which viewers voted for the *Survivor* contestant who exemplified the Tylenol "Push through the Pain" ethic and had an opportunity to enter a sweepstakes that awarded a trip to the *Survivor* location. The process of voting sent viewers to a Tylenol website where they registered for the sweepstakes. In addition to creating the website, Tylenol's agency purchased advertisements in *USA Today*, in magazines (co-branded with the *Survivor* logo), and on radio encouraging viewers to watch *Survivor* and join the sweepstakes. The campaign included in-store promotional displays, commercials in each *Survivor* episode, and a thirty-second vignette at the end of the episode in which the series' announcer revealed the winner of the previous week's vote. In this campaign, then, not only did the product fit "organically" in the show, but also the client spent extensively in other media to support the value that could accrue from the integration deal. Ideally, a well-deployed promotion garners free publicity as well. By 2010, advertisers had social media tools available to gather the kinds of information that sweepstakes entries had previously been valuable for and also provided yet additional ways to reach potential consumers.

New challenges have emerged as these advertising strategies become more central to the industry. Key concerns include establishing norms of pricing and standards of measurement, as well as determining the effectiveness of the various strategies. Even as these experiments began, however, much about the value and effectiveness of new advertising strategies remained uncertain. Substantial change in advertising practices was deterred by advertising clients unwilling or unable to pursue new strategies without proof of their value, while networks and agencies could not offer such certainty without clients willing to engage in the trial and error of initial experiments. Frances Page suggested that in the transitional environment of 2005, clients might place 70 percent of their budget in established and traditional media, 20 percent in more unconventional but still tested media, and allow 10 percent to boldly experiment with new venues and possibilities, though the ability to experiment with even 10 percent was limited to the largest advertisers. The broad economic crisis that disrupted practices a few years later led to scaled-back advertising budgets across the board, and money for experiments was often the first to disappear. This informal 70/20/10 rule indicates the pace of change at which wary clients will finance alternatives to legacy models, particularly when little data exist to indicate the effectiveness of their efforts.

Efforts to Save the Thirty-Second Advertisement

At the same time that the industry has been attempting new strategies in response to the changing technological and economic environment, it has also been pursuing efforts to save or shore up the thirty-second spot. The growth in alternative strategies of advertising should not suggest that any group within the industry sought the elimination of the thirty-second ad. On the contrary, although newly developed control technologies form one of the most substantial threats to it, different groups in the United States and Britain have sought to keep it viable by using emergent technologies. For example, distributors have sought to use set-top and smartphone technology to incentivize viewers to stay with programs through commercials. British Sky Broadcasting satellite service enabled viewers to play along with game show contestants while using interactive features to keep track of the viewer's score. The viewer had to keep the set tuned to that channel for the duration of the program in order for the device to remember the score, which decreased the likelihood of channel changing. Another Sky initiative included the distribution of "loyalty cards" that viewers inserted in their set-top box. The device could monitor the viewer's behavior, and the willingness to stay tuned during commercials yielded rewards for the viewer at area retailers or enabled them to purchase other video content.[71] Significantly, while each of these endeavors offered viewers a service or reward for keeping their sets tuned to channels displaying commercials, none could actually confirm viewership: the messages might have played to empty rooms or to viewers engaged by other media. And these experiments were often fairly content-specific; the reward of playing along with contestants is a motivation specific to game shows. Still, many have seen the introduction of an added-value proposition for the viewer/consumer as a likely trend for the future, especially as new applications and services create new fees.

And despite the radical innovation often perceived of broadband video distributors, most new entrants aspired to network-era fifteen- or thirty-second advertisements as the financial model. Broadband distributors did experiment with the norms for this system—trying shorter commercial pods, less frequent commercial breaks, and more precisely targeted advertisements. In 2012 Hulu announced that it would charge advertisers only for ads that are completely played—its terminology is "viewed"—but that seems impossible to verify, nor could there be certainty that these views were not triggered by bots. Hulu showed more than 1.5 billion video ads in February 2012, which was slightly higher in number than Google's 1.2 billion, but in total time, Hulu offered 650 million minutes in comparison with Google's 119 million.[72] Hulu requires playing of pre-roll advertisements to access video, and reports a 96 percent completion rate. The video-ad serving company Freewheel reported average completion rates closer to 88 percent for long-form content and 54 percent for short clips.[73] A key advantage of broadband-delivered programming was the ability to count only completions, as well as a less cluttered advertising environment and greater demographic precision. But despite the increased certainty about demographic features of audience members and guarantees of completion, the revenue that entities like Hulu and YouTube achieved remains paltry relative to the broader television advertising business.

The early years of the post-network era were spent acculturating audiences with nonlinear experiences and teaching viewers that there were services that could deliver the content they desired. Among the start-up providers, these years featured a sorting out of which companies had the best technology and could build a desirable user experience. By 2014, YouTube and Hulu were certainly dominant, though Hulu's future was constantly in question as sales of the venture were attempted and it remained unclear whether it would hold its value if disconnected from its content-rights–holding owners. Hulu had emerged as an alternative to conventional television and control technologies such as VOD or DVRs. Though both Hulu and YouTube relied on advertiser funding, it was difficult to see them as competitors, except to the degree that any viewer's available time for viewing had constraint. Hulu primarily delivered industry-produced content, while such content was a small component of the YouTube repository. Likely the greatest threat to Hulu would be the assessment by its industry partners that viewers had become accustomed enough with online access that they might more profitably develop proprietary sites that offered the same functionality—which is more or less how the post-network era started.

In 2014 the thirty-second advertisement also was poised as the advertising mechanism that would support VOD. As Michael Bologna, director of emerging communications at WPP PLC's Group M, explains, "VOD was a worthless advertising medium until now," referring to the value added by dynamic ad insertion.[74] The cumbersomeness of VOD advertising—each feed had to be manually uploaded with ads included and could not be changed without re-uploading the entire show—contributed to the slow development of this means of distribution. Further, MVPDs—which tend to be regional—controlled the technology and the initial terms of advertising. This meant that the ad units would need to be sold locally, a more labor-intensive endeavor than national advertising. VOD advertising was still preliminary in 2014, with many norms and practice standards still being worked out. One issue was the C3 advertising currency, which counted views of advertisements only if done within three days of live airing. The audience measurement service Rentrak, which uses set-top box data, reported that 70 percent of advertiser-supported television VOD viewing takes place at least four days after the initial airing—giving it no value in the C3 system.[75] VOD also provided the opportunity for more direct audience targeting than advertisements in linear content, though it was unclear whether the use of thirty-second ads would be restrained enough to maintain a value proposition for viewers who also had other ways of accessing content.

By the mid-2010s, the rapidly changing landscape of television, devices, and audience behaviors made it difficult for advertisers to develop coherent strategies. Different types of advertising messages were more or less suited for the panoply of advertising techniques that seemed to emerge by the day. There would certainly be a place for advertising of all kinds in the post-network era, but the fracturing of audiences—not only across programming but also across ways of accessing it—challenged legacy practices.

In effect, then, none of the many advertising strategies in use by the end of the multi-channel transition replaced the thirty-second advertisement; nor is any likely to become the singular advertising strategy of the post-network era. Instead, the proliferation of strategies at the end of the multi-channel transition suggested that a mix of placement, integration, branded entertainment, sponsorship, and the thirty-second spot will continue to exist in a post-network era in which television encompasses a range of conventional, on-demand, and subscription services.

Cultural Consequences of Post-Network Advertising

At the same time television advertising strategies began to diversify and multiply, a variety of new perspectives in critical thinking about the relationship of culture and consumption emerged. Some scholars construed consumption as an empowering activity expressive of agency. Others, pointing to the construction of the "consumer" in industrialized and postindustrialized countries, questioned the nature of such an identity and whether it could be experienced in meaningful ways.[76] Still others explored issues arising from the increasingly niche quality of contemporary media and looked at the consequences of de-emphasizing mass culture in the advertising of goods.[77] While all register the complexity of consumerism in post-Fordist and postmodernist cultures, the redefinition of television at the beginning of the post-network era poses further challenges for the critical evaluation of emergent advertising strategies. Experimentation with placement and integration in texts that maintain thirty-second advertisements clearly suggests an increasing commercialization of television, but with marketing strategies shifting substantially, their specific cultural effects remain uncertain.

The programming created in service of the industry's commercial goals provides the link between television as a commercial enterprise and television as a social force, and the expansion in strategies used to finance programming affected other production components, just as the shift from sponsorship to magazine-format advertising affected the types of programming supported by those models. My focus here is on the implications of alternative advertising strategies for the creative output of this resolutely commercial medium. Thus, I am specifically concerned with the ways they have affected the transition from mass to niche audience norms, disrupted long-held industry practices, and created new textual possibilities.

Becoming a Niche Medium

Television's transition from broadcasting to narrowcasting has enormous implications for advertising. Narrowcasting enables advertisers to direct their messages to much more specific demographic and psychographic groups, and as a result, design creative messages that are different from those that would be used in broadcast forums. Studies examining advertising in fragmented media spaces have reached disparate conclusions about the sociocultural consequences of this shift. Two of these, quite different in approach, are worth singling out: Joseph Turow's treatment of the historical development of fragmentation across advertising media and the transition from mass to target marketing in various media, and Arlene Dávila's exploration of the creation of a differentiated Hispanic marketing sector.[78] In emphasizing the consequences of fragmentation on dominant white society, Turow finds the "gated communities" of taste and image culture that result to be a particularly negative development for democratic society.[79] Dávila examines the issue from the perspective of Latinos, perhaps the most underrepresented cultural group in U.S. media, and emphasizes the constitutive role of advertising in culture. Instead of viewing Latino-targeted advertising as a means for marginalizing their status in society, she asserts that "the reconstitution of individuals into consumers and populations into markets are central fields of cultural production that reverberate within public understanding of people's place, and hence of their rights and entitlements in a given society."[80] This approach echoes that of Néstor García Canclini, whose work likewise deconstructs the dichotomy between consumers and citizens as an essential aspect of understanding the circulation of culture and capital in a post-Fordist, globalized context.[81]

Many scholars exploring advertising and culture as television transitioned from a mass to niche medium have examined changes in the content of commercial messages. However, advertisements were not the only creative content affected by these adjustments. The transition to niche target marketing bore significant, if subtle, implications for the programming television networks created to collect that audience, as advertisers embraced programs directed to narrower and more specific audiences. Target marketing provides economic support for ways of constructing and viewing television that are emerging in the post-network era, including those that render television a subcultural forum.

Opportunities Arising from the Disruption of the Status Quo

The growing threat to established norms of commercial funding and viewer behavior in the early years of the twenty-first century forced a long-complacent advertising industry to break from conventional practices to a degree absent in the network era and much of the multi-channel transition and created the opportunity to experiment with new strategies, many of which have been chronicled in the preceding pages. Inevitably, some endeavors failed, and though significant change seemed likely in the mid-2000s, by 2014 the exuberance of a decade earlier had nearly died out. Nonetheless, advertisers' support of unconventional programming content and their willingness to explore different strategies to reach audiences have in themselves had an effect on perceptions of what might be possible and viable. Here, too, successful cases create opportunities for programming that is unlikely to be produced under the once-dominant thirty-second advertisement model.

Drawing from the study of organizations and innovation, Turow considers how periods of industrial transition can be linked with the creation of unconventional programming.[82] Such transitional periods, which create "cracks" in established organizational operations, disrupt hegemonic relationships and allow new norms to be established.[83] To be sure, this is not always the case. Those with more power can maintain their status, especially when the hierarchies that privilege them continue to seem natural and a matter of "how it has always been." But the scope of change throughout the production process of commercial television at the beginning of the post-network era has been so substantial that new practices have had to emerge and to do so in a manner that allows an uncommon degree of power redistribution, though mostly among corporate entities.

For example, while the new viability of product placement as an advertising strategy has consequences for programming, it is also shifting power relations behind the scenes. Producers who bring scripts to the networks with advertisers already attached to placement deals not only challenge network sales divisions, but also threaten to usurp their authority and management of advertising dollars. Advertising agencies also gain more control, especially when they develop in-house divisions specifically charged with creating programming into which clients can be integrated. As these new brand integration divisions take on expanded creative roles, previously ancillary jobs such as set dressing also compete for a place in developing shows.

In the post-network environment, advertisers have had the least at stake in continuing to pursue dominant network-era practices, particularly as they steadily came to pay more for less. Frustrated with the status quo and aware of the coming changes resultant from widespread adoption of DVR technologies, advertisers willing to commit part of their budgets to nontraditional strategies pushed the business out of its complacency, but haven't found new norms for television. Advertisers reasserted their status as the economic lifeblood of a commercial media system, but shifts in their behavior enabled significant adjustment throughout many other industrial relationships.

Advantages of Multiple Advertising Strategies

In addition to the experimentation and industrial reconfiguration occurring during the recent period of transition, new opportunities have also arisen with advertisers' and networks' willingness to accept a situation in which multiple advertising strategies coexist. It is too early to know how long this environment will continue or what might ensue, but in the near term, a multiplicity of possible financing models is indeed available, and each model creates more possibilities for variation than could develop in the long-dominant singular system.

To illustrate how the coexistence of multiple advertising strategies might lead to a broader diversity of programming, we might begin by looking at how the differences in programming on broadcast, basic cable, and subscription cable can be linked to their different financing structures. And regardless of economic model, for all the derisive commentary it has garnered, unscripted programming has instigated some valuable innovation in conventional business operations. Since many of these shows incorporate sponsorship or placement fees, they do not require deficit financing. As experiments that originated in unscripted programming come to be repeated in scripted programs, networks are increasingly able to realize value from new and diverse advertising strategies.

Product placement has provided cable networks with additional dollars for production costs that in turn have helped them compete with broadcasters' textual attributes. For cable networks, high-profile programming events have derived value beyond the ratings they achieved. Indeed, by 2005, one breakout-hit series could move a cable network from the tier of relative obscurity to high-profile awareness, as *Trading Spaces* did for TLC, *Queer Eye for the Straight Guy* for Bravo, *The Shield* for FX, and *Mad Men* for AMC. And once a cable network achieves substantial cultural awareness, it is much easier to secure the advertising dollars necessary to maximize its niche status through additional programming, though as the case of TLC now illustrates, cable channels must maintain a subsequent stream of programming to continue their perception of relevance.

Importantly, the highly competitive leisure environment that characterizes twenty-first-century U.S. homes has created a measured negotiation between content and commercialism in television. Here, the situation differs considerably from that of the 1960s. At that time, when television was still the next big thing in home leisure technology, and there were no competitors such as computers, gaming devices, and home video technologies to draw away viewers, these conditions helped enable the resolutely commercial practices of cheap and often uninspired telefilm production of the era. After twenty-five years of audience erosion, broadcast networks were far less cavalier in their attitude toward the audience and recognized that while cheap programming might aid the short-term bottom line, they could not take their viewership for granted.

Conclusion

According to an old advertising industry bromide, the industrialist Lord Leverhulme (or alternately, the department store magnate John Wanamaker) claimed that he knew he wasted half of his advertising budget, he just didn't know which half. The quip acknowledges that advertisers long have been aware of the inefficiency of their promotional endeavors, and yet, history illustrates that they have pursued them nonetheless. The increasing use of product placement and integration, as well as experiments with advertiser-created programming, suggests that the existing model was failing. Advertisers' willingness to devote some of their budgets to strategies other than the thirty-second advertisement indicated the level of crisis they perceived in existing models and the sense that this was not a passing fad, but the beginning of changes that could not be avoided. The early twenty-first-century disruptions to existing television industry norms, the emergence of online means of video distribution, and the quick adoption and abilities of the smartphone—all during a time of general economic crisis and malaise—created a difficult environment for innovation. By 2014, considerable experimentation persisted with little sense of an ultimate rearrangement of norms. Where the online realm was distinguished from "television" as "new media" just a few years ago and "mobile" screens seemed categorically different from domestic sets, these distinctions seem less definitive in 2014, and it instead seems that a more holistic video ecosystem is emerging. Advertisers will follow viewers, but viewer behavior continues to change and take on a greater variety of forms than advertising practices can keep up with. The question, then, is whether there is anything that might make viewers solidify new behaviors more quickly.

Key developments in recent years have been driven by new opportunities in distribution; for example, the experience of Netflix encouraged greater MVPD options, and subscription fatigue encouraged zero-TV homes. Such disruptions will persist, and eventually the forces that allow for the perpetuation of the bundle will be overcome. Or advertisers could start pushing this transition with their dollars. With dynamic ad insertion just becoming widespread as the book goes to press, the space of MVPD VOD is just beginning the negotiation of what advertising experience viewers will find acceptable. In the last year, I found most of the shows I watch on-demand to be delightfully devoid of advertisements, but cases in which VOD fast-forwarding was disabled have become more frequent in recent months and produced a personally unbearable viewing experience that sent me back to my DVR. The "compromise" of such systems might be a lighter advertising load than the linear broadcast and more relevant advertising. But for prized content and for those with leisure dollars, even this might not be an acceptable compromise.

Despite the similarity of current strategies such as placement and sponsorship to those that were dominant before the network era, the renewed use of these methods does not require a return to the power relations that characterized the earlier period. As existing practices eroded, a realignment of power occurred that privileged those best positioned to compete given the composition of the industry at that moment. Advertisers' willingness to explore new financing structures offered an initial salvo to which other players responded. The negotiation of the duties of network sales divisions, advertising agency media divisions, and talent/product placement firms illustrate the degree to which the industry was in flux. Various groups rushed to stake a claim in product placement and integration as the lucrative nature of these practices became evident. Network sales divisions, previously the gatekeepers who determined whose commercial message got on the air and when, saw their role threatened when producers such as the *Survivor* creator Mark Burnett came to networks with programs already integrated with support and financing. Network sales divisions began competing to maintain their status of delivering the golden goose as producers and various agencies attempted to usurp the power afforded to those who control the advertising dollars. Likewise, advertising agencies developed placement specialists, old prop companies morphed into placement contractors, and talent agencies attempted to leverage their existing position in this area of the industry.

Situations such as this represent key moments of crisis when conventional industrial practices and power roles can be challenged. Different workers rarely had the opportunity to redefine their task within the broader production process, as entrenched entities and status quo operations normally prevent change because of the threat of lost power that change poses. As in other sectors of the U.S. television industry, the erosion of norms and practices during the multi-channel transition created a situation in which the dominant processes faltered and created the opportunity for a reallocation of power among entities related to advertising and advertising strategies. Toward the end of the multi-channel transition, such opportunities were particularly important to those whose roles and relative power within the system of production were enhanced or diminished, but these adjustments also affected programming by creating new gatekeepers and allowing for a reallocation of content and audience priorities. The extent to which status quo industrial relationships were disrupted by this environment of innovation should not be underestimated. Where the conventional power relations that had so long stood as barriers to change had come down, fear of being shut out of the newly redefined industry motivated further innovation.

Very little critical scholarship has explored the cultural implications of differences among media that consumers pay for directly versus those that are advertiser-supported, particularly within a single medium. As financing models diversify and coexist, we need a much more comprehensive understanding of the significance of different payment methods to viewers' use and enjoyment of programming. Subscription television—a leading application of the direct-pay model—has defied many foundational assumptions about television's defining attributes. Indeed, it is in some ways more like media such as magazines or even mobile phones than advertiser-supported television, and the video-on-demand environment offers viewers even more opportunities to escape advertising through transactional payment. The availability of such options and patterns of viewer use—what types of viewers choose to pay for what types of content—are important factors to consider in studying the nature and effects of transactional television.

Exploring the differences between television that audiences are willing to pay for and television supported by some advertising means is but one of many new research areas resulting from developments of the multi-channel transition. Many of these developments require substantial research into viewer uses, behaviors, and preferences. They also give rise to many questions about the consequences of the erosion of mass audience advertising opportunities, the social implications of narrowly targeted advertising, and the effects of expansion in placement and branding on consumption. The continued process of transition allows only for conjecture about how the erosion of mass media affects Fordist models of consumption. It is also too soon to know the full range of textual consequences that results from eliminating the thirty-second advertisement norm and the social ramifications of narrowly defining consumer groups not only in programming, but in advertising as well. What we do know is that changes in program financing and in assumptions about the audience scale necessary for commercial viability have played key roles in determining the kind of programming that can be produced.

The diversity in advertising practices that emerged at the beginning of the twenty-first century indicates the extensive changes the television industry was undergoing. While network-era advertising practices had provided advertisers with decreasing returns for a long time, the continued supply of capital from this sector discouraged change. Advertisers' willingness to finance experiments with different advertising strategies was the first domino to fall in the chain of events advancing transformation, and it influenced many subsequent aspects of production and financing. Different advertising strategies led to different business models; different business models led to different funding possibilities; different funding possibilities led to different programming; different programming redefined the medium's relationship with viewers and the culture at large. Changing advertising strategies consequently indicated a vital development in the broader process of redefining television.

6

Recounting the Audience

Measurement in the Age of Broadband

Finding out whether *C.S.I.* beats *Desperate Housewives* is just the
beginning. Change the way you count, for instance, and you can
change where the advertising dollars go, which in turn determines
what shows are made and what shows are then renewed. Change
the way you count, and potentially you change the comparative
value of entire genres (news versus sports, drama versus comedies) as well as entire demographic segments (young versus old,
men versus women, Hispanic versus black). . . . Change the way
you measure America's culture consumption, in other words, and
you change America's culture business. And maybe even the culture itself.
—Jon Gertner, *New York Times Magazine*, 2005[1]

For most of U.S. television history, Nielsen Media Research provided the
common currency supporting the entire economic framework of the U.S.
commercial television industry. The industry trusted Nielsen as an independent player, and remunerated it well for supplying the agreed-upon standard
audience measurement values upon which the industry allocated millions,
and eventually billions, of advertising dollars each year. At first glance, audience measurement may seem a secondary and insignificant business relative
to what is commonly thought of as the "television industry," but as Gertner's
remarks in the epigraph suggest, adjustments in audience measurement and
research norms have the potential to significantly reconfigure a commercial
media system. Audience measurement is also increasingly important for
encouraging and discouraging various innovations during periods of industrial change, and so it is particularly vital for forming an understanding of
the emerging post-network era.

The role of audience measurement became particularly controversial in
the late years of the multi-channel transition, as Nielsen Media Research

endeavored to introduce technological upgrades that contributed to the reallocation of advertising dollars at the same time that new distribution methods and advertising strategies required impartial measurement for validation. The existing paradigm of audience measurement proved increasingly
inadequate for the variation characteristic of post-network television. This
chapter consequently considers the crucial role of audience measurement
and developments in this area during the tumultuous early 2000s, as well
as the effects that adjustments in this sector have had on the production
of television. In a statement that underscores the uncertainty and sense of
transition gripping the long-dominant measurement service as well as the
industry, Nielsen's then CEO, Susan Whiting, commented in 2005 that the
next three to five years would bring more substantial change to the television
industry than had taken place in the previous fifty years.[2]

Arguably, Whiting underestimated the time period just a bit, as it took
until 2010–2012 for broadband video distribution to really emerge and offer
further suggestion of the post-network era. In many ways, the post-network
era requires a complete reconfiguration of measurement systems. The measurement systems that evolved throughout the network era and multi-channel transition worked because the fundamental economic and distribution
mechanisms remained fairly consistent at their core. As broadband-delivered
video debuted, an initial impetus emerged to measure computer, tablet, and
mobile phone screens in the same way as the domestic screen because the
decades-old experience of advertisers trying to reach a mass audience with
thirty-second spots was immediately assumed to continue as the dominant
advertising practice.

The chaos of the television landscape evident by 2010 made charting the
future of measurement difficult. Certainly, technologies and methodolo-
gies could be developed to evaluate audiences wherever they might view.
The challenge of this era resulted from uncertainty about how to value
different types of advertisements relative to others and to existing norms.
Various technological mechanisms could allow much greater information
about viewers and their behavior if they viewed programming delivered by
broadband connection. Moreover, notable demographic differences began
to emerge between those who viewed linear television on a living room
screen and those who used control devices, with the latter tending to be
both younger and more affluent—two characteristics particularly important
to many advertisers. Perhaps one of the elements most effectively retarding
change was the paradox of traditional television spending throughout the
multi-channel transition. Though broadcasters offered smaller and smaller
audiences, their rates continued to climb because their audiences—though

significantly diminished—remained the largest audiences available. If advertisers instead had paid proportionally less for the audiences broadcasters now delivered, they would have had far more substantial budgets available for experiments in funding various types of broadband video distribution, which would have had the effect of pushing these industries into maturity more quickly.

The differences between advertiser- and subscriber-supported services detailed in the last chapter are also a crucial starting point for understanding the issue of measurement considered here. Measurement is most relevant to advertiser-supported media. Subscription services may be curious about how particular fare performs—how many people watch *Game of Thrones* versus *Girls*—but HBO executives are just as avidly watching subscription rates, DVD sell-through, iTunes sales, and international distribution, as well as whether programs are generating "buzz" in the culture that might push new subscribers to HBO. The multiplicity of models for funding television leads to substantially different concerns in assessing the value of advertiser-supported entities such as broadcast networks, Hulu, and YouTube, subscription services such as HBO and Netflix, direct-pay transaction sales of iTunes, Amazon, and DVD sell-through, and the advertising/subscription hybrid of cable and Hulu Plus. Indeed, part of the reason audiences were reasonably confused by the relative success of different shows across these distribution outlets resulted from network-era paradigms of measuring success that proved decreasingly relevant or only offered part of the picture: everyone I know is watching *Community*, it must be a huge hit; the *New York Times* covers *Mad Men* extensively, it must be a popular show. But in fact, the long-standard measure of a Nielsen rating explained less and less about the emerging environment for commercial television viewing.

This chapter deals with the issue of measurement for advertiser-supported television. The early part of the twenty-first century featured measurement strategies that attempted to keep up with the changing ways audiences viewed content on living room screens. As is the case in many developments explored in this book, advertisers and measurement services waited for the other to make a first move. Advertisers were wary of moving too much of their television spending into broadband-delivered programming until there was more information about who was watching, whether they watched the ads, and the effectiveness of these messages. Likewise, the measurement industry was waiting for there to be something someone would pay them to measure before devoting too many resources to measuring broadband viewing. In addition to this standoff, reliable measurement practices require careful planning and testing and precisely selected samples to guarantee that results are generalizable. To add to the challenge, the user base of broadband-delivered video remained in flux, as did the advertising practices that supported them. All of which well explains the slow growth and the uncertain future for measurement.

Many possible futures exist for advertiser-supported video. Live sporting events and contests are still useful for this advertising strategy, but also require innovation. As broadcasters endeavor upon more experiments that include a common package of content and commercials across devices—as in CBS's coverage of the NCAA men's basketball tournament—measurement practices that count audiences across screens will emerge. The work is already being done, but proprietary technologies in the mobile and tablet arenas make this task cumbersome. It is more commonly the case that linear, broadband, and VOD versions of the same content have different advertising, and that will require new agreements between providers and advertisers. Until the holy grail of a mobile/social advertising method emerges, both advertisers and measurement companies have more reason to proceed conservatively than to speed ahead. Measurement services really can't know what to measure so long as such a haphazard array of distribution and economic models persists, and it seems unlikely the post-network era will ever provide the consistency of the eras that preceded it. Broadcast networks continue to seek higher rates for declining audiences, cable channels seek greater parity in rates, advertisers' accounting departments want a clear return on investment for every ad dollar they spend, and media buyers, paid a percentage of the buy they make, have no incentive to try to rein in the advertising fees. Uncertainty over new or replacement measures has enabled the persistence of the network-era paradigm of valuing measures such as frequency (how many times a viewer has seen an ad) and reach (how many people see an ad), even if these were imperfect tools for media of another era that prove even more substandard relative to present and coming measures. As Gertner notes, changing the way you count may change the culture business and the culture, but changing the way you count is a feat unto itself.

Audience Measurement in the Network Era and Multi-Channel Transition

The shift to magazine-format advertising in the days of early television generated a need for audience measurement and spurred the development of the related yet independent audience measurement business. As advertising agencies became less involved in program production and more involved in determining ideal locations for a sponsor's message early in the network era, they carefully evaluated the audience data to determine whether they achieved value in their purchases.

U.S. television history features a range of systems, companies, and methods that provided increased precision as technological tools and the viable approaches to measurement expanded. From the advent of audience measurement in the 1920s, when announcers requested that audience members send letters and postcards, through Hooperatings, Audimeters, and People Meters, the techniques have grown increasingly sophisticated and yielded ever more information about the habits, behaviors, and characteristics of the viewers at home. U.S. television audience measurement has relied mainly on sampling in order to derive the audience size estimates upon which advertisers base the value of their purchases. During the 1960s and 1970s, Nielsen introduced the Storage Instantaneous Audimeter, a device that daily sent viewing information to the company's computers using phone lines and made national daily ratings available by 1973.[3] At this point, audimeters offered no information about the demographic attributes of the audience, but Nielsen could triangulate the audimeters' information about the stations sets were tuned to with diary reports that provided some sense of the audience composition. By the early 1980s, the Nielsen sample included approximately 1,700 audimeter homes and a rotating panel of approximately 850 diary respondents.[4] Nielsen introduced its Nielsen Homevideo Index (NHI) in 1980 to provide measurement of cable, pay cable, and VCRs, and the NHI began offering daily cable ratings in 1982. Nielsen dominated the measurement of national network television and competed with Arbitron in measuring local markets at the beginning of the multichannel transition.

Nielsen made a substantial technological advance before network-era norms entered crisis with its transition to the national People Meter sample in 1987. The initiative was a competitive move required by the entry of Audits of Great Britain (AGB) into the U.S. market and its implementation of a similar technology. AGB helped create a competitive environment that allowed for innovation, but the industry was unwilling to pay for two sets of numbers, and AGB's competitive efforts were short-lived. People Meters represented significant advancement over the preceding technique, but no adjustment in audience measurement norms occurs without substantial controversy. Any change in method—with its attendant change in results—clearly costs those whose audience had been overestimated. This reality leads to protest even if the new data provide more precise results. The nearly constant changes in technology and distribution that characterize the end of the multi-channel transition were thus especially troubling for Nielsen, as they required oner-ous adjustments in measurement techniques. At the same time, though, the industry's uncertainty about emerging advertising strategies, distribution windows, and ways people were using television increased Nielsen's central-ity, because all sectors of the industry were eager for information about audi-ence behavior in the new context.

The implementation of the People Meter at the start of the multi-channel transition resulted in growing pains, as the networks and advertisers knew that such a significant methodological shift would likely indicate some dis-parity from the established system. The networks did not offer audience guarantees on advertising purchased in the upfront for the 1987–1988 season because of the uncertainty of the new measurement system, and CBS in par-ticular experienced significant audience loss because the previous methods tended to overcount older viewers, who disproportionately favored CBS.[5] The People Meters reported that fewer audience members were watching broadcast networks, but did not find these audience members to be watch-ing something else instead. People Meters reported a 5 percent drop in the number of female daytime viewers, a 3.7 percent decrease in broadcasters' prime-time viewing, significantly smaller audiences for many popular pro-grams such as The Cosby Show (10 percent in the case of Cosby), and higher ratings for some late-night shows.[6]

Most of the industry-shifting features characteristic of the multi-channel transition were fairly rudimentary at the time of the introduction of the People Meter, and audience erosion due to cable and VCR penetration had not yet significantly reconfigured audience distribution norms. For the most part, VCRs were used only occasionally and cable programming consisted mostly of "reruns" at this time, enabling the persistence of linear viewing through the multi-channel transition. The People Meters arrived the year after advertising spending dropped for the first time in fifteen years, which compounded anxiety about the measurement switch. Broadcasters were gen-erally nonplussed about the results, but there was little recourse available. CBS dropped Nielsen and briefly opted for the meter service of its competi-tor AGB, which did not show as significant audience decreases; ABC and NBC threatened to do the same, and ABC signed only monthly contracts with Nielsen instead of yearlong commitments during the People Meter implementation.[7]

The backlash against the introduction of the People Meter was mini-mal compared to the industry response to measurement adjustments once the multi-channel transition was more definitively established. The result-ing audience fragmentation created a competitive environment in which fractions of ratings points meant the difference between a network ranking first or fourth and affected the allocation of millions of dollars in advertising. According to an estimate produced by the Broadcast Cable Financial Management Association, by 2005 a prime-time rating point on a Big Three network was worth $400 million per year.[8] By the beginning of the twenty-first century, researchers simply could not test new methods and implement them quickly enough to keep up with industrial changes.

One of the most challenging aspects of audience measurement during the multi-channel transition resulted from the intermediate nature of new technologies and distribution systems. The sampling techniques that most audience research relied upon were based on a fairly uniform nationwide availability of technologies and programming, and thus reflected a network-era experience with television. The arrival of varied programming tiers of cable channels challenged this system, as U.S. television homes began having highly discrepant access to technology and programming and consequently began using television in significantly different ways. The influx of new technologies such as DVRs complicated established in-home measurement systems and required the development of entirely new protocols in order for DVR-owning households to be included among sample homes. Historically, the Nielsen box derived its measurement by registering the frequency of the television signal in order to determine the channel being viewed. This technology does not work in an era of DVRs because even when a viewer watches content live, the signal still goes through the DVR, which constantly emits the same frequency. Nielsen consequently needed to develop an entirely new device, the A/P (active/passive) meter, which reads a code embedded in the audio track of programming rather than the tuning frequency.

Any one change in the measurement environment represents an enormous challenge to audience research norms and requires exceptional resources in response. But as these challenges confounded research firms' current plans, more threatening forces gathered on the horizon with the emergence of the post-network era. The arrival of cross-platform media delivery—"television" content delivered via the Internet, by mobile phone, and so on—would render television-only measurement technologies obsolete, especially once audiences embraced the platform agnosticism many predicted. Data about use of these new technologies and the effectiveness of commercial messages transmitted through them were vital to engender the confidence of advertisers to leave behind legacy models of commercial message delivery. As the measurement incumbent and monopolist, Nielsen was set to play a pivotal role in charting the future of television.

Audience Measurement in the Post-Network Era
A Lesson in the Politics of Measurement: Introducing the Local People Meter

Many accused Nielsen of complacency as adjustments in the industry com-pounded in the mid-2000s, but Nielsen had been making steady changes since the beginning of the decade. Nielsen changed counting methods and began weighting its sample in the fall of 2003 in response to census shifts and requests from some sectors of the industry. But then the industry responded with a torrent of criticism when it appeared that young men had all but dis-appeared from the television audience that autumn. While the networks cried foul, media buyers noted that the season featured few new programs likely to attract men and that the count might address the suspected move of the demographic to cable and video games. Eventually the young men were found—after many new programs in 2004 distinctly targeted the demo-graphic—but this public attention to Nielsen was just the beginning of the firestorm that continued as Nielsen implemented its automated Local People Meter (LPM) technology in New York and Los Angeles in 2003.[9]

Nielsen introduced the Local People Meter at the same time that evi-dence of the fracturing network-era business model became widely apparent and anxiety about the future of the industry raged in all sectors. The LPM marked the shift from active, diary-based local measurement to more pas-sive, meter-monitored measurement of local markets. Technologically, the LPM is similar to the People Meter Nielsen had used for its national sample since the late 1980s. The key advance of the LPM is that, in contrast to the People Meter, which dealt with a sample that was representative nationally, the new device provides accurate measurements of particular local markets. The mechanized LPM system has also pushed the industry further toward year-round measurement, as opposed to focusing on the quarterly "sweeps" periods used in diary-based surveys of local markets. Nielsen began a test of the technology in Boston in 2002 and completed the measure in 2003, at which point it announced plans to roll out the technology in the top ten Des-ignated Market Areas (DMAs) over the next two years.[10]

The initial rollout, particularly in New York, garnered substantial negative publicity and even congressional inquiry as a result of differences in audience behavior reported by the LPMs. Much of the concern centered on the dis-crepancy between diaries and LPMs in reporting minority (African Ameri-can and Hispanic) viewing as broadcasters that traditionally dominated in reaching these audiences, such as FOX and UPN, earned substantially lower ratings. In response, some industry players, specifically News Corp, engaged in a particularly belligerent public relations campaign aimed at discrediting

Nielsen. The creation of the News Corp–funded "Don't Count Us Out" coalition made the new measurement technology a hot-button political issue by framing it in terms of racial disenfranchisement. Nielsen also received substantial complaints and faced a lawsuit from Univision for adjustments in its procedures for counting Hispanic households in Los Angeles.

Such developments make it very apparent that advances in measurement can be politically precarious. Nonetheless, researchers believed that in comparison with the diary method it replaced, the LPM more accurately reported the full range of programming viewers watched, including that observed while viewers channel-surfed. Many speculated that diary writers underestimated the viewing done while surfing and were more likely to remember to write down well-known shows that were more commonly found on broadcast networks, thereby increasing the audience estimates of broadcast shows and decreasing those of cable.[11] The new methods generally indicated greater viewing of cable and decreased viewing of broadcast—results broadcasters were neither interested in knowing about or funding. For example, while LPMs found fewer African American audience members watching FOX than the diaries had, the LPMs also reported a 180 percent increase in total day viewing of BET and reported more than 100 percent increases in the number of African Americans viewing the cable networks ESPN, LMN, Telefutura, and Starz.[12]

Despite the increased precision of LPMs, it is unlikely that Nielsen will utilize them beyond the top 25 markets—out of 210 nationwide—because of the cost of this measurement system and the inability of smaller market stations to afford the resulting expense of an LPM-based report. In July 2010, Nielsen received accreditation from the Media Ratings Council for its LPM measurements in the top twenty-five markets, which account for 49 percent of the population and 64 percent of local television commercial advertising spending.[13] In 2013 Nielsen began testing its "Code Reader" box as a diary replacement in other markets. These light boxes that communicate using cellular technologies can be sent to homes by mail and do not need to be physically connected to the set or a phone line. This eliminates the extensive recruitment and installation costs of traditional meters, though they also do not have a way to register demographic information about the viewer— merely to what channel the set is tuned. Still, this meter overcomes many of the limits of diary measurement.

For the most part, members of the Madison Avenue advertising community remained out of the political fray and focused more on measurement issues arising from DVR adoption; the simultaneity of the LPM controversy and the need for DVR innovation illustrates the array of major challenges that bedeviled the industry even before the emergence of other screens for viewing. Advertisers have allocated their most substantial budgets to national buys guaranteed on national audiences, so the shift in measuring local markets was a less pressing issue for them than for the networks. Importantly, the networks derive most of their profits from the local advertising time sold by stations they owned and operated in the nation's largest markets, which is why they registered such concern about the local transitions.

The amount of public criticism Nielsen endured as a result of the LPM seems curious, considering how small an improvement it represented relative to more radical developments in measurement technology needed to address the new ways of accessing and viewing that emerged in the post-network era. In effect, the LPM involved a linear advancement from norms established during the network era. In other words, the LPM allowed Nielsen to maintain established measurement practices, but to do them better. At the same time that Nielsen pushed ahead with the LPM in a way that did little to account for the chaos increasingly challenging the television business model, it and other companies also attempted to account for various technological and programming developments that were transforming the industry. The turmoil over the LPM offers an important lesson on the challenge of changing measurement norms and the unexpected outcomes of improving measurement techniques. Nielsen faced congressional inquiry and accusations of disenfranchising black and Hispanic audiences because it refined its measurement techniques in a way likely to more accurately measure viewership. In the land of audience measurement, companies don't always get rewarded for doing their job better.

Extended Home Measurement and Panel Expansion

For Nielsen and its sampling-based measurement techniques, the most significant consequence of the multi-channel transition was the fracturing of the audience across tens, then hundreds of new channels. In order to maintain accuracy, Nielsen needed to significantly expand its panel. Nielsen tripled the size of its National People Meter survey between 2007 and 2011, increasing it to 37,000 homes and 100,000 people. The panel had already been expanded from 5,000 households in 2002.[14]

In addition to expanding the base of its panel, Nielsen also added some of the viewing of those panel members that existing norms did not count. Termed "extended home viewing," this initiative, which became part of the panel in 2007, sought to incorporate viewing that occurred in domestic settings not currently measured by the service. One component of the initiative measured the viewing of college-age members of Nielsen families who resided in college housing during much of the year. As of 2003, census figures estimated that 7 percent of the eighteen- to twenty-four-year-old population lived in dormitories, and Nielsen had never included this viewing in its sample.[15] The outcome of the preliminary study found significantly higher viewing among this group than expected, with an average of 221 minutes per day, and notably, this group often viewed with visitors. Additionally, the college students watched ad-supported cable over broadcast networks by nearly a two-to-one margin.[16] Although college students had been counted in the Nielsen sample previously, no viewing was attributed to them when away from the family home, which depressed the usage levels of this particularly advertiser-coveted group. Nielsen began regularly including college students living away from home in February 2007.[17]

This strategy of identifying viewing missed by established protocols was repeated in 2013 as Nielsen redefined the "television home." For all of measurement history, Nielsen defined a television home based on requirements that it have a set and means for the set to receive signals: either broadcast reception or a subscription to cable, telco, or satellite service. In 2013 Nielsen identified the category of "zero-TV" households, which watched "television" that did not meet this definition, a group that had grown from 2.01 million homes in 2007 to 5.01 million homes in 2013. Zero-TV homes use a broadband or wireless connection to access video, and Nielsen added this viewing to its definition of a television household in measurements beginning in 2013.

A full year's worth of data with this new household definition was not available as the book went to press, though it seemed likely to generate controversy; indeed, concerns voiced by local broadcasters forced Nielsen to delay release of the first year of data. Adding the viewing done by zero-TV homes provided a five-million-home shift in the denominator, which would have the effect of decreasing the rating of any content not watched as much in zero-TV homes as those long part of the panel. Given the technological competency required in 2013 to access online video as a household's sole video source, it was reasonable to surmise that the viewing habits of this group of five million likely did not mirror the tastes of the general population and the new results would result in critique of Nielsen. But this too represented a refinement in measurement technique aimed at yielding more accurate data.

The idea for adding zero-TV homes partially came from Nielsen's struggles to make sense of data that indicated a strong downward trend in the viewing of young people, but notable expansions in use of gaming consoles, particularly among those outside traditional gaming demographics, such as women ages thirty-four to forty-nine. Nielsen's measurement technologies cannot discern what a gaming system is doing, but the spikes in use of devices connected to the Internet seemed likely to correlate with use of gaming systems such as the Xbox to stream content from broadband programming services such as Netflix. Though Netflix is not an advertiser-supported entity in need of measurement by an external party, having data about these emerging forms of television viewing was a crucial part of understanding the present and future uses of the medium and for combating the picture suggested by the existing data, which showed suppressed viewing among particular demographic groups.

DVRs and C3

The methodological issues that resulted from audience fragmentation during the multi-channel transition offered only a slight suggestion of the scale of difficulties yet to come. As noted already, DVRs affected a wide range of industry norms, and audience measurement is among them. The devices required new practices for reporting audience viewership, and their capabilities for commercial skipping diminished the significance of series' viewership to the economics of television.

As U.S. homes added new technologies, Nielsen reconfigured its electronic measurement systems in order to adapt them to the hundreds of different models of televisions and then VCRs that became available. Nielsen disqualified DVR homes from their sample for the first eight years that the technology was in use until it could find a way to integrate it into the panel in a way acceptable to advertisers and networks. DVRs and VCRs may appear to offer similar technological functionality, but Nielsen's preliminary DVR use studies found that DVR households recorded a daily average of 30 percent of programming and 46 percent of prime-time viewing.[18] By contrast, viewers used VCRs in more extraordinary circumstances, such as when they needed to be out at the time of a particularly favored program—as in the case of regular time shifting of soap operas—or for the creation of archives. Upon the widespread adoption of the VCR, Nielsen reporting practices changed little, largely because the service lacked a way to measure whether viewers played back the content they recorded. But the situation changed with DVRs. Including DVR use in its sample required Nielsen to create multiple reports in response to different requests from clients. Advertisers wanted to negotiate pricing based on live viewing or at most that done the same day. Networks requested reports that gave viewers a longer opportunity to view, such as that done within a week of the recording.[19] Nielsen offered various reports as the advertisers and networks negotiated their way toward standards on which they could agree. The industry analyst Jack Myers opined in 2005 that "It will realistically be another three to ten years before new technologies such as the DVR have sufficient penetration to upset traditional viewing behavior and before new metrics are fully developed, tested, successfully modeled, and syndicated."[20]

It required four years to adopt a new standard, though nearly ten years later it is difficult to say whether traditional viewing behavior has been "upset" by DVR use alone. Myers, like almost everyone else, did not anticipate broadband delivery, which exponentially complicates the measurement of post-network control technologies. Nielsen began reporting time-shifted viewing in 2005, but the industry did not approve use of a common standard, what would be called the C3 measure, until 2009. The C3 measure counts live viewing plus viewing recorded on a DVR if it is viewed within three days of airing and if the commercials are played at regular speed. C3 represents a compromise between advertisers who wanted only live measures and networks who sought a C7 measurement that would include a full week of playback viewing. By 2013, calls to expand to a C7 rating seemed to be getting louder, and some 2014 upfront purchases were reportedly based on C7 viewing.

The slow initial adoption of DVRs aided measurement services struggling to keep up with the pace of technology: the devices were in just 32 percent of homes when the C3 standard began and nearly 50 percent in 2014. However, the lack of measurement also slowed the adoption of video on demand (VOD), which was included in the C3 measurement only if it had the exact same commercials and commercial load as the live airing. Experiments emerged in 2013 that embedded the commercials of the most recent episode in any episode viewed using VOD in hopes of expanding the monetization of VOD, an initiative Nielsen called "on demand commercial ratings" (ODCR). Video on demand continued measurement dilemmas begun by the DVR, but a whole other way to view emerged coterminously. Just as C3 solved the quandary of DVR measurement in 2009, broadband-delivered video took off and proved a more technologically sophisticated challenge.

Cross-Platform and Extended Screen Initiative

Nielsen initially identified efforts to measure viewing done on new screens and through new technologies as part of "cross-platform" viewership, an initiative begun in 2006, but not an official panel until 2010. These early reports, which Nielsen shared beyond its subscribers—unlike its proprietary ratings data—were valuable to those outside its client base for indicating the precise scale of technological diffusion and insight into how emerging viewing technologies were being used. Though among some enclaves it might have seemed that "everyone" used a DVR, accessed content "over the top," or had an iPad, Nielsen aggregate use data helped moderate the rhetoric that resulted from the tendency of journalists and other opinion leaders to be early adopters and nonlinear television users.

In response to the availability of full-length recent television episodes on outlets such as network websites, Nielsen developed its "Extended Screen Panel" in 2011, which added viewing done on computers, so long as the video viewed on these devices had the same commercial load as the original airing, and Nielsen began developing technology to measure mobile streaming apps for tablets and mobile phone screens as well. As of early 2014, neither viewing on tablets nor viewing on mobile phones was included in the panel, though Nielsen announced target dates of the fall of 2014 for preliminary reporting of these data.

Measuring these new screens presented myriad challenges. Including devices in the Extended Screen Panel required panelists to allow Nielsen to attach a device to computers, which was a more difficult permission to secure than gaining assent for television measurement. Still, much computer-based video use was likely to go unmeasured; at the 2013 National Cable Show, a network sales executive expressed belief that 50 percent of streaming is done in the workplace, which is a computer that would not be included in a Nielsen household panel. In 2014, each different brand and model of phone and tablet required a different technological solution to be included in the sample, and many of the manufacturers made it difficult to devise technological measures of the devices. For many reasons, then, the Extended Screen Panel was a logistical nightmare that still couldn't account for all viewing.

Though the Extended Screen Panel best approximates a combined measure of the same content across devices, in 2014 this wasn't likely to include much viewing. Online viewers were far more likely to go to aggregators such as Hulu that did not have the network-aired ad load, and thus were not included in this measure. Of course various other services, Nielsen among them, were also measuring online behavior and use more generally, but it was unclear whether the business of advertising would adapt to post-network distribution possibilities. Nielsen's next endeavor was the measurement of Digital Program Ratings, which used embedded watermarks to count programs viewed online regardless of the advertising load. Such viewing might not count for the ratings measure—which was primarily of interest to networks as the basis of advertisers' spending—but it enabled Nielsen to report to content creators about what aggregators and technologies viewers were using to watch programs. Nielsen first made Digital Program Ratings available in November 2013.

New Measurement Currencies

As explored in chapter 5, the early years of the post-network era featured experimentation with an array of advertising strategies and a gradual inclusion of placement and integration in some program forms. There was also considerable experimentation with different measurement currencies in response to these advertising strategies, the rapidly expanding data about audiences that new technologies increasingly made available, and the general uncertainty of the era.

"Engagement" was one of the first currencies to receive a good bit of discussion and consideration within the industry. As the post-network era began, it became apparent that regardless of audience size, not all audiences were the same. Those trying to craft an engagement metric argued that viewers watched certain programs more intently or had greater passion for some shows, and that advertising messages placed in such content should be more highly valued as a result. The logic of the argument surmised that in the network era of linear television, a viewer watched whatever was "on," even if that show might have not been desired and may have just been the "least objectionable." As cable channels expanded their original program offerings to more precisely target specific audience interests in the late multi-channel transition, viewers increasingly found programs they were excited to watch and engaged them more closely. Such focused attention to content might reasonably correspond to closer attention to advertising, and some cable channels commissioned or conducted research to make this point.

Engagement was widely discussed by 2005, particularly as cable channels sought to increase their share of national television advertising buys, and as recently as June 2013, the sound recognition company Shazam released a service called the Shazam Engagement Rate aimed at tallying how frequently viewers tagged products after seeing television ads. Engagement was centrally relevant to those seeking to monetize social media, and thus it circulated as an advertising industry buzzword and an idea with considerable conceptual purchase. Still, it did little to displace the conventional metric of ratings for television advertising sales.

Considerable attention to "buzz" also developed as many sought to understand connections between social media mentions and viewership. Various measurement currencies circulated that tried to predict ratings based on social media mentions, and by 2013, Nielsen found that Twitter mentions, *along with past performance of a show,* were a strong predictor of ratings. But this was still very early days for innovative intersections between social media and television, and it was unclear how use of Twitter might change over time, how the body of Twitter users might change to make it more or less representative of the viewing audience, and whether certain types of programming were more likely to be present in social media conversations.

As the post-network era adjusted the U.S. television industry, another revolution casually referred to as that of "big data" simultaneously swept through just about every facet of life, as once unimaginable tools for gathering and making sense of data about Americans led industries to consider how they could use this information to improve their businesses. The television and advertising industries were no different, but as of 2014, the data revolution remained mostly on the horizon. It was certainly clear that a television rating was but one small point of information and that it was now possible to know much more. What remained most unclear, though, was how to value that information or what measurement currencies could be agreed upon as a basis for advertising transactions.

Competitors

Due to its domination of U.S. ratings, Nielsen was a frequent whipping post for networks and advertisers whenever the economics of their businesses shifted in negative ways. Toward the end of the multi-channel transition, the further changes facing the industry and their consequences for media measurement services were difficult to anticipate, but it is likely that the monopolization of U.S. ratings by a single company contributed to its lack of preparedness at the dawn of a new age of television. The absence of competition within the field of audience measurement discouraged innovation until innovation became imperative, yet the clients who fund audience research did not want to face the additional costs of experimental new services; moreover, the networks and channels did not want a new measurement system that suggested a decrease in the value of their content.

The challenges in audience measurement faced by the television industry during the shift to the post-network era are not so much the result of Nielsen's decisions as they are a consequence of the industry's historic willingness to accept a ratings service monopoly. But Nielsen's status is even more complicated. At times it has seemed to exist as a joint industry committee—a collaboratively developed body tasked with industry oversight—an option preferred in European media markets and enabled by their different approach to antitrust and collusion laws.[21] This awkward status of the company contributed to the politicization of the LPM controversy and to heightening agency and network frustration with not being able to keep up with changing audience uses of television that resulted from the new technologies and distribution possibilities by the mid-2000s.

As further breakdown of linear viewing norms threatened to fragment audiences temporally as well as among broad content options, measurement companies have been challenged by the fact that the user characteristics of each of the various delivery systems differ, as does the blend of advertiser and subscriber support. Such variation requires audience measurement mechanisms for each form of delivery, and each faces the lag between invention, adoption, and refinement of measurement methodology that the industry experienced with the DVR. Though long-form commercially produced content is now widely available, viewers remained uncertain of where to find content and what value proposition they might receive from different services, while advertisers and service providers remained locked in a frustrating battle over how to value and measure content across screens and advertising experiences. It is little wonder that subscriber-based services such as HBO and Netflix have been able to innovate so extensively in recent years in comparison with the muddled advertiser-supported space.

Even as new technologies create new challenges for audience measurement and research, they also provide new tools. All of the technologies and devices that have been introduced in the digital era are significantly "smarter" than those they replaced. Digital set-top boxes, DVRs, and devices that deliver broadband services not only keep track of what viewers do but are—or can be—linked to communication systems through which activity is reported to central servers and mined in various ways. The possibility of replacing the sample-based data-gathering methods that have figured audience behavior since radio with actual data of real set use represents the most significant advance in audience measurement to result from the introduction of digital technologies. Though many refinements in data are now possible, the variety of stakeholders and range of proprietary technologies have slowed development of measurements that match the sophistication of contemporary technology.

One possibility for refined measurement of living room set viewing is data from set-top boxes, and the measurement and research company Rentrak began using these data to compete with Nielsen. By August 2013, 231 local stations in 54 markets subscribed to Rentrak's TV Essentials service, though a 2011 report noted that two-thirds of stations used it as a supplement rather than a replacement.[22] Many of the stations are in smaller markets that Nielsen measures only with viewer diaries. Rentrak's ratings are based on set-top data from Dish Network, Charter cable, AT&T's U-verse, and as of 2013, DirecTV services. It collects only passive data—meaning that it can report only what the set-top box does, so it cannot offer demographic information about who is in the room, nor can it be certain that there is anyone watching and the set-top box has not just been left on. Where Nielsen creates a representative sample, Rentrak does not account for how those homes in its census vary from the population as a whole; for example, over-the-air viewers and viewers of all but the four providers listed above are not included in its measure at all.

As the Rentrak methodology somewhat illustrates, the panacea of set-top box data is confounded by the regionality of cable providers and the multiplicity of set-top box manufacturers. Even if all providers agreed to sell their data to a single service, managing all the different data formats would be unwieldy, the active component of who is in the audience is missing, and over-the-air viewers would be entirely disregarded. Some have suggested that the future of measurement will be in technologies embedded in devices—televisions, phones, tablets, and computers—which would require cooperation from consumer electronics manufacturers, but that too is still more idea than practice. Though digital technologies have provided tools to gather data about technologies that rectify limitations of existing measurement protocols, they do so while introducing yet other limitations. The shared and agreed-upon standard of the Nielsen rating has been crucial to a business as vast as national advertising. Though data collection technologies now reveal much more and varied information, these data are likely to remain supplemental until a new standard of sale is agreed upon.

Despite considerable advances, the only certainty in audience measurement at the beginning of the post-network era seems to be that the days of a single measurement service with a standard currency are over. Much of the need for a plurality of measurement protocols has resulted from the adoption of the plurality of advertising strategies, technologies, and means of distribution explored in preceding chapters. As in other sectors of the industry, the multi-channel transition led to a fracturing of a single dominant standard that had been characteristic of the network era and the creation of a more diverse and complicated industrial field. Developing measurement capabilities for new technologies such as VOD has been crucial to attracting advertiser support, yet measurement services for unproven technologies have been slow to develop, creating a classic catch-22 situation that has slowed innovation. Still, the ability to measure audiences and gauge the effectiveness of new advertising strategies for television plays a critical role in determining the forms and content of television in the post-network era.

A fully developed post-network era will be in place once advertisers and networks agree upon ways to harness the exceptional data now available. In addition to changes within the traditional television industry, the continued growth of Internet advertising and the reconfiguring of advertising and marketing by opportunities offered by social media have suggested that a broad-scale change will disrupt these industries further. In 2006, Jon Mandel, then chairman of MediaCom US, claimed, "The research has finally gotten to the point where we can do deals that are based on the advertising actually working. The television industry has woken up to 'the way to beat [new media] is you prove it works as well if not better than other media.' We have finally been able to hit a new level of advertising measurement."[23] When I followed up with Mandel about these remarks in 2013, he laughed, and acknowledged the Sisyphean nature of this career. "I thought it was possible," he explained as CEO of Precision Dynamics, a start-up trying to use science and big data to identify better ways to buy television advertising. Despite the new endeavor, Mandel was less sanguine about the future, acknowledging the inertia preventing change and the extent of vested interests among agencies, networks/channels, and measurement entities to keep the status quo in place.

Although changes in audience measurement have occurred throughout television history and the data supplied by measurement services have often been disputed, by the early 2000s, the scope of changes within the television industry and throughout the broader media environment earned audience measurement an even more central and contested status in the industry's evolution. The technological and industrial shifts of the multi-channel transition led to substantial adjustments in both the type of audience research available and the role of such information in the economic transactions of the industry. Changes in the advertising business, including the conglomeration of its players, resulted in an increase of proprietary research and created greater economies of scale that enabled media research divisions to spread their costs across more clients, and thus fund increasingly detailed research in-house or through commission. Joe Uva, president of the media buying agency Omnicom, explains, "We're developing more customized and proprietary measures on a client-by-client, case-by-case, and even medium-by-medium basis. We're less focused on comparing media choices than ever before and more interested in understanding what consumers are doing and how we can use proprietary insights to help drive our clients' business."[24] Reflecting on the changing environment in 2004, the former FCC commissioner Dennis Patrick acknowledged that television programs might reach fewer audience members, but that advertisers and marketers know more about those viewers than ever before, and that that knowledge increased the value of those audiences.[25] Less optimistic about the expanding availability of research, the veteran market researcher Leo Bogart countered that "More research data and more complex ways of manipulating them on the computer will make media buyers better informed but won't necessarily make them more intelligent."[26]

Advertisers have spent billions of dollars each year despite the lack of definitive data to explain why and how advertising influences buying behavior. The existing system of evaluating advertising based on the number of people watching the show in which it was embedded developed because of the technological possibilities and limitations of the network era. Since then, new technologies enabled more direct and precise measurements, such as that of audience size and composition during commercials, but advertisers have continued to seek measurements that directly link their dollars with the number of goods moved off the shelf or by expanded brand awareness. Experiments with "single-source measurement services" that—through a single provider—aggregate information such as consumer behavior that is gathered from credit card purchases or loyalty card swipes and known media consumption, such as set-top box data, were emerging and creating a much richer picture of advertising exposure and purchasing behavior for particular households over time. The desire for such detailed pictures of behavior and advertising effects was by no means new or particular to the arrival of a post-network era, but changes throughout the television industry had begun to increase the likelihood of attaining its fulfillment.

The Future of Audience Measurement

You can't monetize what you can't measure.
—Stacey Lynn Schulman, Television Bureau of Advertising, 2014

All of the anxiety and uncertainty about the future of the television industry created a degree of stasis in audience measurement in 2013, but it seemed likely the turmoil would quickly resume. The upfront buy of 2013 was notable for the restraint advertisers engaged, refusing the types of fee increases typical in a year that the economy seemed otherwise healthy. What would happen with actual payments for the 2013–2014 buys remained to be seen, as the upfronts are merely a suggestion of the market; yet it seemed the industry had reached some sort of turning point in acknowledging that linear broadcast viewing would not dominate forever. Many of the changes in measurement practices chronicled here emerged out of a need to respond to the fracturing of the network-era television audience and the many additional sources of content that audiences might seek. But these changes largely presumed the perpetuation of linear television, and by 2013 it was clear that two just as sizable disruptions would also be a part of establishing post-network conditions.

The first was adjusting to evidence of decreased linear viewing. The Nielsen zero-TV household adjustment was a preliminary acknowledgment that will better reveal the different ways and places audiences view that are no longer in synch with network and advertiser expectations. Advertisers must subsequently create ads and buy time in a manner consistent with the growing reality that at least some viewers are reorganizing their viewing. Advertising messages can still be valuable, advertisers just need to evolve from a paradigm that expects that message to be seen at a particular time. And again, if that timeliness is of high value, those advertising dollars need to compete for the limited spots available in live sports and contests most likely to be viewed at a predictable time.

The data on ratings using the revised definition of the television household were not yet available as I completed the manuscript, but it seemed impossible that the new data would be welcome. Nielsen's revision—though likely an adjustment that would make its ratings more accurate—would expand the number of television households, and thus decrease many ratings in comparison with previous years. Just as in the case of the LPM, where new viewing of cable was uncovered, the counting of viewing in which audiences use broadband providers such as Netflix or Hulu to manage their viewing in lieu of a DVR will reveal previously unmeasured viewing. And again, this is most likely to diminish ratings of linear viewing and increase the ratings of shows watched by the most tech-savvy demographics.

The other adjustment—one likely to become more pronounced after the industry moves further to accept and encourage nonlinear viewing—is the adjustment to more expansive, and perhaps eventually dominant, use of dynamic ad insertion. The expanded choice and control characteristic of post-network television do not eliminate the utility of the thirty-second advertising model, and dynamic ad insertion was arguably a tool that could beneficially revolutionize television's advertising practices. But the change it introduced was extraordinary enough to make it difficult to merely transition existing practices. The industry had been stymied for a decade trying to move the arbitrary but existing valuations of gross ratings points from broadcast and cable advertising to the broadband world. Dynamic ad insertion would require entirely new evaluations, and in many ways warrants an entirely new language, though one that might be used both in the context of living room VOD and content delivered to broadband-connected screens.

Though audience measurement developed in the network era and multi-channel transition as a service for monetizing audiences through advertising, the sophistication of post-network-era technologies also suggested new businesses for this industry. "Nobody knows" has long served as a maxim of creative industries that references the unpredictability of audience reaction, and a key feature distinguishing creative industries from most others. Audience measurement of shows developed as a proxy for measuring the audience of commercials, but in an era of big data—where much more about viewers could be known—new applications of program-viewing data might have new value.

Netflix offered the first sense of this. Though Netflix was notoriously guarded about its data as of 2013, executives often implied they knew a great deal about the viewing behavior of their subscribers. Unlike other types of programming delivered through an array of distributors, broadband-delivered programmers—particularly those requiring subscriber authentication—knew what went where, to whom, and for how long. Netflix executives suggested that these data enabled them to better understand viewers' preferences and to value content. For example, executives defended the high cost of *House of Cards* on the grounds that they that knew their subscribers particularly enjoyed complex, serialized drama. Given the limited availability of such content in the secondary market, they chose to produce an original show in this form. The wealth of Netflix data was similarly referenced in the decision to produce additional episodes of *Arrested Development*.

Such information isn't limited to broadband providers. Cable distributors too could build algorithms and develop teams to understand viewer behavior, but there just hasn't been much incentive given the existing organization of the industry, which has largely separated distribution from content creation. Nor have they shared this information with the middlemen such as networks and channels or the studios, which might use it to make the intelligence-based development choices Netflix claims to use. Richer data about viewers may help studios better value content as they license it in secondary windows—or such data may help licensors make better purchases.

Additionally, more sophisticated measures of the audience are emerging that will aid television advertisers in targeting their messages. Though basic demographic features such as age and gender have been used as the basis of audience profiles in the past, more detailed information may better predict success in particular cases and decrease some of the significant uncertainty in these transactions. Many of these profiles remain proprietary and are used internally by content companies in an effort to understand the differences emerging in how viewers use television and different types of television. For example, in concert with the Cambridge Group, CBS's research division, CBS Vision, began dividing the audience into more sophisticated categories to better capture the changing ways different audiences had begun to engage with television.[27] With categories such as TV Moderators, TV Companions, Sports Enthusiasts, Program Passionates, Media Trendsetters, and Streamers, CBS is able to identify the variant behaviors of audiences relative to its programming to help advertisers make more efficient buys.

Conclusion

Audience measurement and research services expanded their techniques and technologies throughout the multi-channel transition because there was an economic imperative for them to do so. Advertisers would willingly underwrite endeavors to know more about the viewing and purchasing behavior of the audience members that they paid so dearly to reach. Yet these refinements in audience research had broader consequences than the bottom line of advertising agencies and ad-supported networks. They also substantially affected the speed at which further changes in television could occur. Advertisers' interest in placement techniques, and their shifting of budget allocations from thirty-second spots to placement or branded entertainment, encouraged the development of new measurement protocols. Measurement techniques for other delivery systems developed more slowly because advertisers invested less in these areas. Cable providers certainly desired a faster adoption of VOD use, but they could only make this form of distribution possible—they could not also finance all of the programming available on demand. Consequently, this use developed more slowly.

One way to understand the role of research in industry innovation is to imagine where U.S. television might be had AGB not threatened Nielsen and led the corporation to substantially enhance its service with the People Meter. In the late 1980s, at the same time People Meters launched, cable approached the 50 percent penetration rate, and the average number of channels received in U.S. households had already risen to 18.8.[28] Previous measurement techniques undercounted cable viewing, which made it more difficult for these networks to secure the advertising support they needed to finance innovative and distinctive programming. If Nielsen had not updated its methodology in a manner more likely to accurately value cable viewing, the development of cable might have been much slower. Likewise, fifteen years after the introduction of People Meters, the introduction of the LPM provided a similarly instructive lesson. Here too, a reallocation of viewing occurred alongside a refinement in research technology. Again, it was the emergent segment of the industry (still cable) that the old method undercounted. The more precise data helped cable networks continue to argue their worth to advertisers, and consequently expand programming budgets and the range of new content included on their schedules.

Nielsen's independence and lack of corporate affiliation with any of the buyers or sellers of television programming have been key advantages for the company throughout television history. By contrast, research derived from digital set-top boxes, mobile phones, or Internet streaming data will probably be proprietary—and closely guarded, as in the case of Netflix—and lack external verification. Verizon may report that sixty thousand users watched a sponsored clip last month, but what guarantee do advertisers have that Verizon properly represents data available only to the corporation? The revolution in technology has solved some of the problems inherent in existing research protocols, but it has also introduced new challenges and dilemmas.

Hollywood's creative voices and network executive suites have therefore regarded the potential of new research tools very differently. The network executives who receive training in business schools identified the ability to know how audiences behave minute-by-minute—if not second-by-second—as a tremendous advantage. But telling a good story differs significantly from building a better widget, and the industry's storytellers have argued that the contemporary surplus of data threatens to numb the creativity of those who already work within narrow parameters. Some of the television programming most valued for its creativity has emerged from networks that either opted against extensive testing, such as HBO, or could not afford it, as has been the case for many basic cable networks. The technological possibility may exist to use refined viewer data for content creation, but executives will be wise to open this Pandora's box carefully.

Another aspect of the new measurement and research capabilities that has extensive implications is the increased surveillance of viewers inherent in research such as census data collection. Although news stories regarding the use of the data available to our many service providers appear on occasion, viewers do not seem to be concerned about the amount of information media services gather about their preferences and behaviors. Though bubbles of outrage develop from time to time—usually in relation to government surveillance—most willfully accept that giving over all manner of personal behavior data is a condition of living in contemporary society. The contemporary reaction to such developments has not been like that which gave rise to Orwell's *1984*; rather, the forces of consumerism and experience of audience and market construction have led viewers to accept and even support surveillance. Joseph Turow identifies the acceptance of surveillance as an outcome of multiple shifts in U.S. consumer society and the development of interactive and digital technologies such as those discussed here.[29] Turow explores how the growth of target marketing in the post–World War II era contributed to creating a contemporary society composed of individuals who understand their role and power as those of consumers more than as citizens. Turow even acknowledges the degree to which audience members support surveillance if it can aid them in receiving access to products with increased customization or allow them to use services at a decreased price. Services that use artificial intelligence applications to suggest programming based on what a viewer previously had selected illustrate this application and to some degree, its advantages: we are not so much in Bruce Springsteen's world of "fifty-seven channels and nothin' on" than we are in one of thousands of shows and no hope of finding what we want to watch. Viewers can and do accept monitoring as the means to the end of the greater availability of desirable products.

Despite such acceptance, the transformation of television from a window on the world to a two-way mirror deserves more interrogation and critique, especially insofar as television-related surveillance has not been a topic of much concern. Viewers may willingly allow aggregation and individuation of their data, but most are probably unaware of the type of information that is gathered or what is done with it: service-provider disclosures tend to be exceedingly long documents of fine print. Acceptance of surveillance is effectively a condition of participation in digital societies. Concern or outrage is not likely to be marshaled until a situation widely acknowledged as an abuse of information gathering becomes well known—perhaps a newspaper report of a public official's pay-per-view porn habits? For the time being, though, viewers appear unaware of disadvantages to sharing information, which makes their revolt unlikely.

Although audience research has rarely been part of the discussion of the monumental revolution in television under way since the mid-2000s, the stealth status of this topic should not belie its importance. The outcome of endeavors to develop viable and accurate research will ultimately determine the winners and losers of coming struggles over technologies, distribution platforms, and programming. Media executives and the advertisers that support them have not forgotten the lessons of the dot-com boom in the 1990s, when millions of dollars were lost as a result of money spent without understanding the fundamental attributes of a new technology. Now, with so many more new media technologies, advertisers have become increasingly suspicious of the value of their existing ways of doing business and have demanded that networks prove the value of their audience delivery in thirty-second commercials as well as in the new developing advertising strategies. All of the distribution applications and technologies that rely on advertiser support require tools for counting and valuing audience members. The network-era norm of a singular ratings service and standard exchange of advertising dollars for gross ratings points might remain part of the business, but many other advertising techniques and currencies will coexist as a result of the increasing ways to use and pay for television.

As Gertner notes in the epigraph that begins this chapter, changing the way the industry measures audiences will change the business and the culture. However, changes in audience behavior will precede many of the other adjustments. Different segments of the television industry have invested billions of dollars hoping to identify the technology or use of television that will be central to the new era. Those at home, and those who have begun taking television on the go, will determine the winners and losers—that is, if the measurement services can find a way to count them.

Television Storytelling Possibilities at the Beginning of the Post-Network Era

Five Cases

At the summer 2003 Television Critics Association tour, the CBS researcher David Poltrack presented data reporting the comparatively minuscule amount of attention paid to CBS series by the nation's journalists relative to the network's substantial audiences. The shows, including *JAG*, *The Guardian*, and *Judging Amy*, were among the most watched each week, yet, Poltrack complained, the critics insisted on devoting expansive column space to relatively obscure shows on the WB and HBO. Ten years later, the situation was largely the same. CBS shows such as *NCIS* and *The Big Bang Theory* continued to gather the largest regular audiences, but pop culture chatter about television focused on shows with a fifth or less of their audience, shows such as AMC's *Mad Men* and NBC's *Community*.

What Poltrack's research did not address, however, was the fact that there just wasn't much to say about many of his network's programs. CBS had successfully decreased the median age of its audience and moved into first place for household viewing, but there was minimal innovation in much of its programming—although exceptions such as *Survivor* and early seasons of *CSI* could be found. CBS became the most-watched network by developing formulaic crime and forensic dramas that were exceptional only in their focus on the most uncommon scenarios for death and mayhem. For critics, who watch much more television than the average viewer, many of these successful shows were tried, trite, and predictable, notwithstanding their popularity.

So then, perhaps much to the chagrin of broadcasters like Poltrack, I also do not focus on the biggest blockbuster hits in this chapter. My selection of case studies does not result from how many people watched these shows, but from the lessons these shows provide about the industry's changes. Each of the five cases explored here—presented chronologically—tells a distinctive tale about production on the cusp of the post-network era. I include them here at the end of the book instead of locating one in each of the preceding chapters because of how they illustrate the interconnections among changes in multiple production components. The cases include various genres (comedy, drama, and unscripted) and address shows on major and minor broadcast networks and on basic and premium cable channels. Despite the variety of shows and their distinctive economic situations, some aspects of their development, production, and distribution stories repeat in important ways. None of these series could have existed on network-era television; each illustrates not only changes in the production process, but also how these changes have created opportunities for stories much different from those of the network era.

Sex and the City

HBO's *Sex and the City*, which debuted in June 1998, was one of the earliest series to feature a production process that substantially differed from network-era norms. The series is important to the questions examined here because of its status as an early original cable success that defied many expectations about the popularity and commercial viability of a show produced by a subscription cable network. *Sex and the City* debuted before *The Sopranos* and was the network's first series to achieve considerable popular awareness. *Sex and the City* told a story unlikely to be found on broadcast networks, or at least unlikely to be told elsewhere in the manner that it was on HBO. Because of the distinctive economic and regulatory context of the channel, *Sex and the City* was able to push boundaries even further than its basic cable counterparts that had likewise begun to produce series differentiated from broadcast by their niche address and unconventional themes; Lifetime's *Any Day Now* serves as the clearest coterminous illustration. Much journalistic attention to the first few seasons of *Sex and the City* focused on the explicit sexual content and frank conversations about sex among its four characters. Although such forthright depictions and expressions were unlikely to be found outside the premium cable television space, the series also derived significance from its uncommon focus on female friendships and its examination of the negotiations made by a group of women previously absent from television.

Despite its unconventionality as a series produced for an original run on a subscription cable network, *Sex and the City* followed a fairly conventional development history. The series was adapted from a book written by Candace Bushnell and based on her *New York Observer* newspaper column. Bushnell was actually the "sexual anthropologist" who wrote stories about sex in New York City and served as the model for Carrie's character; the series reduced the much larger collection of people inhabiting Bushnell's book to Carrie's three female friends. The executive producer, creator, and occasional series writer, Darren Star (*Beverly Hills, 90210, Melrose Place*), chose to produce the series for HBO despite an offer from ABC.[1] The premium cable channel was necessary in Star's mind in order to maintain elements of "eliteness" in writing and production, as well as a budget—initially $900,000 an episode—on which he could afford the caliber of directors and writers he desired. The series' home on HBO also provided Star with considerable content freedom. This enabled the series to derive its humor from the sexual adventures of the four characters, rather than from the double entendres and hidden discussion about sex typical of broadcast sitcoms at the time.

Many focus on the freedom from advertiser influence when considering the distinction of subscription television networks, but the difference in these networks' economic processes is more fundamental. Broadcast and basic cable networks are primarily concerned with how many people tune in to their programming. Since advertisers evaluate the networks they support in each hour of each day, networks must select programming likely to reach the broadest audience possible and allocate their programming budget accordingly—for example, by creating or purchasing new programs for each hour of almost every day. In contrast, subscription cable services rely on viewers desiring to watch their programming so much that they are willing to pay for it. They do not worry about how advertisers evaluate programming—there are no advertisers. This difference in subscription networks' financing model changes the conditions of production, but more significantly, provides these networks with a mandate for selecting programs that differs vastly from advertiser-supported networks. Subscription networks care much less about who watches and when; rather, they are chiefly concerned with providing subscribers with a service with enough value that they continue to subscribe.

Of course, willingness to subscribe is closely related to what the network programs, but this mandate leads the network to construct its schedule and select programming much differently than advertiser-supported networks.

Subscription networks seek to provide as many viewers as possible with adequate reason to keep paying their monthly fees. HBO achieved considerable success by offering a wide range of programming that very specifically targeted the tastes of a broad array of audience members, but did not provide an exceptional amount of programming for any one group. It matters little to the network whether a viewer subscribes because of the networks' films, sports, or original series. Rather, its economic model requires that viewers find some aspect of such value—ideally something unavailable on advertiser-supported networks—that they maintain their subscription.[2] A show such as *Sex and the City*, with its specific address to young, upscale, professional women, provided an ideal addition to the HBO schedule. No series on television so precisely addressed this group, and much of HBO's other programming, such as the original series *Oz*, boxing, and *Inside the NFL*, more emphatically sought to entertain male viewers.

The first season of *Sex and the City* consisted of only twelve episodes, followed by eighteen in each of the subsequent five seasons. This shortened and flexible schedule became characteristic of premium cable originals, and then those of basic cable, and has significant artistic and economic consequences.[3] Artistically, the shortened season decreases the burden of originating so much new content each year, which is one of the key aspects of television production about which creative personnel often complain. In addition, HBO's flexible and commercial-free schedule allowed creators to develop episodes at a length determined by the story rather than according to the strict twenty-two-minute format of broadcast comedies designed with narrative climaxes prescribed to occur at regular intervals to allow for commercial breaks.

The substantial syndication success enjoyed by *Sex and the City* could not have been predicted when it began production in 1998. Although the series produced substantial "buzz" within the culture, actual audience size for the original run on HBO was quite small by television standards. At the time the series aired, only about 30 percent of U.S. homes subscribed to HBO, and many of these homes did not view the series. HBO reported an average audience of 6.6 million homes in the series' second season for the premiere play of each episode.[4] This limited audience size in the original run added to the series' syndication value because there remained so many new potential audience members for the series. Those who did not subscribe to HBO first had an initial opportunity to view the show on DVD, which gave viewers an uncut version of the series as it had appeared on HBO. In the first half of 2003, the series earned more than $65 million just from DVD sales.[5] The series also enjoyed immediate revenue from international syndication; it would eventually be syndicated in more than forty-eight countries.

More significantly, the series was among the first original cable series, basic or premium, to earn revenue from other distribution windows.[6] The series was sold for a first run to the cable channel Turner Broadcasting System (TBS) and Tribune-owned local stations in deals that involved a combination of cash and advertising time. TBS paid $450,000 for each of the ninety-four episodes. The terms of the Tribune deal were not disclosed, but a month later HBO sold the series to the independent San Francisco station KRON for $10.4 million plus advertising time.[7] Unlike most broadcast series that could be sold in the same form in which they aired on the network, *Sex and the City* episodes had to be reedited by distributors to cut them to a conventional length, allow for commercial insertion, and make them acceptable for broadcast content standards. Although this reediting incurred costs, they were not as high as they might have been, since the producers had shot additional scenes during the original production to replace scenes that might be troublesome for international syndication or a non-premium domestic outlet. This lengthened original production minimally and proved invaluable once domestic buyers such as TBS and Tribune emerged. HBO also sold a second cable run of the series to Comcast for its E! and Style Network channels beginning in January 2011, and HBO continued to air the series long after it finished production and included it among its offerings on HBO GO. Interest in the series was also reanimated by two subsequent *Sex and the City* theatrical films.

It is nearly impossible to estimate the ultimate reach of this show, as I can't imagine a methodology that could meaningfully aggregate all of the places it has aired and account for repeat viewing that likely occurred as well. Though it began as a niche, premium cable series, the audience available to this original licensor has been surpassed many, many times over, suggesting the challenge of trying to assess the cultural importance of programming in the post-network era. Though television has always had a long revenue arc, we increasingly need to theorize the cultural relevance of programming in a way that places less priority on performance at debut, especially if it premieres in a window not widely available, as was the case of Netflix's *House of Cards*. Or, as VOD viewing allows viewers to discover shows seasons after their original airing and then view them systematically in the manner of prized content, the importance of a show might not be clear until well after its production has concluded, as could be argued of a show such as

The Wire. There is much to be learned about how and why different cultural properties build audience interest over time and as their availability expands across distribution platforms.

This case also indicates the importance of the availability of a programmer such as HBO that operates with a distinctive and differentiated set of production norms. *Sex and the City* eventually defied many conventional norms of distribution and proved surprisingly valuable in many secondary markets, but it could never have been created outside a subscription network. Certainly, the version of the show Star might have produced for ABC would not have had nearly the same tone or storytelling emphasis, as ABC executives would have insisted on making the content more accessible to a broader audience. The programming HBO began producing in the late 1990s was among the most widely hailed by critics and in award competitions, and it indicated previously unimagined possibilities for television as a creative medium. While HBO's particular industrial situation enabled much of its programming, HBO's efforts had effects across the television spectrum—and not just by increasing sexual and violent content.

In another strategy particular to HBO, the channel reported that it accepted no paid product placement despite what many have noted to be an exceptional amount of clearly named commercial goods in shows such as *Sex and the City* and *The Sopranos.*[8] Producers do accept products for use in shows, but HBO has repeatedly denied that any money changes hands for them, including the oft-noted use of Absolut vodka in an episode of *Sex and the City* that required the creation of a fake ad including one of the series' actors as a model.[9] Many in the marketing community are skeptical of HBO's claims of refusing payment, especially considering one marketer's 2002 estimation that *The Sopranos* could have earned $6.8 million per episode if it accepted placement payments.[10] Nonetheless, the channel emphasizes its particular industrial status as a subscription service. As the HBO spokesman Jeff Cusson explained, "We're not a network that accepts advertising. And product placement is a form of advertising."[11]

The unusual economic mandate of subscription networks provided an incubator for creative television that expanded program content through their niche address and innovative production techniques. Contrary to its late-1990s slogan—"It's Not TV, It's HBO"—HBO programming was indeed television, and the network's original series had significant implications for both subsequent shows on all television channels and on how others approached, produced, and viewed television. Though HBO's strategy and success inspired competitors that diminished HBO's hold on critically hailed television, it largely maintained its strategy as the post-network era developed further. The 2012 debut of *Girls* both harkened back to *Sex and the City* topically and revealed how much more narrow the niche segmentations of audiences had become. Where *Sex and the City* seemed narrowly targeted in the late 1990s with its emphasis on late-thirty- and forty-something single career women who might as well have been in their twenties and early thirties, *Girls* presented an experience largely particular to single, urban, career women in their early twenties with particular educational and class privilege. Though both shows obviously drew audiences much broader than the demographics of their characters, the stories in *Girls* were far more specific than those told by *Sex and the City*.

Survivor

One of the most profound adjustments in programming during the multichannel transition resulted from the unexpected success of unscripted, or "reality," programming. Not only did these shows alter expectations about norms of program content, but they also introduced a vast variety of production practices. Just as the original production of *Sex and the City* involved the unique economic arrangements of a subscription network, so, too, do most unscripted series use economic practices fundamentally unlike those common to most scripted television programming. The success of unscripted programming led producers to break from status-quo assumptions about how shows could be financed, and this in turn disrupted many residual network-era norms of industrial practice and programming.

Although unscripted television has often been haphazardly amalgamated into a single category of "reality" television, substantial differentiation in both program content and production norms had arisen by the mid-2000s. Indeed, production practices for unscripted series vary significantly—so much so as to support the contention that no single one serves as a norm. This variation makes it difficult to select a meaningful case that might loosely characterize this genre. CBS's 2000 breakout hit, *Survivor*, provides an informative case, although its exceptional success made it unlike most other unscripted shows produced in this era. Nonetheless, *Survivor* illustrates how unconventional formats can revolutionize the business and programming of major broadcasters and yield *broad*cast success in an increasingly narrowcast competitive environment.

Debuting in May 2000, *Survivor* was one of the earliest unscripted competition series to disrupt the balance broadcasters had established in the 1990s of schedules filled mostly with scripted programming and some cheaper newsmagazines to help compensate for rising scripted programming costs.

The newsmagazines and various early unscripted shows such as *Unsolved Mysteries* and *Cops* had low production costs that could be covered by license fees that were even lower than those for scripted programs.[12] The frugal production expenditures eliminated the need to syndicate this programming, although many shows, including *Unsolved Mysteries*, *Cops*, and *America's Funniest Home Videos* still earned substantial profits in both international and domestic syndication.

As shows such as *Survivor* entered homes in the early 2000s, many believed that "reality" programming represented something entirely new. These shows developed not so much out of innovation as out of necessity, as integrating low-cost shows simply became an essential part of networks' programming cost equation in the mid-1980s. The development of competitive reality shows such as *Survivor*, *American Idol*, and *The Amazing Race* made sense for networks that had saturated viewer interest in newsmagazines but needed to maintain some lower-cost shows. Despite the sense that reality television was taking over in the early 2000s, consistent reports indicated an increase in dramatic programming and only a slight decrease in comedy, which was mainly due to difficulties in developing successful comedies. The unscripted series primarily replaced newsmagazines and theatrical films, and drew larger or at least younger audiences than other lower-cost programming forms. "Reality" never really threatened to take over network schedules—viewers showed interest in the novelty of the programs, and younger audiences have watched in greater numbers than viewed newsmagazines, but scripted series remain central to network identities and overall strategies. It is the case, however, that some sort of lower-cost programming is likely to play a role in network programming norms as long as a linear system remains in place.

Mark Burnett, a newcomer to television whose reputation was mostly based on the *Eco-Challenge* race he produced for various cable channels including MTV, ESPN, Discovery Channel, and USA, acquired rights to the concept that would become *Survivor* in the mid-1990s, when U.S. "reality" television consisted of *Cops* and *America's Funniest Home Videos*. As he pitched *Survivor* to various networks, he received rejection after rejection in a manner that retrospectively seems inconceivable (although significantly, *American Idol* received similar treatment at the time). Burnett spent years trying to sell a network on *Survivor* and succeeded only when a junior member of the CBS programming department became an advocate for the show and persistently argued his case to upper-level executives.[13]

Few suspected that unscripted programming would succeed, and certainly no one believed that these shows would attract blockbuster audiences in the manner that came to be regularly achieved by *Survivor* and *American Idol*. Many did perceive, however, that the contest format of these series would prevent them from performing well as reruns or in syndication of any kind. Without the syndication revenues to support deficit financing, these prime-time "game" shows would have to find an alternative financing system that did not require distributing them through additional windows to recoup costs. Most unscripted series could be produced more cheaply than scripted series because they required no actors or writers who must be paid at guild rates—a considerable advantage at a time when actors' fees were particularly blamed for the "skyrocketing" costs of scripted series and when networks were casting major stars to help distinguish their programs in the increasingly competitive environment.[14] Yet some of these shows, such as those produced by Mark Burnett, were known for "putting money on the screen" rather than in the producer's pocket, and had production costs similar to those of scripted series. Such shows faced the dilemma of lacking syndication value, which made deficit financing a poor option, yet were unable to finance production from license fees alone. *Survivor* was one of these series.

Mark Burnett and the CBS executive Les Moonves tell somewhat different stories about the arrangement negotiated to get *Survivor* on the air. According to Burnett, the network was uncertain about the completely untested nature of the series and was wary of losing money. CBS consequently entrusted Burnett with the responsibility of selling the advertising time in the series and agreed to split the profits 50–50 if the show actually succeeded.[15] CBS, wary that any producer think it would agree to such a deal, has denied Burnett's claim repeatedly and asserted that Burnett received only 50 percent of the advertising revenue earned from expanding the *Survivor* finale—which drew an unthinkable fifty-one million viewers—from one hour to two. Regardless of where the precise truth of this deal might be found, Burnett's unconventional approach to selling advertising time is not disputed. Rather than trying to convince advertisers to simply buy thirty-second commercials, Burnett sought for companies to buy in to the show through what he termed "associative marketing," but has since come to be known as product placement or integration. The fees paid by the eight sponsors Burnett attracted covered the show's production costs, and many received both placement in the program and a package of thirty-second commercials.

The reason the truth of the agreement between Burnett and CBS is so elusive is that the stakes are so high. The 50–50 split, or even sharing any revenue with a producer, would involve an unfathomable reallocation of economic norms that could wreak havoc for the networks. If highly sought-after producers thought they might negotiate for a share of the initial advertising revenue from programs, the networks would relinquish additional control and revenues in a way that would further require massive adjustments throughout all their financial arrangements. For his part, Burnett became an instant sensation, and networks quickly began bidding for him to develop subsequent projects. It served him well for the other networks to believe that CBS had been willing to offer such a grand deal.

Regardless of what CBS paid, few could argue that they overcompensated given the key role *Survivor* played in bringing more viewers to the network, decreasing the network's average audience age, and wresting control of the lucrative Thursday night from NBC's long-held dominance. As long as a linear model continues, Thursday-night advertising is in particular demand due to the night's proximity to the weekend; this makes it especially key for advertising films opening on Friday and for weekend events and sales. Indeed, the viability of the show was not the only expectation that proved faulty. The exceptional popularity of *Survivor* helped fuel DVD sales of the series, and the popularity of other unscripted series led to the creation of windows for subsequent syndication. Some unscripted series were distributed on the short-lived FOX Reality Channel launched by News Corp in May 2005, which was primarily available to the eighteen million households subscribing to then News Corp–owned DirecTV, though it ceased operations in early 2010. In the case of *Survivor*, Burnett and CBS sold the rights to air seasons 1 through 10 to the Outdoor Life Network (OLN; which changed its name to Versus in September 2006 and NBC Sports Network in January 2012), defying expectations that the show had no subsequent value. CBS also established unconventional distribution deals for new episodes of *Survivor* in various transactional payment markets. The network included *Survivor* in its limited initial on-demand offerings available for purchase in certain local markets with Comcast cable service and experimented as the distributor by making downloadable episodes available for purchase on the CBS website, in both cases for ninety-nine cents each. Such sales further upset the notion that unscripted series have no value after their initial airing, particularly those shows with stories or contests confined to specific episodes, such as *Fear Factor*, which earned $250,000 per episode in its syndication sale to FX.[16] Despite these expanding opportunities to earn subsequent revenues, deals for new unscripted series rarely relied on deficit financing. Notably, all unscripted series are not the same. Some, such as *American Idol*, *Dancing with the Stars*, and arguably *Survivor*, create a viewing pleasure related to seeing what contestant is eliminated next, much like live sports. Others hew more closely to the narrative pleasures of scripted series and likewise have diminished viewing urgency.

Another key source of revenue for unscripted programming developed through international format sales. This means that templates of these relatively cheap to produce contests are sold in markets around the world, rather than episodes of U.S. contestants competing; thus, each country can reproduce the show with local contestants and in response to cultural particularities. Such flexibility has proven to be especially advantageous. For example, *Who Wants to Be a Millionaire?* was produced in 130 different territories at one point. *The Weakest Link* sold similarly well, but the brash and rude persona of the female questioner and her winking had to be adjusted—replaced with a male host or the winking eliminated—in some cultures that responded negatively to the initial importation of the style and gender performance of the original British host, Anne Robinson.[17]

The sale of formats was a phenomenal business in the early 2000s and suggested very different global relations than the cultural imperialism thesis in which Western countries, and the United States in particular, were believed to impose their culture around the globe by dominating network-era program exports.[18] Indeed, many of the biggest successes were imported to the United States, and format trade was so vibrant as to lead to the sale of scripted formats, as in the BBC's sale of *The Office* to NBC. Scripted formats had been sold before and sometimes with considerable success—a case in point is Norman Lear's *All in the Family*, which was based on the British series *'Till Death Us Do Part*. But the format business and especially international format sales of unscripted series helped enable culturally specific production and decreased U.S. dominance in the international market.[19] In fact, few of the unscripted series that succeeded in the United States originated as domestic productions, and the country purchased more formats than it sold. The case of *Survivor* discussed here is one instance: Burnett purchased the series format from creators in Sweden.

Burnett's *Survivor* continued to perform well into 2013, defying much conventional wisdom. The financing arrangements Burnett constructed after *Survivor*'s success indicated what he learned from his early endeavors and the changing nature of the business. Despite some stumbles with subsequent series such as *The Contender* and *Rock Star*, Burnett's more recent productions such as *The Voice*, *Shark Tank*, and even the scripted cable miniseries *The Bible* have drawn large audiences. Burnett continued to require unconventional deals with networks and increasingly sought and achieved greater control of international distribution, sponsorship, and product integration.[20] From the beginning of *Survivor*, Burnett rightly insisted that the show's innovation was not in the programming form, but in the business model, and he is lauded in the industry for understanding that different shows require different financing arrangements.[21] Shows such as *Survivor* changed broadcasters' thinking about how shows could be financed and the audience sizes that lower-cost formats could attract. Out of sheer necessity, *Survivor* introduced a major shift in advertising norms with its sponsorship funding and skillful integration of products into the story.

Significantly, though, even after thirteen years and nearly four hundred episodes, many executives would look at the chances of a concept like *Survivor* succeeding with great skepticism. The success of the show and its consequences for television's production process illustrate the profound uncertainty that sometimes characterizes this industry as well as the highly haphazard process of change. Burnett's persistence helped CBS assuage its fear of risk, and *Survivor* opened the industry's eyes to the variety of programming and industrial practices available through its success in captivating audiences with its competitive narrative and by deviating from conventional program economics. But for anyone looking at the series at its inception, these were hardly predictable or even imaginable outcomes.

Burnett's willingness to take a risk on the show's economic model is also an illustration of a strategy useful for broadband-delivered programming attempting to establish new distribution and economic models. Advertisers desire proven concepts, so the ability to self-fund or establish a track record on a minimal budget might pave the way for transitioning into a model in which studios and producers rather than networks secure advertiser funding. Notably, experiments in original series produced by broadband-delivered programmers such as Yahoo's *Burning Love* and Hulu's *Battleground* haven't borne out the viewer interest Burnett captured—though in these cases, Yahoo and Hulu functioned very similarly to networks. But smaller-scale examples of content with strong producer control and vision, such as Freddie Wong's YouTube-originated *Video Game High School*, exemplify the strategy of content finding an audience and then advertisers coming to the content in order to reach that audience.

The Shield

The Shield debuted on the FX cable network in March 2002, where it aired for seven seasons. The series provided the network's most high-profile attempt at original series production, and the network supported its premiere with an extensive promotional campaign. Textually, *The Shield* reinvigorated the police drama genre by centering its narrative on a rogue police detective clearly playing outside the bounds of proper procedure. As is true of all of the shows discussed in this chapter, *The Shield* is unlike any show created or financed during the network era.

The series achieved a variety of "firsts." *The Shield*'s central protagonist, Vic Mackey, is neither hero nor antihero, and the series' writers artfully compel the audience to be alternately drawn in and repulsed by him. Although some successful original series had emerged on advertiser-supported cable networks by this point, the extensive promotional campaign FX used to announce the show, cultural debate and discussion about the series' content, and early advertiser defections generated more expansive cultural awareness than other basic cable series had. The series won a Peabody Award in 2006 and a Golden Globe Award for best drama in 2003, and the actor Michael Chiklis won both a Golden Globe and an Emmy for his portrayal of Vic Mackey following the show's first season. Not only was Chiklis's Emmy the first won by a basic cable program in a major series category, so too was the series' Peabody the first earned by a basic cable live-action original series, and all these awards were significant to establishing the legitimacy of basic cable original series.[22]

The story of how FX came to purchase *The Shield* is as unusual as the series' content and is symptomatic of the uncertainty and unpredictability of production practices at the beginning of the post-network era. The show's creator, Shawn Ryan, began his career as an intern on the comedy *My Two Dads* before suffering five years of unemployment and writing comedy spec scripts (scripts for shows already on the air, which writers use to find jobs). In 1997 he began writing for the CBS action/buddy show *Nash Bridges* and then moved to the more character-driven WB series *Angel*. Ryan also had been hired to write a sitcom script for Fox TV Studios, but the idea Ryan had at the time was for a dark police drama. The Fox TV executives liked the script, although it was clearly not the comedy they needed. A few months later Kevin Reilly, FX entertainment president, had an idea for a show, but needed a writer. The Fox TV executives sent Ryan's script to Reilly as a sample of his work, and Reilly liked the script so much that he decided to pursue making *The Shield* rather than his original idea. [23]

It is significant that Ryan had not bothered trying to sell the show, as it indicates how strong and pervasive ideas about what shows networks will and won't make may be. The story also illustrates the chance that sometimes characterizes the operations of this industry. Ryan could have been pitching the series all over town and not have achieved a network deal, simply because he might not have thought to pitch to a small cable network that had not previously produced original drama—up until then, FX had originated only the mildly successful *Son of the Beach*, sort of a parody of *Baywatch* created by Howard Stern. If happenstance had a role in the development of the show, so too did conglomeration. The FX cable channel and Fox TV Studios are both owned by News Corp, and Reilly may never have received this particular script were it not for inter-conglomerate awareness that the network was seeking to produce an edgy drama.

The particular circumstance of producing an original police drama for a basic cable network while maintaining "broadcast-quality" production values presented other challenges. The risk that production deficits could not be recouped made any single studio hesitant to produce the show, so Fox TV Studios and Sony Pictures Television (formerly Columbia TriStar) agreed to coproduce, thereby reducing risk. Such arrangements have become less common, and cable channels have also adopted the broadcast norm of producing original series through a conglomerate-owned studio. The studios produced the show for $1.3 million per episode—which compared with $1.8–2.2 million for an average broadcast drama at the time—and divided the various syndication rights; Fox Home Video distributed the DVD, while Sony syndicated the show to broadcast and cable networks.[24] To save money, *The Shield* shot each episode in seven days instead of the more standard nine of broadcast series, and the producers were able to pay a lower union-scale rate in the early seasons because the series aired on a basic cable channel.[25] The producers were able to find enough cost savings in these production adjustments to keep production in Los Angeles, unlike many other original cable shows that produced in Canada in order to avoid union costs, developed international coproduction deals to spread costs (such as SciFi's *Battlestar Galactica*), or partnered with a foreign distributor to share costs (as was the case for USA's *The 4400*).

The limited budget of *The Shield* had consequences for how the show looked. The show utilized a lot of handheld camera shooting, which is more time-efficient and created a distinctive, frenetic visual style. Budget limitations also probably played a role in the relative inexperience of the writing team Ryan assembled and the series' use of primarily unknown actors. Although the inexperience of the writers could be regarded as a drawback, it also helped the series in its pursuit of originality. The writers were not burdened with the acculturation of "how things are done" that prevents deviation from established norms. The police drama has an elaborate history, and Ryan sought to tell different stories or at least tell them in a different way than those that had come before. The selection of Chiklis for the role of Mackey initially broke the series' budget, but Chiklis's audition persuaded the network and studios that the actor was worth the cost.[26] The budget flexibility that allowed the hiring of Chiklis arguably paid great dividends as the attention to the series that resulted from his Emmy nomination and win provided invaluable promotion and legitimacy.

One way *The Shield* tried to draw audiences to basic cable was by pushing the boundaries of established norms of acceptable use of violence and coarse language. Although basic cable networks were not prohibited from airing such content, no network had aired such graphic material in a series. This is not to suggest that *The Shield* pushed these boundaries gratuitously; the violence and language were narratively motivated, and its visual sophistication and high stylization that harkened back to the innovation of the 1980s police drama *Hill Street Blues* also helped establish it as a show that would defy expectations. Critics—who tended to pay little attention to basic cable originals at the time—appropriately hailed the series as the most HBO-influenced show to air outside a subscription network.

It was consequently unsurprising that *The Shield* immediately became a target of public advocacy groups concerned about violence and adult themes on television. Trade magazine articles recounted the increasing number of advertisers who withdrew from the series as early episodes of *The Shield* aired. To be sure, the series' location on FX and its post–10:00 p.m. time slot reasonably freed it from fears that it would come under government regulation, but advocacy groups such as the Parents Television Council pursued a successful strategy of exerting pressure on advertisers through threats of boycotts and negative publicity in order to starve the series of the commercial support necessary for survival. Many advertisers did exit during early weeks of the series, but as FX continued to air new episodes and audience numbers not only remained steady but continued to grow, this trend reversed. FX Networks' president of ad sales, Bruce Lefkowitz, explained that despite the polarizing tendency of shows like *The Shield* to alienate some audiences, they also delivered a "demographic that is often MIA. With *The Shield*, *Nip/Tuck* and *Rescue Me*, we can reach an underserved audience that isn't generally catered to by the mass market. Advertisers see that FX has those early adopters, the trend-setters—people with spending power who tend to get drawn to our authentic, unique programming."[27] The series proved successful in drawing otherwise difficult-to-reach upscale male audiences, and by the end of the first season new advertisers stepped in to replace those who had exited. *The Shield* and other FX shows did remain on many advertisers' "do not buy" lists, but the network was able to fully sell its inventory despite the unabashedly mature content of its original series.[28] *The Shield* was not the first or only series that violated norms of acceptable content, but it is meaningful in the degree to which it did so and survived—and even thrived.

The Shield thus indicated some advertisers' desire to be associated with distinctive content, as well as their willingness to support programming that willfully offended some viewers. Like much of the programming produced in a niche-focused media environment, *The Shield* exhibited a substantial amount of "edge," meaning that it clearly defined the boundaries of its intended audiences and deliberately excluded some tastes and sensibilities.[29] Such a strategy is clearly the opposite of programs that seek the broadest possible audience through the least objectionable programming, as was characteristic of the network era. The commercial viability of the newer strategy—in such an extreme case—is a significant indicator of the changing dynamics of advertising at the beginning of the post-network era. Whereas conventional wisdom suggested that the protest of advocacy groups would result not only in advertisers pulling support from the series but also the end of the show, as had been the case in the network era and would likely have been the case if the show aired on a broadcast network, the old rules have been rewritten. In the new environment, consisting of fragmented audiences and niche-programming strategies, edgy programming produced in clear affront to some viewers can do more than succeed, it can become particularly attractive to certain advertisers and accrue value from distinguishing itself so clearly in the cluttered and intensely competitive programming field.

By the beginning of the second season, thirty-second advertisements in the show were selling for 30–50 percent more than the median price of $40,000 from season 1, which made them among the most expensive in cable.[30] These rates hardly compared with those of broadcast networks, but were significant enough for the series to be an economic success for FX. *The Shield*'s audience was also much smaller than that of most broadcast shows: it averaged only 2.8 million viewers in its fifth season, but then 1.8 million of these were in the key eighteen- to forty-nine-year-old demographic.[31] Importantly, the value of a series like *The Shield* extended well beyond the advertising rates the series earned. When *The Shield* launched, FX lacked a channel brand identity and was not well known. The publicity *The Shield* garnered in awards and critical praise raised the profile of the channel and probably contributed to adding viewers throughout the programming day. Of cable channels in prime time, FX was ranked twelfth among eighteen- to forty-nine-year-olds in 2001; it jumped to fifth place in 2006, in large part due to its development of original series like *The Shield*, *Nip/Tuck*, and *Rescue Me*.[32]

The particular industrial positioning of FX also contributed to the making of *The Shield*. FX launched in June 1994, but had a fairly low subscriber base until the late 1990s. In 2003 an industry analyst noted that the channel had fairly low CPMs (cost-per-thousand advertising rates), but that it benefited from high license fees because it operated as FOX's "ransom channel."[33] Throughout the multi-channel transition, the broadcast networks and cable systems reached a relative détente over the issue of whether cable systems would compensate broadcasters for carrying their networks on cable systems. The broadcasters generally did not require payment from the cable systems—which they had a right to under the must-carry rules—but instead leveraged this nonpayment in order to gain better carriage or license fees for commonly owned cable channels. So, for example, instead of ABC negotiating a fee to carry the broadcast network, ABC convinced cable systems to launch ESPN2. Likewise, News Corp allowed systems to carry FOX for free, but negotiated a license fee for FX estimated to be around $.30 per subscriber in the early 2000s, which was three times greater that what its competitors earned.[34] By 2011, however, FX seemed undervalued, earning $.48 per subscriber in comparison with $1.18 for TNT or $.68 for USA.[35] In 2013, a point by which broadcasters had changed strategies and were now demanding retransmission fees from cable operators, FX "split" into two channels, spinning off its comedy series such as *Archer*, *It's Always Sunny in Philadelphia*, and *The League* to FXX, a channel focused on targeting men through irreverent comedy.[36] The success of *The Shield* and subsequent FX programs that built on its efforts to expand the norms of television storytelling was important for positioning FX as a must-have offering for MVPDs.

This particular situation of FX revealed the complexity of production processes that we must consider when we assess the consequences of shifting industrial practices on the types of programming that are produced. The survival and success of *The Shield* as a boundary-defying program relied upon a shift in the tastes of advertisers: by 2002, there were some who were willing and eager to be associated with a show otherwise experiencing advertising boycott. The industrial position of FX as a "ransom channel" that earned substantial fees from cable system operators is likely to have played an important role in both financing the license fee for the series and allowing the network to be patient after the initial advertiser defections. These industrial practices also greatly contributed to enabling Ryan to produce a series of exceptional quality and unlike that found on broadcast and other advertiser-supported cable networks. A difference in any one of these industrial practices could have yielded a very different outcome for the series.

The potential syndication life for *The Shield* was highly uncertain when the series began production. At that time, it would have been reasonable to assume that the international market would provide the only additional revenue for the series. Many original cable series had successfully sold international syndication rights, but not found buyers in the domestic market. By the time the series finished its first season in the United States, it had already been sold in forty countries, which helped repay the production deficit.[37]

But the international market proved to be just one source of potential revenue. *The Shield* premiered just as DVD sales of television shows began to escalate. As was the case for *Sex and the City*, this distribution format was particularly valuable for *The Shield*, a show that aired on a more obscure cable network to a smaller audience. The first-season DVD sale enabled new viewers to join the audience before the second season, which was particularly important because many may not have heard about the show until the press coverage related to its awards. Like many broadcast shows that have sold particularly well on DVD, *The Shield* featured a fairly serialized storyline that made DVD viewing especially attractive, and by 2006, the series had sold roughly 250,000 units of every season.[38] *The Shield* became available for iTunes transactional purchase in 2012 and was licensed to the subscription libraries of Amazon's Prime service, to Netflix, and to Sony's ad-supported Crackle, all providing additional revenue for the series.

In addition to the DVD distribution, a more surprising secondary revenue window emerged when Sony was able to sell cable syndication rights to the U.S. cable channel Spike. No series produced for a basic cable network had been purchased in syndication by another basic cable network, and the boundary-pushing content of *The Shield* also could have served as a deterrent to buyers because it would require many outlets to reedit it or only air the show at certain times of the day.[39] The cable network Spike actually outbid FX in this first case of a signature series of a basic cable network being syndicated by a competing basic cable network, and Spike paid an estimated $300,000 to $350,000 per episode.[40] Although somewhat odd in terms of cable competition, this purchase made great sense creatively. Spike had been explicitly promoting itself as a network for men since 2003, while FX more successfully established this brand through its original programming. Sony also sold *The Shield* to Tribune and The WB 100+ station groups in an all-barter sale that made the series available on local broadcast stations across the country in a deal expected to earn at least $30 million.[41] (An all-barter sale means the stations pay no cash, but give the studios half of the advertising slots to sell to national advertisers.) Significantly, there is no hint of conglomerate self-interest in either deal, as Spike is part of the Viacom conglomerate and has no links to either News Corp or Sony. The serialized nature of *The Shield*'s story led Sony to encourage buyers to break from the conventional strategy of "stripping" the show, or running it Monday through Friday in the same time slot, and instead to "stack" it by choosing one night and airing three episodes consecutively. Though Spike followed this strategy, it remains the case that serialized shows continue to struggle to find audiences when embedded in a linear schedule. As Netflix streaming and VOD access has suggested, viewers appreciate the ability to self-determine serial viewing, but the scheduling of such viewing needs to be determined by the viewer.

The economic success of *The Shield* beyond its original airing on FX provided an important lesson for the industry about the ability of original cable series to return profits. Although the earnings from *The Shield* may not come close to those of a broadcast success, it has been more profitable than most moderate broadcast hits. And this achievement has had additional consequences. The FX entertainment president, Kevin Reilly, noted that writing submissions to FX increased tenfold in the fifteen months after *The Shield*'s premiere, indicating shifting perceptions about where to sell series, and the network followed *The Shield* with the similarly edgy, critically lauded dramas *Nip/Tuck, Rescue Me, Over There, Thief, Damages,* and *Sons of Anarchy* that made FX a leader in original cable series production.[42] By 2014, many other case studies of original cable series could be written, though none that reveal new lessons. Shows such as *Mad Men* and *Breaking Bad* have built AMC's reputation and enabled it to increase its license fee even though its shows are viewed by a narrow audience, even by cable standards. On the opposite end of the basic cable spectrum, channels such as USA and TNT have developed original series more similar to broadcast fare.

Just as the differentiated economics of HBO and Showtime allowed the production of unconventional television and radically adjusted notions of television's storytelling capabilities, basic cable networks such as FX, AMC, and USA also contributed to expanding television storytelling in important ways. Before *The Shield*, both audiences and producers had good cause to be wary of original cable productions, but cases such as *The Shield* illustrated the different possibilities available to producers willing to forgo the conventions of broadcast production.

Arrested Development

An Emmy would be nice, but I'd settle for an audience.
—Mitch Hurwitz, creator and executive producer of
Arrested Development, 2004[43]

When *Arrested Development* premiered in the fall of 2003, no one could have predicted its fate. On one hand, the series had been the subject of an intense bidding dispute between FOX and NBC that earned it the status of a "put pilot" a year before it was to go on air—that is, it was effectively guaranteed to have a place on the programming schedule. There is often a substantial financial penalty that the network must pay if it chooses not to air put pilots—a high six-figure penalty in the case of *Arrested Development*.[44] It was also being produced by an established studio, Imagine TV, the studio of Brian Grazer and Ron Howard that was aligned with 20th Century Fox TV Studios. Moreover, not only did the show have the support of the successful director and producer Ron Howard behind it, but he was actually involved in production, providing the series' voice-over.

On the other hand, there was no doubt that the series was unconventional. Producers planned to develop an unusual film style, the show used a somewhat sprawling cast, and none of the characters was particularly likable. Yet the early 2000s were dark days for television comedy, with no breakout hits emerging after *Everybody Loves Raymond*'s debut in 1996 until *Two and a Half Men* in 2005. There were comedies that lasted many seasons and were sold in syndication (*King of Queens*), but this was largely a function of the required programming economics of television, and none achieved the riches or cultural currency of *Seinfeld*, *Friends*, or their many predecessors; and though *Two and a Half Men* came to be the most popular comedy on television in the late 2000s, it still drew audiences with roughly five million fewer viewers than the top comedies of a decade earlier. Each year the number of comedies rating among the top ten and twenty programs dwindled. Even using a proven comedic formula no longer predicted hits, so networks increasingly considered that being unconventional might provide the key to success.

The paradox of expectations bore out as a paradoxical reality for *Arrested Development*. The series was nominated for an unimaginable seven Emmys in its first season and won five, including best comedy, best writing, and best direction; it also won the Television Critics Award for best new show. But it limped through the first season in the ratings competition, finishing the year as the 120th most viewed show among households and 88th among eighteen-to forty-nine-year-olds.[45] Set amongst the other case studies examined here, *Arrested Development* illustrates the situation of a comedy with too niche a tone to succeed on a Big Four network in the early 2000s. Although the innovation of *Arrested Development*'s tone and visual style would have been impossible a few years earlier, even consistent and uniform praise and a cult following could not keep the show on the air beyond three seasons.

Arrested Development proved to be too narrow a hit for the original-run distribution possibilities of its time. Its edgy comedy earned it a devoted fan base, but not one large enough to support broadcast economics. Its audience was more comparable to that of basic cable original series, but these channels also struggled with original comedy development and, as of 2006, had yet to produce a successful original narrative comedy series.[46] The uncertainty of televising comedy at the beginning of the post-network era was apparent beyond the circumstances of this particular failing sitcom. Growing acculturation with narrowcast strategies and niche comedy yielded uncomfortable results when audiences with diverse comedic tastes reconvened for what had once been media events—such as the Academy Awards—as it seemed, by the mid-2000s, that no comedian could bridge the country's varied comedic tastes. This was a complicated environment in which to produce any comedy series, especially one needing to attract broadcast-sized audiences.

Arrested Development first marked its difference from other contemporary comedies through its visual style. It did not use the multiple-camera, fixed-set, studio-audience, laugh-track style long dominant in comedy production, but it also deviated from the emerging single-camera style used in shows such as *Sex and the City* and *Scrubs*. Instead, it blended the two, shooting on location and on film, but with multiple cameras—in a manner similar to some unscripted series.[47] The use of natural light and handheld cameras further contributed to its documentary-like visual effect and helped facilitate the show's rapid pace, which was not primarily the result of economic considerations, as was the case for *The Shield*. On the contrary, the distinct stylistic attributes of *Arrested Development* complemented the particular tone and depiction of family life the comedy sought to offer. In addition to these visual innovations, the use of voice-over also contributed to the series' singular style.

Most basically, the series told the story of a wealthy disconnected family that must unite when its patriarch goes to jail. The show presented family as neither an idyllic space nor a dystopia, as many previous sitcoms had offered. Rather, much of the comedy developed from the distinct characters and their absurd reactions to absurd situations. Fast-paced, the show also offered jokes that were neither obvious nor particularly complicated (a character named Bob Loblaw, which is pronounced blah, blah, blah) but that rewarded regular viewers, whose familiarity with the characters and their backstories added a layer of meaning from which much of the comedy evolved. This self-referentiality and use of in-jokes evolved over multiple episodes; old jokes returned unexpectedly and passed by quickly in a manner that added to the pleasure of longtime viewers but made it difficult to begin watching the series after its first season. The show demanded close attention from viewers, but also compensated them—perhaps making its cult fandom the least surprising aspect of its story.

Another somewhat astonishing component of *Arrested Development*'s history is the consistent support FOX gave the series despite its poor ratings performance. The series received a full-season order for season 2 despite lackluster ratings. The decision to schedule the series for a third season seemed more labored, as the crew did not learn that the show would return until thirty-six hours before it was announced at the network's May upfront presentation. But as ratings continued to falter in the third season—especially after it failed to gain after being scheduled in a prime post-*Simpsons* time slot during the second season—it seemed FOX had done all it could to support the series. News of cancellation never came, despite repeated cutbacks in the original order of twenty-two episodes and the network's intermittent airing of remaining episodes in poor positions in the schedule. Unwavering fans held out hope that rumors of another network buying the series would materialize, and in early 2006 the subscription cable network Showtime reportedly offered a deal for twenty-six episodes over two seasons, so long as the creator and executive producer, Mitch Hurwitz, remained at the helm. Fans' dreams were dashed when Hurwitz declined the offer, with many speculating that he found the compensation package unacceptable.[48] In any event, he and the writers had clearly anticipated the end on FOX and had produced final episodes that adequately closed the story, allowing the series a creative completion many do not receive.

The series produced only fifty-three episodes, which in the network era would have made opportunities for subsequent distribution most unlikely; in that era, the key syndication markets were local stations and international channels, both of which required many more episodes in order to construct weeklong strips of episodes. But post-network distribution created other means for subsequent revenue; as has become increasingly standard, the series was released on DVD and sold internationally. In addition, the negligible HDNet cable network (renamed AXS in July 2012)—a high-definition network then available in only three million homes—purchased a syndication run for undisclosed terms; MSN bought exclusive portal rights to syndicate the show online for three years; and the cable network G4 also purchased the basic cable rights for a three-year run.[49] This deal marked the first time a show was simultaneously syndicated on three platforms, but because of its unprecedented nature and the small audiences of each venue, the deal gave little indication of whether it might be successful for any of the distributors or the studio, or what it might mean for subsequent programs with strong fan interest, but small overall audiences.

Rumors of an *Arrested Development* movie persisted for years after the series' end, and in late 2011 Netflix announced that the cast and creative team would reunite for what became fifteen episodes released to Netflix subscribers on May 26, 2013. These episodes weren't precisely a continuation of the series (a result of the inability to resume full production with the entire cast at the same time), but they did make full use of the viewing control a nonlinear distributor allowed. Though creating new episodes of a show that garnered only four or five million viewers in its original airing may seem ill-considered, Netflix knew how many of its subscribers had watched the older seasons in its library as well as how frequently they viewed and reviewed and their pace of viewing, which minimized the risk of producing the new episodes.

The story of *Arrested Development* leaves as many questions as answers. Although it is unlikely that viewers would have been able to enjoy this series in a previous era, the competitive environment of the late multi-channel transition was ultimately unavailing as well. The coda of the Netflix revival is more interesting than truly innovative. In the end, the Netflix deal only yielded an additional season of episodes, which were certainly appreciated by audiences, but not significantly disruptive to industrial practices. In the first edition of this book, written well before Netflix streamed (let alone created) television, I worked out a thought experiment regarding how *Arrested Development* might be indicative of a new economic model for television. Though I didn't name it then, *Arrested Development* was an early example of what I categorize here as prized content. As prized content with high value to a passionate—even if small—fan base, *Arrested Development* illustrates how the ability to distribute and receive payment from niche audiences might enable alternative economic models for the industry.

A key part of the niche-hit equation is not only the smaller audience, but also the intensity of feeling for a show, which in turn makes viewers more willing to pay directly for content. Niche hits have an advantage over general interest successes—for example, linear television—that draws audiences who are seated in front of the television at the right time, but take little notice of missing an episode. *Arrested Development* faced cancellation by FOX once its audience diminished to an estimated 4.3 million each week. Assuming standard mid-2000s production costs of $1.8 million per episode, an audience of just 1.25 million paying $2 per episode could finance production costs, even after an approximately 20 percent payment to the distribution service, which was Apple's estimated cut of the early iTunes deals (though now distributors take more). By creating the new episodes, Netflix avoided external distribution costs, though its subscription economic model also prevented it from earning revenue specifically for the show.

In a truly post-network era of all nonlinear content in which viewers deliberately select programs and the intensity of feeling for a show characteristic of cult hits creates greater economic value, a series like *Arrested Development* would be more likely to succeed than a long-lived yet low-interest comedy. Yet *Arrested Development* cultivated its loyal fans through "free" viewing on FOX, and how to create initial support for new series remains an uncertain component of such a nonlinear and transactional economic system. Will people be willing to pay to sample content? What economic model might introduce new programming? Some forecast a scenario in which broadcast network airtime could serve as a promotional vehicle—FOX might air the first season of a show, but continued production would depend its gathering enough viewers to support subsequent seasons through some combination of advertising and subscription. Such a transaction model might not produce the billions of dollars in profits some producers have enjoyed over the years, but it could provide a real opportunity for those who seek to use the medium to tell a story regardless of whether it might bring exceptional fame and fortune. With economic models continuing to develop, it may be possible to find ways to support niche hits other than through direct transaction. After all, advertisers very much want to remain part of viewers' television experience—they're just wary of supporting unproven methods of reaching audiences.

Off to War

Of the five cases under consideration here, *Off to War* arguably deviates most significantly from any norm of network-era commercial television. In a stunning piece of documentary, the filmmakers Craig Renaud and Brent Renaud embedded themselves with the men and families of the 239th National Guard Infantry during the eighteen months of their deployment to Iraq in 2004 and 2005 in order to tell a rich and detailed story about the war and its effects upon the part-time soldiers. One of the filmmakers followed the Guard unit to Iraq, while the other stayed behind in the small town of Clarksville, Arkansas, to record the experiences of their families, thereby allowing viewers to see how war affects both those who go and those who remain at home.

The limited-run series, which aired on the Discovery Times channel in 2004 and 2005, began as a documentary with funding from the Japanese broadcaster NHK. As the Renauds prepared to join the unit, the Pentagon mandated that they find a U.S. network or channel to support their project as a condition of being embedded. The filmmakers subsequently reached a deal with the digital cable channel Discovery Times for three one-hour specials that went inside the world of the unit preparing for deployment and then going to Iraq. The response to the specials, aired in the fall of 2004, led the network to commission another seven hours, which aired in the fall of 2005. Significantly, the unit was called up and began training before the insurgency changed the nature of the war into a more fraught and extended military occupation. The additional episodes consequently allowed the filmmakers to explore the situation of part-time soldiers with limited preparation and equipment for the task they ultimately faced.

Off to War first aired just before the 2004 U.S. presidential election, although most episodes aired during the period in which support for the president and the war eroded under the mounting evidence of falsehoods circulated by the administration in the run-up to the war. Under the circumstances, the series' honest look at how the war affected the soldiers and their families in profound, unfair, and largely unconsidered ways made it an important contribution to the culture's (re)assessment of the war. The series also featured working-class, rural, southern Americans—that is, individuals who were most unlikely to appear on television screens, despite the way the 2004 campaign often invoked this population as the heart of the nation.

For all of its cultural centrality, however, *Off to War* existed in comparative obscurity on what might be described as an ultra-niche cable network. Discovery Times launched in March 2003 and reached only 37.3 million homes by November 2005. At the time, Discovery Communications and the New York Times Company jointly owned the network, which serves as a prime example of the additional providers made possible by the efficiency of digital cable transmission. Many of the networks launched after 2000 received carriage only on digital cable systems, as by that point, space in analog transmissions systems had become too crowded for new entrants, especially those without some sort of leverage—the "ransom" channel phenomenon discussed in relation to FX. The Discovery Times channel would thus never have existed without digital cable, and its part ownership by an established cable entity also greatly contributed to creating an opportunity for such a niche venture.

New and niche channels such as Discovery Times struggle financially, as their limited distribution makes it difficult to draw advertisers, yet without substantial advertiser support their budgets remain inadequate for developing programming that might grow viewership. Rather than purchasing off-network series such as the various iterations of *Law & Order* and *CSI*, as has been the strategy of many cable channels that attempt to use known content to draw viewers to an unknown channel, Discovery Times attempted to develop limited original content that it promoted as extensively as it could afford outside the network in hopes of finding viewers. *Off to War* provided a timely project for the channel, and Discovery Times's relative newness enabled it to schedule the series as soon as it was complete—as opposed to an established network like HBO, which often had its documentary schedule set at least a year in advance.[50] Although the commission Discovery Times could afford was not as substantial as some other networks might have paid, the production costs of the series were limited enough to enable the filmmakers to cover them. (Because the Renauds were embedded, their housing, food, and flights to Iraq were paid by the military.) They also retained the rights for international and DVD distribution in order to profit from the series. Just as a marquee program can be invaluable for establishing the identity of a cable channel, so too can an opportunity that might be small by broadcast network standards be extremely valuable for independent filmmakers. *Off to War* earned the Renauds a nomination for outstanding directorial achievement in documentary from the Directors Guild of America, an Overseas Press Club award, and significant exposure at a number of international film festivals that likewise provided more value than just the tangible revenue from the series.

Discovery Times economized the purchase of the series by replaying it many, many times. A review of the channel in June 2005 found that it produced approximately thirty hours of new programming each month, filling much of the rest of its schedule with programming from other channels in the Discovery family.[51] The same article asserted that about 75 percent of Discovery Times's content was unique to the channel, which suggests how many times original content such as *Off to War* might be replayed. Due to its small audience size, the network sold its advertising time as a companion to cable news channels, which have a similar demographic composition. At the time *Off to War* aired, the channel targeted men and adults aged twenty-five to fifty-four; its viewers had a median age of forty-seven, compared with fifty-seven for the cable news channels.[52]

From the channel's perspective, *Off to War* was a big success and catapulted it to where its staffers hoped it would be in five or six years in just a matter of one or two.[53] In one weekend that aired an *Off to War* marathon, the network ranked among the top ten of cable channels, which is significant relative to the competition, but still negligible compared with broadcast audiences.[54] Despite this success, the New York Times Company announced plans to sell its stake in the channel in April 2006, and in 2008 Discovery rebranded the channel as Investigation Discovery, which came to focus on real and fictionalized stories of murder. The rationale is perhaps indicative of the arriving post-network era: the New York Times Company reportedly decided to shift the budget allocated to the channel to developing online video content for the paper's website. In many ways, the Discovery

Times channel itself was a prime candidate for such nonlinear distribution. Although the new content the *Times* endeavor developed was much shorter than the documentaries aired by Discovery Times, the decision underscores a perception of the need for content to be available on demand. An on-demand Discovery Times would allow viewers to more easily access its limited new programming, but before dynamic ad insertion, monetizing VOD wasn't possible.

Since the series reached only thirty-seven million homes and was buried deep in the expanse of rarely viewed cable channels—channel number 111 in my home—few viewers were likely to just happen upon *Off to War*. Discovery Times promoted the show heavily, but the nature of the channel forced it to focus its effort and money outside the channel—spending nearly as much on promotion as production in order to draw viewers to the series. Promotion was one of the key assets of the channel's part ownership by the New York Times Company, with full-page color advertisements often appearing in the *Times* in advance of each episode. The filmmakers also appeared on CNN, FOX News, and radio shows such as *This American Life* and *Weekend Edition*. In terms of more conventional television promotion, the channel also benefited from its co-ownership by Discovery, as it could air promotions on the other four Discovery channels available in most digital cable packages. Still, the series garnered little promotion compared with a "broadcast" show and would probably have escaped even my attention were it not for my regular reading of many of the nation's television critics, particularly those who offer less mainstream recommendations.

The story of *Off to War* may well beg an answer to the question, "If an unconventional show airs in obscurity on an ultra-niche cable network, what contribution can it make to the culture?" Here, there is room for much speculation, but there are other issues to which the series speaks directly. *Off to War* illustrates that commercial television at the beginning of the post-network era could encompass a once inconceivable range. For those who saw the series, this capability mattered a great deal. Indeed, while the other shows considered here derive much of their significance from their unconventional economic and distribution practices, the significance of *Off to War* is in its very existence. *Off to War* also reminds critics that the medium encompasses more than the base motives of commercialism. Alongside those willing to tell whatever story will earn the highest profits are those who have a meaningful story to share and a passion for telling it, undeterred by limited economic reward. The production conditions of the multi-channel transition and post-network era create far greater opportunities for such stories than existed in the network era.

And in many ways, this is where the significance of the emergence of broadband distribution, and particularly a platform for amateur content such as YouTube, becomes clear. During the network era in the United States, there was no means to distribute video content perceived by gatekeepers as noncommercial. Indeed, entities such as premium and basic cable channels were able to expand the array of content that could be commercialized during the multi-channel transition, but these entities too are foremost driven by programming decisions aimed at maximizing profits. The ability to distribute video content online enabled the sharing of noncommercial video and created a space to test assumptions about what audiences will watch and pay for. Though the rhetoric in 2005 proposed that broadband video distributors such as YouTube might overthrow the established television industry, it is now clear this is not the case. What the "democratic" distribution of gatekeeper-free entities such as YouTube have offered is a new way for the established, commercial industries to identify talent and audience interest. Several of those who have emerged as YouTube stars—Lucas Cruikshank, Issa Rae, Freddie Wong—have or are moving from the ranks of YouTube stardom into the established entertainment machinery, and this alone is a significant development. Though many hoped to overthrow the commercial hegemony, the ability for those with both passion and talent to rise to the attention of those in Hollywood and New York is important for expanding the array of voices, content, and ideas communicated by television.

Conclusion

In an era of a couple of hundred channels, I obviously could have selected a group of cases to tell just about any story about the television industry. I could have looked at a stalwart like *Law & Order*, which was one of the most profitable shows in television history regardless of its adherence to conventional production practices. Or I could have considered one of the many programs that survived the grueling gauntlet of development and pilot selection, only to disappear into network oblivion after a handful of episodes. Another story, admittedly a bit harder to tell, is of the shows that still do not achieve distribution on U.S. television. Many of these are no more unconventional than those I consider; they just lacked the right mix of happenstance and opportunity that intervened in these cases.

Each of the cases I did select tells a different story that brings to life the detailed production practices chronicled in the other chapters. I offer them in hopes that the concrete circumstances of specific shows might help illuminate the behind-the-scenes practices that influence the business of television in crucial ways. Program success and failure may be difficult to anticipate, but a confluence of industrial factors contributes significantly to explaining why audiences now choose among a broader array of televised storytelling than at any previous moment in the medium's history. Likewise, understanding the yet-to-be-established practices hinted at in previous chapters prepares us to assess how television is likely to continue to change in the coming months and years.

Of the cases that I present here, it is *Off to War* that gives me the most hope for what the post-network era might mean for television as a cultural institution, but it is also the case that makes me most skeptical of the consequences of what is to come. I could replace *Off to War* with many different series: PBS's *American Family* (2002–2004), Lifetime's *Any Day Now* (1998–2002), even Comedy Central's *The Daily Show with Jon Stewart* (1998–), though this last example became far more mainstream by the late 2000s. All offer stories about people or perspectives not usually found on U.S. television and address important social issues. But these shows typically reached a few hundred thousand people on a medium well known for its ability to gather tens of millions.

The variation in the audiences reached by different programs at the beginning of the post-network era suggests the need for multiple ways of thinking about television as a cultural institution. Those shows that are watched by audiences numbering in the multimillions continue to allow television to operate as an electronic public sphere, as it did in the network era. By and large, these will continue to be shows designed for outlets that require mass audiences, such as the broadcast networks, and will have certain attributes that will aid their broad address. As is the case now, in seeking inclusiveness, they will often "plane" the contentious edges from their programs. With more neutral and mainstream ideas, these shows for mass audiences will do significant work in reaffirming certain ideas and norms within the culture.

By the late multi-channel transition, other categories of programming supported television's more emergent cultural functions, the significance of which we have yet to fully understand. This programming exists on a continuum. At the extreme are programs that address narrow audiences, whether they be Bill O'Reilly's angry conservatives or Bill Maher's liberal hipsters. Somewhere in between are more niche-specific shows and channels for distinctive audiences—MTV's adolescents, Univision's Spanish speakers, Lifetime's traditional women. These are not nearly as exclusive as programming at the extremes, but they also raise the question of how to understand the contribution of exceptional niche shows relative to the mass hits that remain. Interestingly, some of the top YouTube channels feature content related to video gaming, though efforts to put this content on "television," such as the programming on G4, weren't adequately successful and were rebranded. Also, some of the most successful YouTube ventures feature voices and perspectives still largely absent from mainstream media. Issa Rae's "Awkward Black Girl" videos on YouTube expressed a racial and gender experience rarely present on television. Likewise, Asian and Asian American producers and actors remain missing from conventional television, but dominate several of the most-viewed YouTube channels.

Changes in production practices enabled new stories to be told on television, but these industrial processes could not overcome the behavior of viewers who follow niche tastes into demographic and psychographic silos. The new capabilities of the medium to offer diversified stories and perspectives became moot against viewers' inability to find these stories or the disinterest of many in stories not directed to their particularities. Television offers its viewers access to profound and meaningful narratives—admittedly interspersed among ample uninspired and mind-numbing drivel—but few members of the audience venture outside their like-reflecting environs of self-interest. But there is still much more out there, and it is brought to us by television.

Conclusion

Still Watching Television

So this is how we end up alone together. We share a coffee shop, but we are all on wireless laptops. The subway is a symphony of earplugged silence while the family trip has become a time when the kids watch DVDs in the back of the minivan. The water cooler, that nexus of chatter about the show last night, might go silent as we create disparate, customized media environments.
—David Carr, *New York Times*, 2005[1]

Despite the wide-ranging changes in the norms and experience of this technology we have called television, I feel safe in asserting that the verb "watch"—or maybe "view"—will remain the primary word most of us will continue to use to describe our experience. Regardless of the screen size or our location, television fundamentally remains a cultural experience valued for the simultaneous visual and aural glimpses it provides of everything from fictional worlds to breaking news. But otherwise, any further commonality in the experience or the use of television is likely to continue to erode.

We may keep watching television, but the new technologies involve new rituals of use. Few have considered their conventional behavior with regard to network-era television in terms of ritual, probably because there was nothing with which to compare it. Television use typically involved walking into a room, turning on the set, and either turning to specific content or commencing the process of channel surfing; notably, true channel surfing is a behavior that developed early in the multi-channel transition. But integration of post-network technologies into regular television viewing has inaugurated entirely new television-related behaviors and created the possibility of a nonlinear future. In just a few years, early adopters have transitioned from linear viewing to using a DVR to record specific shows weekly or daily and to developing viewing plans based on on-demand availability. Using a DVR or VOD requires a very different and far more deliberate process of content selection than was possible when network programmers made many of these decisions for us. Where the practice of not viewing at appointed times had just begun to trickle beyond the leading edge of adopters as the first edition went to press in 2007, broader asynchronous viewing is emerging now, much as a function of delayed windows that attempt to maintain exclusivity. Netflix viewers may binge on *House of Cards*, but far more may view it once available on DVD and for transactional streaming. HBO subscribers may excitedly view the weekly installments of *Game of Thrones*, while an equally engaged fan base awaits its release beyond the HBO confines or once it can be viewed at a viewer-determined pace. Nonlinear options are increasingly not only a matter of rescheduling viewing to match one's schedule, but reorganizing viewing to accommodate personal priorities related to the desire (or not) to experience culture synchronously, to create greater story continuity, and to complement a viewer's mood.

Although my particular nonlinear viewing behavior is probably not widely representative, it nevertheless illustrates the process of adapting to new modes of decision making regarding television. When I first considered my emerging television ritual in 2006, my DVR was set to automatically record *The Daily Show* Monday through Thursday and keep two episodes. Typically I automatically recorded cable series only because I often forgot when they were on; and I usually took advantage of the late-night replay of episodes so that I could record something on a broadcast network during the prime-time airing because my DVR had only one tuner. I set the broadcast recordings nightly, usually before preparing dinner. I had a better sense of what broadcast shows were on which nights then than I do now, but would scroll through the interactive programming guide making selections based on what shows might be repeats or new episodes and to keep abreast of network scheduling changes. Then I set the VCR on the other television in cases where two shows aired simultaneously. On rare occasions, I set something to record far in advance—typically a show on a rarely viewed cable channel—because of reading a provoking review or due to buzz within the culture or among my students. My DVR held only thirty hours of programming on extended record level, and I recorded almost all shows on the "extended" (poorest) quality level because my hard drive was often nearly full, and this mode allowed me to fit the most shows on the recorder; I also only had a standard-definition television, so the image quality wasn't detrimental. Some exceptional shows—those for which viewing was an event in my household—were recorded in high quality, so long as there was space. Neither then nor now do I watch or record much programming outside prime time, with the exception of limited sports coverage, which is always viewed live. Even though online news availability was pretty nascent in 2006, I was more likely to go online for breaking news coverage than bother with television, but in some cases I might have turned on the television for live updates of a developing story.

By 2014, my practices were much different, both for personal reasons and because of the intervening adjustments in television distribution and control technologies. I purchased my first HD television in the spring of 2007 and had my first child that summer; both have extensively changed the way I use television. I view less television; my prime time is now limited to after kids are in bed and before my own fairly early bedtime. I watch only on the main HD set, though we also have a second, smaller one for the kids, in earshot of my perpetual mom-mode location in the kitchen. The standard-definition set in my bedroom (an old analog set lacking even an IPG) is only used occasionally by the kids in the morning. Though I used to retire to watch TV in bed, the substandard quality of the image and the fact that it is connected to no control device make it pretty unusable; once it breaks I doubt I'll replace it. On rare instances, I'll watch something streamed over wifi on a laptop in bed.

The HD television necessitated an HD DVR, and we added the "whole home" option when Comcast made it available a few years later. Our main set has an HD DVR that records content that can be viewed on either my set or the kids' set. I maintain a couple of episodes of the handful of shows they are most interested in. I view little on broadcast anymore, so my nightly DVR setting ritual has ceased. Almost everything I watch is set for automatic recording. I'll often go through the schedule manually at the beginning of the week to set one-off recordings of new shows or specials. We delayed Netflix subscription until recently because there is always a sizeable queue on the DVR and I don't have time to view more content. I'm an ideal candidate for cord cutting as someone who watches a narrow and particular selection of shows, but it could never work for the rest of my household. The main thing my husband deliberately views is live sports, and my kids' viewing changes often enough that a package of channels and the on-demand service considerably help me manage their viewing. And the truth is, as I've pushed my regular behavior in researching this book, I've been consistently impressed with what my Comcast subscription now offers. Because I mainly view on the television that has an awful VOD interface, I hadn't realized the more expansive content available to me from Comcast online, or the superiority of the online interface. Mind you, I saw Comcast's next-generation interface at a trade show over two years ago. The knowledge that something far superior is coming hasn't made me any more patient and it still hasn't rolled out in my market, probably because there is no telco here competing. Though the trade press reports steady hikes in cable subscription fees, I have to acknowledge that the VOD offerings provide me with a lot more value, and I'd much rather have on-demand options than additional linear channels.

My viewing behavior has changed in the last year as Comcast has expanded its VOD library and original license holders have allowed recent episodes to remain available for a few weeks after airing. I've transitioned to VOD instead of DVR viewing in cases in which content seems reliably available on VOD and if the VOD has a minimal advertising load or allows fast-forwarding. I'm more likely to try something new and watch the first episode because I can find it on VOD than I am to record the most current episode from the linear stream. But I find it unbearable when programs include the full commercial load and fast-forwarding is disabled—that sends me right back to the DVR. I don't watch much online. For me, television viewing happens when all the demands of the day are complete and it is a time to get lost in a story. I sit back and enjoy the visual wonder of my ridiculously large and beautiful screen.

This particular set of practices may be unique to me, but it illustrates the deliberate nature of my viewing and its evolution in response to new technologies, distribution practices, and economic models even over a brief period. I am one of the people who can honestly say that I have not channel surfed since my DVR arrived in my home, and my replaying behavior is equally purposeful. I never watch anything live—except sometimes an HBO show, although I usually watch these on demand—largely because my viewing is pretty much contained to prime time. I can't stand to watch live commercial programming: I've become very impatient with the commercial breaks and often occupy myself with something else—typically a tablet screen—until the show returns, and then am often distracted by whatever I found to entertain me during the break. Sometimes I'll start watching something in progress and time it so that I catch up with the live airing by the end. Yes, I skip nearly all commercials, but I will go back and watch some if they catch my eye during my thirty-second jumps.[2]

I know my use differs considerably from that of others, particularly those who walk into their house or a room and turn the television on regardless of what is on or whether they intend to sit and watch it. Sometimes I feel sandwiched between adopters of more radical "cord-cutting" behavior and the great majority that continue to engage most viewing linearly. In thinking of these things in 2006, I presumed that those long accustomed to linear viewing would never change their behavior, but since the first edition, my parents have acquired a DVR and have come to use it for quite a bit of their viewing, which I wouldn't have predicted. The demise of linear viewing unquestionably has begun; the questions keeping industry executives up at night are how quickly more viewers will join my ranks and just how much linear viewing will remain.

This is all a very long way of illustrating the complicated, deliberate, and individualized nature of television use for those adopting new technologies and shifting into the post-network era. The technologies have been available for some time, but required studios and networks to release content—or for pirates to beat them to it—in order for us to fully experience their capabilities. As I examine my own behavior, it becomes clear to me how complicated negotiating new uses might be for industry decision makers. The industry seeks comprehensive new practices and financial models, while each viewer probably values each piece of content, as well as the opportunity to access it, differently. In 2006 it seemed as though a new story emerged weekly reporting that x percent of viewers prefer free content over paying for it, or y percent watch the commercials in DVR-recorded content—with the values of x and y varying considerably depending on the study. These days such stories tout the fact that x percent of viewers watch y amount online, on a mobile phone, or of late, buzz about "second screens." The problem with such studies that seek aggregate answers is that they reveal little about the intricacies of individual uses and values, and it is only at the individual level that the viewer finds any of these new capabilities meaningful. For example, if I missed an episode of prized content, I'd think nothing of paying five, maybe even ten dollars to download the missed episode—with or without commercials—though this is a problem I rarely encounter now, as almost all shows are available for some slightly extended window. My willingness to pay decreases from there. I might pay a few dollars for some other shows, and I'd make the effort to download or stream a few other shows if there was no cost involved, but I would probably just skip most others and wait until the next episode. Every viewer allocates value differently based on taste, ability to pay, and the technology at hand, all of which makes establishing standard industry pricing very difficult.

This suggests how the experience of television in a post-network era fragments beyond the narrowcasting of the multi-channel transition to personcasting in terms of what is viewed, when, and how, and even in how viewers pay for it. Although such variation in what we watch might not entirely disrupt our network-era understandings of television and culture, the disparity in when we watch programs will make asynchronousness a defining feature of the post-network era. When the industry journalist Diane Mermigas queried in 2006, "Isn't 'primetime' a misnomer in this new 'anytime' era?," the question itself suggested one of many coming adjustments in the very way we talk about television.[3]

Imagining Viewing in the Post-Network Era: Television circa 2020?

Though a range of industrial developments—particularly in technologies used for viewing and in the distribution of television—necessitated a second edition of this book, I maintain that we have not yet achieved a post-network era. More of its features are apparent now, but much more profound change will yet transpire. In my mind, the full arrival of the post-network era will feature the common use of expanded tools that enable easy accessing and ordering of content, the breakdown of bundles—certainly the cable package bundle and maybe even the bundle of programming we call networks or channels—and adjusted economic models that account for these new distribution opportunities and the emergence of advertising mechanisms superior to those of linear television.

First, a key to the future use of television is the development of a nonproprietary tool for searching, organizing, and recommending content. By nonproprietary, I mean a mechanism that searches broadly (not contained to a particular provider such as CBS, Hulu, or Warner Bros. Studio), but also personalized enough to know what a particular user has ready access to; the website locatetv.com provides a most preliminary example of this. I imagine that if my students are buzzing about a new show, let's call it *The Scoop*, I'll be able to quickly enter the title in a smartphone app (or whatever the equivalent of the pocket computer is at that point) that will find the show—in this case, a low-budget feminist talk show posted to YouTube, and a click will add it to my queue. Or if I'm making dinner and I hear an NPR feature on a new documentary about Generation X, I'll quickly enter it as well. Since it is a film, maybe it is being released in theaters in major cities—even here in Ann Arbor—but it doesn't meet my justification of media worth hiring a babysitter. I'll see it when it reaches a more accessible window, but I can enter it in my queue now and it will be delivered to me when its first run is complete (this is a dystopian world where windowing hasn't been fully eradicated). What is important to me is that I don't have to re-remember my interest when the later window begins or search it out a second time, though of course, I may have lost interest by the time it does appear, and can simply delete it from my queue.

Such a tool advances the convenience of accessing desired content for viewers, and from an industry perspective, also could help in reducing marketing costs. When spare time permits, I can peruse my queue and adjust the position of these new entries. When I sit down to view content—whether on a pocket device or on my large living room screen—the content arrives seamlessly and consistently no matter where it originates. By 2014, some of the smarter MVPD interfaces featured some of this functionality. I can access and search my Xfinity offerings from almost anywhere, but I haven't figured out how to help it distinguish between the users in my house, and it isn't helpful for finding content that exists outside the realm of offerings Comcast has negotiated to provide, which means I often search Xfinity, then Netflix, then Amazon Prime, and so on, when looking for a particular title.

In terms of the breakdown of the bundle, if we look at the consequence of digitization on every industry affected so far—newspapers, magazines, music—the consequence of digital distribution has been disaggregation: I don't want the whole *New York Times*, I want articles about television; I don't want the album, I want the songs I like. Bundling was a way to add value in the era of analog technological affordances when the cost of physical creation and distribution needed to be built into pricing, but it is a less valuable strategy in an era in which each audience member desires specific content rather than arbitrary expansion of media goods—I want to watch what I want, when I want, not another "channel." The U.S. television industry must identify an economic model that matches the post-network technological and distribution possibilities that have already emerged. The extraordinary profitability of network-era television has allowed the industry to expand its program offerings to a once unimaginable degree, while pushing the cost for a lot of content that it isn't clear anyone really wants onto all audiences through subscriber fees and advertisers who pay more and more for less. In the process, an illogical system full of inefficiency has become the norm. The future of television is not the persistence of this model, but the establishment of one that matches conditions of the present and future.

It was not coincidental that the over-the-top crisis emerged at the same time retransmission disputes became more frequent and contentious: there might be better content than ever before, but viewers were reaching the point where the incremental cost increases could no longer be borne, and though not a perfect replacement, the Internet provided much to see and do. In the network era, networks were needed to organize and "curate" content; today, many viewers desire to do these tasks themselves. We still need the people who make the content (studios) and someone to deliver it to our houses and mobile technologies (MVPDs; with better compression, maybe cellular providers), but in an economic squeeze, it is unclear whether networks and channels add adequate value to maintain the costs they add. To be clear, I'm not suggesting that all the functions of a channel are redundant, just speculating that many of the costs added by these organizing entities that also need to turn a profit could be eliminated or reallocated if some of the activities currently shouldered by studios and MVPDs were expanded.

The on-demand offerings some MVPDs have built offer a good illustration of this. In large part, my Xfinity on-demand offerings replicate the syndicated linear offerings of the 300-some channels I can choose among, except it often lets me select an episode from the entire library of the series. Building these libraries has required extensive dealings between the MVPDs and the studios; might the MVPDs license new content directly as well? Instead of the current system, where a network is the gatekeeper, selecting among series based on fit with brand, schedule availability, and budget, what if studios just made the shows they believed would attract an audience? Maybe they make the first set of episodes at a deficit and give them to MVPDs for free, and then if people like them, the audience pays a transaction fee for more. Certainly these transaction payments couldn't be added to the already extensive fees viewers pay to MVPDs, but if you take away the cost of the bundles, these fees may not be onerous and may eliminate the broadband-only temptation for light viewers. Or what if the MVPDs, who have the capability to more specifically target advertising, became the sellers of advertising that would subsidize viewers' costs?

In an era of nonlinear television, there are no schedules that need to be filled each and every hour of the day. There may well be a retraction in the amount of programming available, but the programming threatened is that which does not rise to the standard of audience desire—not just a matter of audience size, but audience passion and willingness to pay for it. Such a system instead encourages the production of content that different audiences really do want. There are certainly many dilemmas—mostly related to finding economic models that bridge those who want to pay for the content consumed and those who retain linear viewing—but willfully refusing to consider a radically unbundled future is delusional given the experience of other media to date. The print and music industries have been disrupted, if not decimated, by digitization because they failed to find an economic model that matched the new distribution opportunities. Only time will tell whether the television industry provides a mutually acceptable economic model or faces the crises experienced by these other industries.

As Nicholas Negroponte wrote in his 1995 best seller *Being Digital*, "The future of television is to stop thinking of television as television."[4] My television use now—and that I imagine in the future—takes on more characteristics of the ways I've previously used other media delivery systems. In many ways, I find the novel the best analogy. I've come to view episodes of fictional series in consecutive installments whenever possible, like chapters in a book. Sometimes I'll interrupt the stream if I'm just in a mood for something else, as I would choose to read a magazine instead of continuing with a book from time to time. The greater portability that allows me to start a show on my living room set, then watch an episode on my commute the next morning or in a waiting room also reminds me of how I consume novels.

At the same time, the television industry isn't the only media industry with a role in the television future. I've noted with some curiosity how other media have moved into providing or competing with "television" with little notice. For example, some newspapers, most notably the *New York Times's* TimesCast, began producing videos featuring reporters talking about what the paper is selecting as the news of the day. Sometimes this content is visually dull, but informative in a way most news produced for U.S. television isn't. In an era in which sophisticated video news continues to be absent from U.S. screens, this seems a viable alternative for viewers who are more concerned with news of substance than style. Other segments include video that is rich and expands the visual experience that can be offered in online stills. The emergence of the video commentator—both those independent and those connected to an established media entity—is a phenomenon to watch in coming years. The analog here is the print industry and the emergence of bloggers. Speaking of which, what industry should we consider videos on *Huffington Post* to be a part of?

The Future of Prized Content, Live Sports, and Linear Television

Community lost can be community gained, and as mass culture
weakens, it creates openings for the cohorts that can otherwise get
crowded out. When you meet someone with the same particular
passions and sensibility, the sense of connection can be profound.
Smaller communities of fans, forged from shared perspectives,
offer a more genuine sense of belonging than a national identity
born of geographical happenstance.
—Tim Wu, *New Republic*, 2013[5]

The book to this point should make clear why a uniform future for television cannot be predicted. The future for prized content and live sports appears rich, while linear viewing has begun a downward spiral that may require decades to complete, but seems unlikely to return as a norm of experience.

Just as U.S. television's industrial practices continue to evolve, so too does its role in the culture. Speculating on the precise nature of its future seems even more impossible than forecasting economic models. Communities of viewers sharing interests, favorites, and self-produced content have emerged to organize nascent post-network viewing in the manner previously provided by the networks. As a result, the industry has struggled to restructure its profit and production systems in accord with these new conditions. Here, communities may be more mediated, but thinking that mediated communities are less meaningful than unmediated ones indicates our bias toward previous norms. As Chris Anderson explains, the fact that you and I may have both watched *CSI* last night might start a conversation, but this might indicate only slightly more intimacy than talking about the weather. The investment in stories and ideas that lead to the connections people make online and through social media provides the tools for the beginning of relationships—what he has identified as the creation of "tribes of affinity."[6] What remains difficult to know at this point is the consequence of such tribes if they do emerge. Equally compelling armchair speculation may suppose that this loosens cultural ties or that it creates stronger ties than existed previously. Adjustments on the scale of the post-network era do not introduce change in just one area, but a whole series of unanticipated counter-adjustments that makes this early stage in transition far too preliminary for hand wringing. Indeed, the network-era mass audience has fragmented, but we also now have new tools and technologies with which to connect and engage.

One lesson already apparent from emerging research on new communication technologies is that it is foolhardy to theorize the consequences of technologies based only on how their technological affordances differ from previous technologies. Though many posited that new phone technologies decreased face-to-face communication, research on how these technologies are used finds they help us create stronger ties to our personal communities.[7] Likewise, fragmentation in television viewing audiences across a greater variety of content that is viewed with greater asynchronousness may suggest reduced cultural affect, but we cannot know that is the case until we study use behaviors and consequences. There may be unexpected consequences of the content produced through niche economics—for example, the creation of programs that more regularly move, affect, and provoke us than was the case in the era of programming that tried not to make anyone turn away. In just the past few years, many have explored how social media have been commandeered for the discussion of television. Indeed, Twitter is not the same as the apocryphal watercooler, but the discussion and engagement with television it encourages should not be presumed inferior. New and unanticipated cultural consequences of television may develop in the post-network era; we might not notice them if we are too focused on identifying how they are not precisely like those that came before.

As I completed work on the first edition of this book in early 2007 and still now in early 2014, many uncertainties about the future of television remained, but some new practices and norms had solidified enough to allow us to contemplate the dominant features of the emerging post-network era. The changes in the industrial norms and conditions of production chronicled here yield substantial implications for the creative possibilities of cultural industries. As a result, much of the "conventional wisdom" of industry workers, concerning the type of programs that can be profitably produced and of scholars and analysts regarding how the industry operates, requires reconsideration. The previous pages illustrate many instances in which monolithic network-era norms have been replaced with a variety of practices that are in turn likely to support a further diversification in the content produced by the industry and available to viewers.

As the new competitive environment began to take shape, many industry assessments compared the situation of the television networks with a perhaps apocryphal story of the railroad industry's response to the advent of passenger air travel.[8] The story goes that the downfall of the passenger train service resulted from the industry's inability to recognize air travel as a competitor—that it narrowly viewed itself as engaged in the railroad business rather than the broader transportation business. The oft-made comparison in the rise of the post-network era of television contemplates whether the television networks and channels will choose to narrowly conceive of themselves as networks in the network-era sense—as entities bound to previous norms of program acquisition, distribution, and scheduling—or whether they will recognize that they are in the content aggregation and distribution business and adjust their competitive practices and business models accordingly. Delivery platforms such as HBO GO suggest that HBO understands that it is not in the network-era television business, and MVPDs who've developed rich TV Everywhere offerings also provide evidence of a radically revised future. Entities likewise must consider whether network-era strategies such as bundling and exclusivity make sense in a post-network era, and the role of the aggregator is also due for reconsideration.

I don't mean to downplay the difficulty of this transition; redeveloping industrial practices requires coherent action among studios, channels, networks, MVPDs, and advertisers, and the future status of these entities is not equally rosy. How many executives of channels will eagerly endeavor upon the future imagined here? A forward-thinking one might start developing a relationship with an MVPD or a studio, recognizing that they aren't in the channel business, but the video delivery business. The question now is whether these businesses can adapt their financial models quickly enough to be able to evolve through a more apocalyptic scenario: what happens when mechanisms that can capture intent—other than search—become more mainstream and present a more valuable proposition to advertisers? What happens when such alternative advertising tools unite the presence-at-purchase feature of the pocket computer and the tribes of affinity cultivated by social media that are also accessed on that pocket computer? There's little that I'm certain of, but I'm confident that people won't be standing in the aisle of Target or Kroger watching video ads on those devices.

The changes of the multi-channel transition and post-network era obviously have manifold consequences for the study of media and its role in society, some of which I have raised here and about which I have offered some preliminary ways of thinking. Few observers have offered detailed considerations of the significance and repercussions of the erosion of mass media, or how audiences exercising choice and control require us to revise fundamental ideas about media and culture.[9] Questions such as how cultures and subcultures come to know themselves and each other without widely shared programming and how this affects perceptions of difference in society require new thinking. I'm not proposing that academic researchers can forecast these developments, but we need to be ready to rethink much of what we've known about television that was established during its network era. Many assumptions about the "mass" nature of media undergird theories postulating the emancipatory potential of media. Even as the new norm of niche audiences eliminates some of these imagined possibilities, it may create others.

A Closing Salvo

The post-network era is introducing immense changes to the medium and its role in society, yet television remains every bit as relevant and vital a site for exploring intersections of media and culture as it has ever been. Some prefer identifying this era as one of "convergence" and specifically object to the way "post-network" might suggest the irrelevance of the entities that have long defined the medium. Others are concerned that the "post-network era" might indicate a newer, more improved version of television that disregards its rich past. There is a certain irony that a post-network era comes to characterize television at the same time as theorists of new technology posit the establishment of a network society—a concept that draws on the networked communication systems characteristic of this era.[10] In this case different notions of networks operate on parallel but distinctive trajectories, so that the shifts in industrial practice that inaugurate the post-network era of television are complementary to developments in the network society.

Let me be clear that in using the term "post-network" I do not mean to suggest the death or complete irrelevance of what we have known as television networks or channels. Rather, the term acknowledges the degree to which the centrality they achieved as controllers of distribution and schedulers of programs has diminished. The post-network era will still include some semblance of "networks" and "channels"; however, their fundamental activities and responsibilities will be greatly adjusted. There still will be a need for program aggregation and organization, though I'm less bullish on an entity such as a network serving this function. The same studios, conglomerates, and distributors that dominated the network era and multi-channel transition remain important at the dawn of the post-network era, but their relationships and control of cultural production require significant renegotiation. Ultimately, the relevance of these entities will depend upon how they adapt to viewers' changing desires and expectations. If established interests try too hard to maintain norms, subsequent editions of this book will have to note yet new players, just as this edition has required discussion of Netflix, YouTube, and Hulu. AT&T, Aereo, Roku, and Google (beyond its ownership of YouTube) are mostly absent from the discussion here, but have projects that might radically disrupt the status quo that existing entities continue to try to maintain.

By noting the increasing control that viewers achieve in determining when, where, and how they watch and even participating in it increasingly as creators, I do not mean to indicate that power shifts to the viewers. Viewers have come to enjoy a meaningful increase in and expanded diversity of programming as a result of the industrial changes of the multi-channel transition and emerging post-network era, and the experience of watching television is extraordinarily different from network-era viewing. But commercial interests still control professional video production—that is, the production of video content as a means of economic support. Broadband-delivered programmers such as YouTube provide a venue for distributing and finding an audience for content that is ignored by this commercial system, but

YouTube's move into developing channels suggests that the future for the distribution of such content is uncertain. It even now seems unlikely that YouTube or similar amateur aggregators will become a viable economic alternative for those with narrow audiences, but is more likely to persist as a space for hobbyists to "broadcast" themselves and a means for a very select few to achieve entry into the industry.

The new possibilities in programming that have been achieved are significant, but by no means do they indicate that viewers now control the process or that a democratization of the medium has occurred or will occur. The conditions of the post-network era decrease networks' power as gatekeepers by enabling greater choice—though certainly not unbridled choice. And, as the industry journalist Todd Spangler notes, it remains the case that "nine companies—Disney, Fox, Time Warner, Comcast/NBCUniversal, CBS, Viacom, Discovery, Scripps, and AMC—control about 90 percent of the professionally produced TV content in the U.S."[11] New means of distribution haven't curtailed their dominance—these too are the entities most present on Netflix and Hulu. Finding audiences becomes far more complicated in the contemporary cluttered content space, and those who can marshal vast promotional budgets maintain the status of gatekeepers.

The ways that new television technologies, uses, and programming both separate us and bring us together provide rich new topics for study and interrogation. Acknowledging the fluidity of the medium's use is crucial. We may switch at any time from watching television "alone together" (the increasingly default mode described by David Carr in this chapter's epigraph) to turning our collective eyes to the same content. Although we have little data or experience to explain how modern societies respond to the loss of common culture, Joseph Turow is probably right in noting that our move into individual silos of entertainment and information is significant. Turow, however, does not allow for viewer-driven efforts to reestablish community through the media now used, and this is also an important consideration for understanding these media and new norms of use. If the lack of interactivity inherent in the one-way transmission of television made it difficult for viewers to re-create viewing communities during much of the multi-channel transition, social media and the Internet have since created locations for the development of rich fan cultures and communities. As the "viewsing" of television and the Internet continue to converge, audience members will be better able to participate in communities of fanship, view together virtually, and share their viewing tastes and pleasures with friends, family, and others. Many aspects of the industry are very interested in developing empirically based understandings of how viewers use television differently in the post-network era, and this information is important for reassessments of its cultural role as well. We know a lot about how viewers used to watch television, but those old understandings provide little information about current and coming experiences.

Just as the conditions of the post-network era require consideration of content distinctly—as in the characteristics that differentiate prized content and live sports and contests—the nature of the television audience needs more distinct consideration. Network-era norms allowed very little differentiation in viewer behavior, but constraints of time and money break down what were once standard viewing behaviors. My own use story is one of a viewer with very little time and considerable discretionary income. It makes commercial pods intolerable and weekly intervals between episodes archaic. In contrast to my time shortage, the financial analyst Craig Moffett began calling the industry's attention to what he terms the "poverty problem" in 2011, making the case that the reality of income distribution in the United States makes it unlikely that the network era's mass audience can possibly afford the array of services and devices MVPDs seek to make standard. Moffett opines, "No one would argue that the entertainment choices offered by Netflix are better than what's available on cable, and neither [are] those offered by Hulu, or YouTube. But when faced with a choice of pay TV or a third meal, will some customers choose to make do with a back catalog or off-the-run TV shows and movies? Of course they will."[12] Part of what earned television its network-era status as cultural storyteller was the ubiquity at least partially enabled by its free over-the-air delivery. Though cable subscriptions remain a high value entertainment opportunity relative to much other media, that does not diminish the reality that Moffett's economic analysis finds 40 percent of American households "essentially bereft of discretionary spending power."[13]

And just as the characteristics of programming and audiences must be discerned with greater specificity in the post-network era, so too must television's programming day—at least as long as linear norms persist. This book focuses nearly exclusively on prime-time programming, but other program areas require similar reconsideration. Some of the arguments made here apply to considerations of post-network television news, post-network sports programming, or post-network daytime programming, but each of these areas also has aspects that distinguish it from prime-time shows—and few have the economic mandate of earning revenue in syndication. These other types of programming further indicate the inadequacy of assuming that "television" exists as a coherent entity. I am uncertain whether we should ever have spoken of "television" in a manner suggesting uniformity and cohesiveness, but we most definitely cannot continue to think of it in this way.

In his 1974 book *TV: The Most Popular Art,* Horace Newcomb closed his pioneering critical look at the nation's most derided and engaged media form by proposing characteristics of television aesthetics.[14] Basing his views upon the network-era content and available uses of television in the 1970s, Newcomb argued that intimacy, continuity, and history were the elements that distinguished television and earned its status as popular art. These characteristics particularly differentiated television storytelling from that of cinema, and while he noted that the "smallness of the television screen has always been its most noticeable feature," Newcomb could not have anticipated the roughly two-and-a-half by four-and-a-half-inch display on ubiquitous smartphones some thirty years later. The established norms of visual storytelling common to cinema framed the media context through which we initially began to understand television, but television now has its own visual storytelling history to write.

We no longer need a separate medium to frame our understanding of television, because its own historical features and distinctions now serve that function. I have provided scant consideration of television programming or television as an artistic form beyond my attention to the consequences of production on the stories it tells, and I won't start now. But it is worth noting, despite my attention to the changes, adjustments, and disruptions of network-era television, that the same aesthetic elements Newcomb outlines continue to characterize television storytelling—and in some cases are even more pronounced now.

In many ways the longevity of the multi-channel transition resulted from the slow realization that the future of television, to paraphrase Negroponte, was to stop thinking of television as it existed, as the box introduced half a century earlier. Negroponte's words were prescient in 1995 and continue to be relevant nearly two decades later. Television remains very much alive and an important part of culture. Understanding the variety of industrial practices and multiple functions of the medium provides a first step toward explaining how television has been and will be revolutionized.

NOTES

INTRODUCTION

1. Mark Fischetti, "The Future of TV," *Technology Review*, November 2001, 35–40.
2. IBM Business Consulting Services, "The End of Television as We Know It," 27 March 2006, http://www-1.ibm.com/services/us/index.wss/ibvstudy/imc/a1023172?cntxt=a1000 062&re=endoftv; Adam L. Penenberg, "The Death of Television," *Slate*, 17 October 2005, http://www.slate.com/toolbar.aspx?action=print&id=2128201; Burt Helm, "Why TV Will Never Be the Same," *Business Week* online, 22 November 2004, http://www.businessweek. com/technology/content/nov2004/tc20041123_3292_tc184.htm?chan=search; Brooks Barnes, "How Old Media Can Survive in a New World," *Wall Street Journal*, 23 May 2005, R1.
3. John Borland and Evan Hansen, "The TV Is Dead. Long Live the TV," *Wired*, 6 April 2007, http://www.wired.com/entertainment/hollywood/news/2007/04/tvhistory_0406.
4. Max Fisher, "Cable TV Is Doomed," *Atlantic*, 18 March 2010, http://www.theatlantic.com/ business/archive/2010/03/cable-tv-is-doomed/37675/.
5. Michael Curtin, "On Edge: Culture Industries in the Neo-Network Era," in *Making and Selling Culture*, ed. Richard Ohmann, Gage Averill, Michael Curtin, David Shumway, and Elizabeth Traube (Hanover, NH: Wesleyan University Press, 1996), 181–202, 186.
6. Michele Hilmes, "Cable, Satellite and Digital Technologies," in *The New Media Book*, ed. Dan Harries (London: British Film Institute, 2002), 3–16, 3.
7. Graeme Turner and Jinna Tay, *Television Studies after TV: Understanding Television in the Post-Broadcast Era* (London: Routledge, 2009), 8.
8. The distribution of television content on Apple's iTunes in October 2005 marked a significant turning point, and if pushed to identify the beginning of the post-network era, this is the event I would likely propose.
9. Nielsen Media Research, "Viewing on Demand: The Cross-Platform Report," September 2013. Figures are for people ages two and up, and are based on total users of each medium.
10. Bill Carter, "Reality TV Alters the Way TV Does Business," *New York Times*, 25 January 2003.
11. Examples of early constraining corporate behavior can be seen in early versions of AOL that allowed only paying members to access content and industry efforts to prevent DVR use. Later, AOL shifted to allow anyone access to ad-supported content, and networks made shows available on iTunes for use on personal computers and iPods, and began enabling limited accessibility outside the linear broadcast airing.
12. Brian Winston, *Media Technology and Society: A History; From the Telegraph to the Internet* (London: Routledge, 1998), 6.
13. A practice with extensive historical precedence; see ibid.
14. Emily Nussbaum, "When TV Became Art: Good-Bye Boob Tube, Hello Brain Food," *New York* magazine, 4 December 2009, http://nymag.com/arts/all/aughts/62513/.
15. Jason Jacobs, "Television, Interrupted: Pollution or Aesthetic?," in *Television as Digital Media*, ed. James Bennett and Niki Strange (Durham: Duke University Press, 2011), 255–82.

16. Nielsen Media Research, "Viewing on Demand: The Cross-Platform Report," September 2013.

17. It is not immune, however. Issues of unauthorized access quickly emerge when artificial limits such as geographic blackouts and time zone issues prevent fans from watching when they want.

18. Robert Marich, "TV Distribs Caught in Sports Vise," *Variety*, 1 March 2013, http://variety.com/2013/tv/news/tv-distribs-caught-in-sports-vise-1200001893/.

19. Derek Thompson, "Sports Could Save the TV Business—or Destroy It," *Atlantic*, 17 July 2013, http://www.theatlantic.com/business/archive/2013/07/sports-could-save-the-tv-business-or-destroy-it/277808/.

20. Todd Gitlin, *Inside Prime Time* (New York: Pantheon, 1983).

21. The implications of a post-network era for U.S. television outside prime time are explored in Amanda D. Lotz, ed., *Beyond Prime Time: Television Programming in the Post-Network Era* (New York: Routledge, 2009).

1. UNDERSTANDING TELEVISION AT THE BEGINNING OF THE POST-NETWORK ERA

1. The channel allocation freeze encompasses the four years from 1948 to 1952, during which the FCC granted no new station licenses while it determined a strategy to organize the spectrum. During the same period, CBS and NBC competed to establish the television standard, which included various color and black-and-white standards that were not always interoperable.

2. The exception was RCA's parentage of NBC; CBS also included the CBS Records division, and ABC had links to theatrical exhibition, but this conglomeration is minimal compared with what developed later. See Erik Barnouw, *Tube of Plenty: The Evolution of American Television*, 2nd rev. ed. (New York: Oxford University Press, 1990); Les Brown, *Television: The Business behind the Box* (New York: Harcourt Brace Jovanovich, 1971); Sally Bedell, *Up the Tube: Prime-Time TV and the Silverman Years* (New York: Viking, 1981); Ken Auletta, *Three Blind Mice: How the TV Networks Lost Their Way* (New York: Vintage, 1992).

3. See William Boddy, *Fifties Television: The Industry and Its Critics* (Urbana: University of Illinois Press, 1993); Lynn Spigel, *Make Room for TV: Television and the Family Ideal in Postwar America* (Chicago: University of Chicago Press, 1992); George Lipsitz, "The Meaning of Memory: Family, Class, and Ethnicity in Early Network Television Programs," in *Private Screenings: Television and the Female Consumer*, ed. Lynn Spigel and Denise Mann (Minneapolis: University of Minnesota Press, 1992), 71–110; Mary Beth Haralovich, "Sitcoms and Suburbs: Positioning the 1950s Homemaker," in Spigel and Mann, *Private Screenings*, 111–42.

4. However, as McCarthy notes, tavern viewing was also common, particularly until penetration levels grew to the point that televisions were common in the home. Anna McCarthy, *Ambient Television: Visual Culture and Public Space* (Durham: Duke University Press, 2001). Also, Spigel recounts that Sony launched a portable set in 1967. While technologically possible, portability was not a defining attribute of the medium in the way the advent of a technology such as the Walkman fundamentally redefined music use. As Spigel notes, the "portable" sets of the 1960s were often marketed as "personal TVs" to allow individualized viewing. Despite this marketing rhetoric, purchase of such sets was slow, and when purchased, these portable televisions seldom moved. See Lynn Spigel, "Portable TV: Studies in Domestic Space Travel," in *Welcome to the Dreamhouse: Popular Media and*

Postwar Suburbs (Durham: Duke University Press, 2001), 60–103, 75; and Tracy Stevens, ed., *International Television and Video Almanac*, 44th ed. (La Jolla, CA: Quigley, 2001), 4.

5. James G. Webster, "Television Audience Behavior: Patterns of Exposure in the New Media Environment," in *Media Use in the Information Age: Emerging Patterns of Adoption and Consumer Use*, ed. Jerry L. Salvaggio and Jennings Bryant (Hillsdale, NJ: LEA, 1989), 197–216.

6. Paul Klein, "Why You Watch When You Watch," *TV Guide*, July 1971, reprinted in *TV Guide: The First 25 Years*, ed. Jay S. Harris (New York: New American Library, 1978), 186–88.

7. This particular change actually preceded the others, beginning in the early 1970s, but it continued to affect the industry in crucial ways during the period in which the industry negotiated these other changes.

8. John Thornton Caldwell, *Televisuality: Style, Crisis, and Authority in American Television* (New Brunswick: Rutgers University Press, 1995), 11.

9. FOX averaged a share of 9 percent, and UPN and the WB each drew 4, leaving ABC, CBS, and NBC with an audience share of only 41 in 1999–2000. Aggregate cable first drew more prime-time viewers than aggregate broadcast networks in 2003–2004. John Dempsey, "Cable Aud's Now Bigger Than B'Cast," *Variety*, 20 May 2004, http://www.variety.com/story.asp? l=story&a=VR1117905396&c=14. The 1999–2000 figure is from *Broadcasting & Cable*; the 2004–2005 figure is from Monica Steiner, "Primetime Update (Full Season)," *Media Insights*, 27 May 2005, 2.

10. Initiative Media, "Today in National Television," report, 9 April 2004.

11. James R. Walker and Robert V. Bellamy Jr., "The Remote Control Device: An Overlooked Technology," in *The Remote Control in the New Age of Television*, ed. James R. Walker and Robert V. Bellamy Jr. (Westport, CT: Praeger, 1993), 3–14.

12. Webster, "Television Audience Behavior."

13. Ibid.

14. In his examination of whether new technologies contribute to making "interactive audiences," Henry Jenkins reports Matt Hills's concern about how asynchronous viewing of programming around the globe troubles synchronized fan discussions and interactions with programming online; Henry Jenkins, "Interactive Audiences?," in *The New Media Book*, ed. Dan Harries (London: British Film Institute, 2002), 157–70, 161. Hills contemplates only the beginning of coming issues, as DVRs and VOD technologies as well as the breakdown of network scheduling conventions challenge the likelihood of television viewing cultures experiencing content synchronously even within national or regional contexts. The future model is more likely to be similar to that of film as viewers choose available content on their own schedules, with fans accessing content immediately.

15. In many instances this required coproduction with an international market, as in the case of *La Femme Nikita*, *The 4400*, and *Battlestar Galactica*.

16. "The Macro View: Investment Analysts on Cable's Economic Outlook," presentation at the National Cable Show, Washington, DC, 10 June 2013.

17. Data from 1985, 1995, and 2005 are from Knowledge Networks Statistical Research, "The Home Technology Monitor: Spring 2005 Ownership and Trend Report" (Crawford, NJ: Knowledge Networks SRI, 2005), 40. Data from 2013 are from Nielsen Media Research, "The Digital Consumer," February 2014, 5.

18. Nielsen Media Research, "The Digital Consumer," February 2014, 5. In-home use of tablets for viewing video was even higher, at 82 percent. Wayne Friedman, "Most TV/Video

and Mobile TV Usage Is in Home," *Media Post*, 3 June 2013, http://www.mediapost.com/ publications/article/201657/most-tvvideo-and-mobile-tv-usage-is-in-home.html.

19. Dan Harries, "Watching the Internet," in Harries, *The New Media Book*, 171–82, 172.

20. Issues of generation are also noted by Nicholas Negroponte, *Being Digital* (New York: Vintage, 1995); and Charlotte Brunsdon, "Lifestyling Britain: The 8–9 Slot on British Television," in *Television after TV: Essays on a Medium in Transition*, ed. Lynn Spigel and Jan Olsson (Durham: Duke University Press, 2004), 75–92, 85.

21. Neil Howe and William Strauss, *Millennials Rising: The Next Great Generation* (New York: Vintage, 2000); Sharon Jayson, "Totally Wireless on Campus," *USA Today*, 2 October 2006, http://www.usatoday.com/tech/news/2006-10-02-gennext-tech_x.htm, accessed 6 October 2006. There are unquestionably factors of class that create substantial differences in the technological access available to the members of this generation.

22. Anne Sweeney, remarks made at the National Cable Show, Atlanta, GA, 10 April 2006.

23. IBM Business Consulting Services, "The End of Television as We Know It," 27 March 2006, http://www-1.ibm.com/services/us/index.wss/ibvstudy/imc/a1023172?cntxt=a10000 62&re=endoftv.

24. Jason Mittell, "TiVoing Childhood," *Flow*, 24 February 2006, http://flowtv.org/2006/02/ tivoing-childhood/.

25. Nielsen Media Research, "Viewing on Demand: The Cross-Platform Report," September 2013.

26. Totals come from hits noted on the site, although these figures are far from definitive.

27. Bill Carter, "Here Comes the Judge," *New York Times*, 12 March 2006, sec. 2, p. 1; Ciar Byrne, "And the Real Winner Is . . . ," *Independent*, 16 January 2006, 4; Claire Atkinson, "Marketers Hunger for *Idol* Reprise," *Advertising Age*, 28 May 2006, 1.

28. See http://www.youtube.com/yt/press/statistics.html.

29. Internet World Stats, "World Internet Users Statistics Usage and World Population Stats, June 2012," http://www.internetworldstats.com/stats.htm; Sara Dover, "Study: Number of Smartphone Users Tops 1 Billion," *CBS News* online, 17 October 2012, http://www.cbsnews.com/8301-205_162-57534583/study-number-of-smartphone-users-tops-1-billion/.

30. Lisa Gitelman, *Always Already New: Media, History and the Data of Culture* (Cambridge: MIT Press, 2006), 7.

31. James Bennett, "Introduction: Television as Digital Media," in *Television as Digital Media*, ed. James Bennett and Niki Strange (Durham: Duke University Press, 2011), 2–3.

32. Spigel and Olsson, *Television after TV*, 2.

33. Charlotte Brunsdon, "What Is the 'Television' of Television Studies?," in *The Television Studies Book*, ed. Christine Geraghty and David Lusted (London: Arnold, 1998), 95–113.

34. Michael Curtin, "Feminine Desire in the Age of Satellite Television," *Journal of Communication* 49, no. 2 (1999): 55–70, 59.

35. This section speaks of dominant norms of this time. Of course exceptions existed as early adopters bought early versions of remote control devices and others utilized portable sets.

36. Nielsen Media Research, *2000 Report on Television: The First 50 Years* (New York: Nielsen Media Research, 2000), 13. See McCarthy, *Ambient Television*, for a critical and theoretical examination of this phenomenon.

37. By using the term "electronic public sphere," I do not intend to invoke Jürgen Habermas. My usage is far more descriptive than analytical and emphasizes the way television made content broadly available; it does not speak to issues surrounding the public sphere that are more theoretically complex.

38. This is evident in Horace Newcomb and Paul Hirsch's theorization of television creating a "cultural forum" and John Fiske and John Hartley's explanation of television's "bardic function" in storytelling. Horace Newcomb and Paul Hirsch, "Television as a Cultural Forum," in *Television: A Critical View*, 5th ed., ed. Horace Newcomb (New York: Oxford University Press, 1994), 503–15; John Fiske and John Hartley, *Reading Television* (London: Methuen, 1978).

39. Todd Gitlin, "Prime Time Ideology: The Hegemonic Process in Television Entertainment," in Newcomb, *Television: A Critical View*, 5th ed., 516–36.

40. This use of the idea of the cultural institution is comparable with Louis Althusser's concept of the ideological state apparatus, and I presume them to function similarly. See Louis Althusser, "Ideology and Ideological State Apparatuses," in *Lenin and Philosophy and Other Essays*, trans. Ben Brewster (London: Monthly Review Books, 1971), 127–88.

41. In fact, a considerable amount of hostility existed between those identifying their work as belonging to the field of cultural studies and those seeing their work as political economy. A 1993 International Communication Association session produced a heated forum for this debate that became legendary; the exchange is captured in a colloquy section of *Critical Studies in Mass Communication* 12, no. 1 (1995). In that volume, see Lawrence Grossberg, "Cultural Studies vs. Political Economy: Is Anybody Else Bored with this Debate?," 72–81; Nicholas Garnham, "Political Economy and Cultural Studies: Reconciliation or Divorce?," 62–71; and James Carey, "Abolishing the Old World Spirit," 82–89. For more on this debate, see also Douglas Kellner, "Overcoming the Divide: Cultural Studies and Political Economy," in *Cultural Studies in Question*, ed. Marjorie Ferguson and Peter Golding (London: Sage, 1997), 102–20.

42. Paul du Gay, Stuart Hall, Linda Janes, Hugh Mackay, and Keith Negus, *Doing Cultural Studies: The Story of the Sony Walkman* (London: Sage, 1997); David Hesmondhalgh, *The Cultural Industries* (London: Sage, 2002); Andrew Calabrese and Colin Sparks, eds., *Toward a Political Economy of Culture: Capitalism and Communication in the Twenty-First Century* (Lanham, MD: Rowman and Littlefield, 2004); Andreas Wittel, "Culture, Labor, and Subjectivity: For a Political Economy from Below," *Capital & Class*, no. 84 (Winter 1984): 11–30.

43. Amanda D. Lotz, "Using 'Network' Theory in the Post-Network Era: Fictional 9/11 U.S. Television Discourse as a 'Cultural Forum,'" *Screen* 45, no. 4 (2004): 423–39.

44. Raymond Williams, *Television: Technology and Cultural Form* (New York: Schocken, 1974).

45. Bernard Miège, *The Capitalization of Cultural Production* (New York: International General, 1989), 146–47.

46. Horace Newcomb, "This Is Not Al Dente: *The Sopranos* and the New Meaning of Television," in *Television: The Critical View*, 7th ed., ed. Horace Newcomb (New York: Oxford University Press, 2006), 561–78; Horace Newcomb, "Studying Television: Same Questions, Different Contexts," *Cinema Journal* 45, no. 1 (2005): 107–11.

47. Tom Scocca, "The YouTube Devolution," *New York Observer*, 31 July 2006, 1.

48. James Poniewozik, "Here's to the Death of Broadcast," *Time*, 26 March 2009, http://www.time.com/time/magazine/article/0,9171,1887840,00.html.

49. Michael Curtin argues that this is at least the case with television. Curtin, "Feminine Desire in the Age of Satellite Television."

50. See Anna Gough-Yates, *Understanding Women's Magazines: Publishing, Markets and Readerships* (New York: Routledge, 2003); Janice Winship, *Inside Women's Magazines* (London:

Pandora, 1987); Ros Ballaster, Margaret Beetham, Elizabeth Frazier, and Sandra Hebron, *Women's Worlds: Ideology, Femininity, and Women's Magazines* (London: Macmillan, 1991).

51. Joseph Turow, *Breaking Up America: Advertisers and the New Media World* (Chicago: University of Chicago Press, 1997).

52. This is a derivation of what Turow describes as "electronic equivalents of gated communities." Ibid., 2.

53. Subcultural concerns are also important, but need theory distinct from that which treats television as a cultural institution.

54. Lotz, "Using 'Network' Theory."

55. Fiske and Hartley, *Reading Television*, 66.

56. Austan Goolsbee, "The BOOB Tube Won't Make Your Kid a Boob," *Chicago Sun Times*, 5 March 2006, B3. To be fair to the authors, however, I should point out that their method looked at network-era television, which may make their universalizing use of television more defensible.

57. Newcomb and Hirsch, "Television as a Cultural Forum," 503–15; Gitlin, "Prime Time Ideology"; Fiske and Hartley, *Reading Television*.

58. In 1970–1971, *Marcus Welby, MD* was the top show, with a rating of 29.6; at the time the U.S. television universe was estimated at 60.1 million; hence the show reached 49.2 percent of households. By 1980–1981, *Dallas* was the most-watched show, with a rating of 34.5, but since the television universe had increased to 76.3 million, it reached 45.2 percent of households. Data on top shows are drawn from Tim Brooks and Earle Marsh, *The Complete Directory to Prime Time Network and Cable TV Shows, 1946–Present*, 8th rev. ed. (New York: Ballantine, 2003). Television universe figures from Nielsen Media Research, *2000 Report on Television*.

59. Kathryn C. Montgomery, *Target: Prime Time; Advocacy Groups and the Struggle over Television Entertainment* (New York: Oxford University Press, 1989).

60. Du Gay et al., *Doing Cultural Studies*; Julie d'Acci, "Cultural Studies, Television Studies, and the Crisis in the Humanities," in Spigel and Olsson, *Television after TV*, 418–45.

61. See Patricia Aufderheide, *Communications Policy and the Public Interest* (New York: Guilford, 1999), for an account of the regulatory, industrial, and public interest machinations and struggles over the Telecommunications Act of 1996; and Joel Brinkley, *Defining Vision: The Battle for the Future of Television* (San Diego: Harcourt Brace, 1997), on the switch to high-definition television.

2. TELEVISION OUTSIDE THE BOX

1. "Convergence Fulfilled," *Hollywood Reporter* online, 13 September 2005.

2. John Borland and Evan Hansen, "The TV Is Dead. Long Live the TV," *Wired*, 6 April 2007, http://www.wired.com/entertainment/hollywood/news/2007/04/tvhistory_0406.

3. Louise Benjamin, "At the Touch of a Button: A Brief History of Remote Control Devices," in *The Remote Control in the New Age of Television*, ed. James R. Walker and Robert V. Bellamy Jr. (Westport, CT: Praeger, 1993), 15–22.

4. Bruce C. Klopfenstein, "From Gadget to Necessity: The Diffusion of Remote Control Technology," in Walker and Bellamy, *The Remote Control in the New Age of Television*, 23–39. It should be noted that definitive penetration rates for remote controls are uncertain because the primary data collected counted set sales and consequently did not account for homes with more than one set or homes replacing one remote-equipped set with another.

5. Jackie Byars and Eileen R. Meehan, "Once in a Lifetime: Constructing the 'Working Woman' through Cable Narrowcasting," in "Lifetime: A Cable Network for Women," ed. Julie d'Acci, special issue, *Camera Obscura* 33–34 (1994): 12–41, 23.

6. William J. Quigley, ed., *International Television and Video Almanac*, 50th ed. (New York: Quigley, 2005), 15.

7. See Byars and Meehan, "Once in a Lifetime."

8. The launch of some of the most competitive cable networks preceded the mid-1980s, with HBO (1972), CNN (1980), and MTV (1981) seeking audiences years before the industry reached the 50 percent penetration mark. Major events also followed; DBS systems offered another method of program delivery (although they arguably offered no more additional program providers than those available through cable, especially since the rollout of digital cable systems). Tracy Stevens, ed., *International Television and Video Almanac*, 45th ed. (La Jolla, CA: Quigley, 2000), 9.

9. Most recently, 52 percent are reached by digital cable and 33 percent by satellite. Nielsen, "The Media Universe: Device Ownership among Americans within TV Homes," September 2012.

10. Meg James, "TV in Your Pocket Is the Next Small Thing," *Los Angeles Times*, 1 November 2005, http://www.latimes/news/printedition/front/la-fi-mobile1nov01,1,1902147,print. story?coll=la-headlines-frontpage.

11. Knowledge Networks Statistical Research, "The Home Technology Monitor: Spring 2005 Ownership and Trend Report" (Crawford, NJ: SRI, 2005), 11. Much of these data are drawn from Knowledge Networks' *Home Technology Monitor*, a report of home technology ownership and trends the company has produced for twenty-five years. The survey relies on calling respondents using a random national sample. Importantly, these data are limited because they survey only households with telephones.

12. Ibid., 17.

13. Ibid., 27.

14. Ibid., 14.

15. The study considered English-language programming only. Nielsen Media Research, "High Definition Is the New Normal," 17 October 2012, http://www.nielsen.com/us/en/newswire/2012/high-definition-is-the-new-normal.html.

16. Ibid., 18.

17. Figures are based on data from International TV and Video Almanac.

18. Nielsen Media Research: "Changing Channels: Americans View Just 17 Channels Despite Record Number to Choose From," 6 May 2014, http://www.nielsen.com/us/en/newswire/2014/changing-channels-americans-view-just-17-channels-despite-record-number-to-choose-from.html.

19. Knowledge Networks, "The Home Technology Monitor: Spring 2005 Ownership and Trend Report," 30.

20. Ibid., 31.

21. Nielsen Media Research, "Viewing on Demand: The Cross-Platform Report," September 2013, 3.

22. Shalini Ramachandran and Amol Sharma, "Cable Fights to Feed 'Binge' TV Viewers," *Wall Street Journal*, 20 September 2013, http://online.wsj.com/news/articles/SB10001424127887324 80770457908317099619059 0#project%3DBINGETV0921%26articleTabs%3Dinteractive.

23. Knowledge Networks, "The Home Technology Monitor: Spring 2005 Ownership and Trend Report," 32.

24. Home computer ownership was 11 percent in 1985; 18 percent in 1990; 30 percent in 1995; and 52 percent in 2000; ibid., 39–40.

25. Ibid., 46.

26. Pew Internet and American Life Project, "Internet Use and Home Broadband Connections, 24 July 2012, http://pewinternet.org/Infographics/2012/Internet-Use-and-Home-Broadband-Connections.aspx.

27. Knowledge Networks, "The Home Technology Monitor: Spring 2005 Ownership and Trend Report," 57.

28. This capability was not introduced in the United States until 2005. Ibid., 56, 59.

29. Nielsen Media Research, "An Era of Growth: The Cross-Platform Report," March 2014, 11. The figure of 235 million is extrapolated from research noting 133.7 million smartphones at 57 percent of the U.S. mobile market. ComScore, "ComScore Reports February 2013 U.S. Smartphone Subscriber Market Share," 4 April 2013, http://www.comscore.com/Insights/Press_Releases/2013/4/comScore_Reports_February_2013_U.S._Smartphone_Subscriber_Market_Share.

30. Nielsen Media Research; "Changing Channels."

31. Nielsen Media Research, "State of the Media: The Cross-Platform Report: Quarter 3, 2012," 7.

32. Steve Sternberg, "Television Insights: A Publication of Magna Global USA," 8 November 2005, 1. Notably, Nielsen was technologically incapable of measuring how many of the 6 percent of viewers ever actually viewed their tapes.

33. Stuart Elliot, "Watching a Show Live, with 72 Hours to Do It," *New York Times*, 17 May 2007, http://www.nytimes.com/2007/05/17/business/media/17adco.html?_r=1&,a. Of all the data I report here, this remains the most mind-boggling; with the exception of live sports, no live viewing takes place in my DVR household.

34. In 1985 Nielsen reported that soap operas constituted seven of the ten most frequently recorded programs. James Traub, "The World according to Nielsen," *Channels* 4, no. 1 (January–February 1985): 26–32, 70–71, 70; Josh Bernoff, "The Mind of the DVR User: Acquisition and Features," Forrester Research, 31 August 2004; Steve Hoffenberg, "DVR Love: A Survey of Digital Video Recorder Users," DTV View Lyra Research, May 2004; Brian Hughes, "Nielsen's DVR Impact Assessment Study," *Media Insights: A Publication of Magna Global*, September 2005.

35. Though playback of DVRs is nonlinear, the original recording of content is based on a linear television schedule.

36. John M. Higgins, "Empty Screens: If Cable's Video-on-Demand Is So Hot, Where Are All the Good Shows?," *Broadcasting & Cable*, 19 September 2005, 14.

37. Knowledge Networks, "The Home Technology Monitor: Spring 2005 Ownership and Trend Report," 31.

38. Andrew Wallenstein, "Study: Cable Industry Fumbled VOD," *Variety*, 5 March 2012, http://variety.com/2012/tv/news/study-cable-industry-fumbled-vod-1118051094/.

39. Eighty percent of Americans buying a wired high-speed connection sign up with their local cable incumbent; more than 90 percent of new wired Internet access subscriptions now go to the local cable incumbent. Susan Crawford, *Captive Audience: The Telecom Industry and Monopoly Power in the New Gilded Age* (New Haven: Yale University Press, 2013), 64, 53.

40. Bernard Miège, *The Capitalization of Cultural Production* (New York: Information General, 1989).

41. Devindra Hardawar, "Netflix Now Accounts for 25% of North American Internet Traffic," *Venture Beat*, 17 May 2011, http://venturebeat.com/2011/05/17/netflix-north-america-traffi/.

42. The term MVPD is new—multi-channel video programming distributor is now the more precise term the industry uses to account for what used to be called MSOs.

43. Cliff Edwards, "Netflix Declines Most since 2004 after Losing 800,000 U.S. Subscribers," *Bloomberg News*, 25 October 2011, http://www.bloomberg.com/news/2011-10-24/netflix-3q-subscriber-losses-worse-than-forecast.html.

44. Ibid.

45. Diane Werts, "A New Way to Watch TV," *Columbus Dispatch*, 21 January 2004, F1, 6; John Maynard, "With DVD, TV Viewers Can Channel Their Choices," *Washington Post*, 30 January 2004, C1.

46. Maynard, "With DVD, TV Viewers Can Channel Their Choices."

47. Werts, "A New Way to Watch TV."

48. Stephanie Rosenbloom, "*Lost* Weekend: A Season in One Sitting," *New York Times*, 27 October 2005, http://www.nytimes.com/2005/10/27/fashion/thursdaystyles/27dvd.html?ex=1288065600&en=808bc3d02751557&ei=5090; Sam Anderson, "The Joys of Rising from the Cultural Dead," *Slate*, 6 April 2006, http://www.slate.com/toolbar.aspx?action=print&id=2139457.

49. See Anderson, "The Joys of Rising from the Cultural Dead."

50. Horace Newcomb, "Magnum: The Champagne of TV," *Channels of Communication*, May–June 1985.

51. Thomas Goetz, "Reinventing Television," *Wired* 13, no. 9 (September 2005), http://wired-vig.wired.com/wired/archive/13.09/stewart.html; Sue McDonald, "Jon Stewart's *Crossfire* Transcript Most Blogged News Item of 2004, Intelliseek Finds," *Business Wire*, 15 December 2004.

52. Mike Reynolds, "TV Guide Net Fills Out Its Full-Screen Roster," *Multichannel News*, 17 December 2012, 9.

53. Anna McCarthy, *Ambient Television: Visual Culture and Public Space* (Durham: Duke University Press, 2001).

54. See Anna McCarthy's work on tavern culture in particular. Ibid.

55. Lynn Spigel, "Portable TV: Studies in Domestic Space Travel," in *Welcome to the Dreamhouse: Popular Media and Postwar Suburbs* (Durham: Duke University Press, 2001), 60–103, 71.

56. Shortly after mobile phone providers began making television available on phones, the Slingbox further disrupted place-based limitations of television. This shoebox-size device plugs into the home cable or satellite feed and Ethernet line and enables the viewer to watch the content currently available on the home television screen or stored on the home DVR on any broadband-connected computer, PDA, or smartphone. Eliminating many of the negative features of viewing on a mobile phone screen, Slingbox technology vastly enhances the convenience of mobile television use. It also challenges the geographic specificity central to a network system that has relied on local affiliates. Slingboxes offer viewers a technological solution to sporting event blackouts—such as when local games are unavailable from local broadcasters because they did not sell out. They also compromise local affiliates' promises to advertisers of "local" audiences. As a result, the industry has begun erecting electronic fences in an attempt to restrict video

content to certain geographic areas—what the industry terms "geofiltering," although the primary use of geofiltering involved limiting online downloading to specific national regions to preserve the profitability of international distribution. Amy Schatz and Brooks Barnes, "To Blunt Web's Impact, TV Tries Building Online Fences," *Wall Street Journal*, 16 March 2006, A1.

57. James, "TV in Your Pocket."

58. In an experiment involving streaming forty-eight games of the NCAA basketball tournament live in March 2006, CBS found that even as the online games drew four million visitors, there was no erosion in their conventional broadcast viewing. Such data underscore the manner in which the technologies and distribution methods that make television more convenient also allow networks to expand their audiences.

59. Jim Wu, "Baseball Officials Plan Live Video Streaming," CNET News, 30 October 2001, http://news.cnet.com/2100-1023-275123.html.

60. Devindra Hardawar, "The Magic Moment: Smartphones Now Half of All U.S. Mobiles," *Venture Beat*, 29 March 2012, http://venturebeat.com/2012/03/29/the-magic-moment-smartphones-now-half-of-all-u-s-mobiles/.

61. Kathryn Zickuhr and Aaron Smith, "Digital Differences," Pew Internet and American Life Project, 13 April 2012, 19, http://pewinternet.org/Reports/2012/Digital-differences.aspx.

62. Wayne Friedman, "Most TV/Video and Mobile TV Usage Is in Home," *Media Post*, 3 June 2013, http://www.mediapost.com/publications/article/201657/most-tvvideo-and-mobile-tv-usage-is-in-home.html.

63. Katherine Rosman, "In Digital Era, What Does 'Watching TV' Even Mean?," *Wall Street Journal*, 8 October 2013, http://online.wsj.com/news/articles/SB10001424052702303442004579123423303797850.

64. Faith Popcorn, *The Popcorn Report: Faith Popcorn on the Future of Your Company, Your World, Your Life* (New York: Doubleday, 1991), 207–33; Faith Popcorn and Lys Marigold, *Clicking: 16 Trends to Future Fit Your Life, Your Work, and Your Business* (New York: HarperCollins, 1996), 51–64. This trend continued through the 1990s, and then the terrorism of September 11 exacerbated it with fears that public venues such as malls, movie theaters, and theme parks might become subsequent targets.

65. Spigel, "Portable TV."

66. Barbara Klinger, *Beyond the Multiplex: Cinema, New Technologies, and the Home* (Berkeley: University of California Press, 2006), 22–23.

67. Nielsen, "The Media Universe."

68. This discussion of resolution, however, shifts as constantly as the technology, CNET.com has consistently provided the clearest explanations. See David Katzmaier, "Quick Guide: HDTV Resolution Explained," CNET, 12 September 2006, http://www.cnet.com/4520-7874_1-5137915-1.html?tag=tnav.

69. For a detailed explanation of the complicated development of HD, see Joel Brinkley, *Defining Vision: The Battle for the Future of Television* (San Diego: Harcourt Brace, 1997).

70. James Hibberd, "Bridging the HD Gap," *TVWeek*, 17 August 2006, http://www.tvweek.com/page.cms?pageId=238.

71. Nielsen Media Research, "High Definition Is the New Normal."

72. High-definition and digital transmission are difficult to extricate because high definition functionally requires digital transmission. At an operational level, digital transmission most profoundly changes television distribution capacity, although it provides

limited enhancement of the image; by contrast, high definition markedly improves image quality.

73. Some broadcasters decided to use the efficiency of the digital spectrum to "multicast," meaning fit more than one channel in the spectrum. Others "leased" extra spectrum to other service providers. As broadcasters were driven by hopes of maximizing available profit from the spectrum, many of their actions deviated significantly from what regulators hoped the spectrum would provide.

74. Consumer Electronics Association, "One in Five Owners Bought Their HDTV to Watch the Super Bowl," 31 January 2013, http://www.ce.org/News/News-Releases/Press-Releases/2013-Press-Releases/One-in-Five-Owners-Bought-Their-HDTV-to-Watch-the. aspx.

75. Boutique strategies of differentiation are also important in the 1980s, as argued by John Thornton Caldwell in *Televisuality: Style, Crisis, and Authority in American Television* (New Brunswick: Rutgers University Press, 1995).

76. Daniel Dayan and Elihu Katz, *Media Events: The Live Broadcasting of History* (Cambridge: Harvard University Press, 1992), 1.

77. Jodi Kantor, "The Extra-Large, Ultra-Small Medium," *New York Times*, 30 October 2005, http://www.nytimes.com/2005/10/30/arts/television/30kant.html.

78. Claire Atkinson, "TV Ad Effectiveness Drops 7 Percent in Non-DVR Households," *Advertising Age*, 16 March 2006, http://www.adage.com/news.cms?newsid=48317.

79. Steve Sternberg, "Most Homes Have Only One Set On during Prime-Time," *Television Insights: A Publication of Magna Global*, 12 September 2006, 1.

80. Nicholas Negroponte, *Being Digital* (New York: Vintage, 1995).

3. MAKING TELEVISION

1. Quoted in Brian Steinberg, "*Law & Order* Boss Dick Wolf Ponders the Future of TV Ads (Doink, Doink)," *Wall Street Journal*, 18 October 2006, B1.

2. International syndication revenues can be earned simultaneously.

3. Joe Flint, "*I Love Lucy* Still a Cash Cow for CBS," *Los Angeles Times*, 20 September 2012, http://articles.latimes.com/2012/sep/20/entertainment/la-et-ct-cbslucy-20120920.

4. The syndicator or distributor, however, commonly receives 35 percent of gross revenues as a sales commission and an additional 10 to 15 percent to cover costs incurred for marketing, distribution, editing, and so forth. Howard J. Blumenthal and Oliver R. Goodenough, *This Business of Television*, 2nd ed. (New York: Billboard, 1998), 39.

5. By 2003, budgets had increased to $1.6–$2.3 million, while license fees averaged $1–$1.6 million. Paige Albiniak, "Deficit Disorder: Why Some Good Pilots Never Get Produced," *Broadcasting & Cable*, 5 May 2003, 1, 27.

6. Bill Carter, *Desperate Networks* (New York: Doubleday, 2006), 130.

7. Paige Albiniak, "License to Thrill," *Broadcasting & Cable*, 8 November 2004, 28; Ray Richmond, "*CSI* 100th: Scene of the Crime," *Hollywood Reporter*, 18 November 2004; Tamsen Tillson and Elizabeth Guider, "CBS Eyes a *CSI* Buyout," *Variety*, 20 December 2006, http://www.variety.com/index.asp?layout=print_story&articleid=VR1117956158&categoryid=14.

8. Maria Elena Fernandez, "*CSI: Miami* Goes Global," *Los Angeles Times*, 18 September 2006, 12; Daniel Frankel, "Ancillary Waters Run Deep," *Variety*, 9 July 2006, http://www.variety.com/index.asp?layout=print_story&articleid=VR1117946444&categoryid=2162.

9. Katie Benner, "Meet Content Partners: The Investors Who Bought *CSI*," *CNN Money* online, 11 March 2013, http://finance.fortune.cnn.com/2013/03/11/content-partners-csi/.

10. Mark Alvey, "The Independents: Rethinking the Television Studio System," in *Television: The Critical View*, 6th ed., ed. Horace Newcomb (New York: Oxford University Press, 2000), 34–51.

11. See Alvey, "The Independents"; J. Fred MacDonald, *One Nation under Television: The Rise and Decline of Network TV* (New York: Pantheon, 1990); Ken Auletta, *Three Blind Mice: How the TV Networks Lost Their Way* (New York: Vintage, 1992), 31; James Walker and Douglas Ferguson, *The Broadcast Television Industry* (Boston: Allyn and Bacon, 1998).

12. Warren Littlefield with T. R. Pearson, *Top of the Rock: Inside the Rise and Fall of Must See TV* (New York: Anchor, 2012), 229.

13. John M. Higgins, "It's Not All in the Family," *Broadcasting & Cable*, 22 May 2005, 8.

14. Robert Fidgeon, "FOX on the Run," *Herald Sun* (Sydney), 26 July 2000, H05.

15. The complexity seen by scholars such as David Hesmondhalgh remains, though, as the leading producer in this era was Warner Bros., which funneled only a limited amount of its programming to the like-owned weblet the WB. David Hesmondhalgh, *The Cultural Industries* (London: Sage, 2002).

16. Robert W. McChesney, *The Problem of the Media: U.S. Communication Politics in the 21st Century* (New York: Monthly Review Press, 2004); Ben H. Bagdikian, *The New Media Monopoly* (Boston: Beacon, 2004).

17. Many proclaimed the shuttering of Carsey-Werner in 2005 as the death of the last independent (Mandabach had left the group a year earlier).

18. The panel included Marc Graboff, president, NBC Universal Television, West Coast; Gary Newman, president, 20th Century Fox Television; Mark Pedowitz, president, Touchstone Television, and executive vice president, ABC Entertainment Television Group; and Bruce Rosenblum, president, Warner Bros. Television Group. Las Vegas, 17 January 2007.

19. Conglomerate-owned studios do not exclusively produce for the commonly owned network, however, and common ownership can lead to complications when a studio such as NBC Universal Television Studio (NUTS) attempts to sell a program to a non–commonly owned network. Marc Graboff, then president of NBC Universal, West Coast, and responsible for both NBC network and NUTS, noted that being on both sides of deals also means that the studio has to give in to the same concessions when producing for other networks as he demands of the non-NBCU studios that license their shows to NBC. For example, if he insists on obtaining rights to stream a Touchstone-produced show aired by NBC on the NBC website, then NUTS would have to be willing to give ABC those rights if it licensed a show to ABC.

20. Many argued that there were no remaining independent producers of scripted series, but definitions of independent varied considerably. Some argued that Warner Bros. was an independent once it no longer had the WB to buy its programming, but it is difficult to categorize a major studio as independent even if it did not have a commonly owned broadcast network.

21. Jason Mittell, *Complex Television* (New York: New York University Press, forthcoming).

22. Jason Katims, quoted in Alan Sepinwall, *The Revolution Was Televised: The Cops, Crooks, Slingers, and Slayers Who Changed TV Drama Forever* (privately printed, 2013), 280.

23. "William H. Macy Is *Shameless* on Showtime," *Fresh Air with Terri Gross*, 30 January 2013, transcript available from http://www.npr.org/templates/transcript/transcript.php?storyId=170463939.

24. The originator of a format might profit immensely through sales of the format to markets around the globe.

25. Bill Carter, "NBC to Pay Outsiders for Blocks of Programs," *New York Times*, 3 December 2007, http://www.nytimes.com/2007/12/03/business/media/03nbc.html?_r=0.

26. Ibid.

27. Ted Sarandos, "Ted Sarandos, Chief Content Officer, Netflix." Interview. Carsey Wolf Center, Media Industries Project, http://www.carseywolf.ucsb.edu/mip/article/ted-sarandos.

28. Andrew Wallenstein, "Netflix Series Spending Revealed," *Variety*, 8 March 2013, http://variety.com/2013/digital/news/caa-agent-discloses-netflix-series-spending-1200006100/.

29. Jim Benson, "Debmar-Mercury Secures $200 Million Distribution Deal for *Tyler Perry's House of Payne*," *Broadcasting & Cable*, 23 August 2006, http://www.broadcastingcable.com/article/105512-Debmar_Mercury_Secures_200_Million_Distribution_Deal_for_Tyler_Perry_s_House_of_Payne.php.

30. Cynthia Littleton, "Fast-Tracked Sitcom May Be Way of Future," *Variety*, 26 June 2012, http://variety.com/2012/tv/news/fast-tracked-sitcom-may-be-way-of-future-1118055951/.

31. Helford, cited in ibid.

32. Charlie Warzel, "YouTube Phenoms Raise Record Cash," *Adweek*, 14 February 2013, http://www.adweek.com/news/advertising-branding/youtube-phenoms-raise-record-cash-147287.

33. Cited in Hayley Tsukayama, "YouTube Channel YOMYOMF Launches, Focus on Asian-American Pop Culture," *Washington Post*, 15 July 2012, http://articles.washingtonpost.com/2012-06-15/business/35462753_1_ryan-higa-youtube-channel-yomyomf.

34. Tessa Stuart, "YouTube Stars Fight Back," *LA Weekly*, 10 January 2013, http://www.laweekly.com/2013-01-10/news/machinima-maker-studios-YouTube/full/.

35. Louis C.K., "A Statement from Louis C.K.," 13 December 2011, https://buy.louisck.net/news.

36. See Hesmondhalgh, *The Cultural Industries*, for more on creative labor.

37. Chad Raphael, "The Political Economic Origins of Reali-TV," in *Reality TV: Remaking Television Culture*, ed. Susan Murray and Laurie Ouellette (New York: New York University Press, 2004), 119–36.

38. Filmscape News Center, "1988 Hollywood Writers Strike," 30 April 2001, http://www.filmscape.co.uk/news/fullnews.cgi?newsid988621880,79143.

39. Dave McNary and Ben Fritz, "Guilds Out of the Online Loop," *Variety*, 15 August 2006, http://www.variety.com/index.asp?layout=print_story&articleid=VR1117948511&categoryid=18.

40. Film shooting in Los Angeles peaked in 1996 and had fallen 38 percent by 2005. Richard Verrier, "Movies, Schmovies—TV's Taking Over L.A.," *Los Angeles Times*, 19 August 2005, A1.

41. Sean J. Miller, "'Tragic' Drop in L.A.'s TV Production in 2012," *Backstage*, 8 January 2013, http://www.backstage.com/news/tragic-drop-ls-tv-production-2012/.

42. The shifting management of the networks that resulted from the purchase of all three networks also contributed to the demand for cost savings. Raphael, "The Political Economic Origins of Reali-TV," 122–23.

43. NBC, FOX, and the WB all announced variations in January 2003, while FOX made changes in its development calendar to enable the "fifty-two-week" season that many suggested should become the new industrial norm.

44. Paige Albiniak, "Sunburned: Cable, Broadcast Each Had Summers That Hurt," *Broadcasting & Cable*, 1 September 2003, 1.

45. James G. Webster, Patricia F. Phalen, and Lawrence W. Lichty, *Ratings Analysis: The Theory and Practice of Audience Research*, 2nd ed. (Mahwah, NJ: Lawrence Erlbaum Associates, 2000).

46. For example, NBC's experiment with new, non-narrative programming during the summer months of 2003 led it to lose fewer audience members than it had in previous summers, and it was successful in drawing the largest audiences among broadcast networks.

47. Stuart Levine, "*Sons of Anarchy* Stretches Episode Length," *Variety*, 26 November 2012, http://variety.com/2012/tv/news/sons-of-anarchy-stretches-episode-length-1118062663/, accessed 4 June 2013.

48. "Paris Barclay: Director, President of the Directors Guild of America," interview, Media Industries Project, 27 June 2013, http://www.carseywolf.ucsb.edu/mip/article/paris-barclay.

49. Micelli, quoted in Wallenstein, "Netflix Series Spending Revealed."

50. David Bauder, "Study: Clutter of Advertising Soaring on Prime-Time Television," *Associated Press*, 12 April 1999; Susan T. Eastman and Gregory D. Newton, "The Impact of Structural Salience within On-Air Promotion," *Journal of Broadcasting & Electronic Media* 42 (1998): 50–79.

51. Eastman and Newton, "The Impact of Structural Salience within On-Air Promotion."

52. Laura Martin and Dan Medina, "The Future of TV," *Needham Insights*, 11 July 2013, http://capknowledge.com/research_reports/disclaimer_thm_future_of_tv.html.

53. Jon Fine and Tobi Elkin, "On the House," *Advertising Age*, 18 March 2002, 1.

54. The WB also distributed coffee cup sleeves with an image that changed based on temperature to further promote the supernatural aspect of the show. Chris Blackledge, "They Deserve a Fat Promotion: Nets Angle for Fall Viewers," *NATPE News*, September 2005, http://www.natpe.org/memberresources/natpenews/articles/story.jsp?id_string= 200023:X58IMn7ClV8Ry8yPMfzucQ**; Meg James, "TV Networks Pursue the 'Super Fan,'" *Los Angeles Times*, 19 September 2005, http://www.latimes.com/business/la-fi-tvbuzz19sep19,1,1350987.story?coll=la-headlines-business&ctrack=1&cset=true.

55. Christopher Lisotta, "WB, Yahoo! Offer *Super* Preview," *Television Week*, 5 September 2005, 1.

56. Among other experiments, the SciFi network loaded a special recap episode of *Battlestar Galactica* onto the online gaming service for Microsoft's Xbox, while NBC reran the pilot episodes of its new series on the various cable networks owned by NBC Universal and allowed free iTunes downloads. CBS embraced cutting-edge technology with billboards advertising shows that allowed commuters with Bluetooth-enabled mobile devices to download a thirty-second clip of the show to their device.

57. James, "TV Networks Pursue the 'Super Fan.'"

58. See the five separate stories in the 10 March 2006 issue, including a front-page story about the "real" mafia relative to the HBO series.

59. Brian Stelter, "A Long Wait Stirs Enthusiasm for Fox Show *Glee*," *New York Times*, 1 September 2009, http://www.nytimes.com/2009/09/02/business/media/02adco.html?_r=0.

60. Ann Oldenberg, "TV Goes to Blogs: Shows Add Extra Information as a Treat for Fans," *USA Today*, 5 April 2006, 1D.

61. Daisy Whitney, "MySpace Video a Boon to *Mother*," *Television Week*, 4 December 2006, 6, 40.

62. Preliminary findings of research by Vicki Mayer, personal communication, 9 September 2006.

63. Quoted in Sally Bedell, *Up the Tube: Prime-Time TV and the Silverman Years* (New York: Viking, 1981), 141.

64. See Susan T. Eastman, "Orientation to Promotion and Research," in *Research in Media Promotion*, ed. Susan T. Eastman (Mahwah, NJ: Lawrence Erlbaum Associates, 2000), 3–18.

65. Jim Bennett, "The Cinematch System: Operation Scale, Coverage, Accuracy, Impact," presentation made at "The Present and Future of Recommender Systems" conference, Bilbao, Spain, 13 September 2006, http://blog. recommenders06.com/wp-content/uploads/2006/09/bennett.pdf.

66. Chris Pursell and Michael Freeman, "Studios USA Will Stay the Course," *Electronic Media*, 29 October 2001, 1.

67. This situation is contemplated further in Jason Fry, "O Pioneers! Watching a Child Who's Grown Up with TiVo Leads to Questions about the Future of TV," *Wall Street Journal*, 12 December 2005, http://online. wsj.com/public/article_print/SB113398924199016540.html.

4. REVOLUTIONIZING DISTRIBUTION

1. Quoted in Phil Rosenthal, "On-Demand Deals a New Dawn for TV," *Chicago Tribune*, 9 November 2005, www.chicagotribune.com/business/columnists/chi-0511090177nov09,1,4745093.column?coll=chi-business-hed.

2. The term "original run" (or "first run") is commonly used to refer to programming produced originally for syndication (sale to individual stations rather than networks), but I use it here to include the original run of shows on networks as well.

3. Full-service networks typically provided more daily content. For example, while FOX supplied stations with only two hours of nightly prime-time content, NBC supplied three hours of prime time, three hours of the *Today Show*, *NBC Nightly News*, and the *Tonight Show*, and two hours of soap operas. Affiliates of FOX, the WB, and UPN still purchased many hours of programming because the network supplied so little. Jonathan Levy, Marcelino Ford-Livene, and Anne Levine, "Broadcast Television: Survivor in a Sea of Competition," Federal Communications Commission Office of Plans and Policy, September 2002, p. 19, http://hraunfoss.fcc.gov/edocs_public/attachmatch/DOC-226838A22.doc.

4. Practices of exclusivity and windowing have been declining in the film industry as well. A 2006 report noted an 11 percent drop in the number of days between the theatrical and DVD release of films. This was the same year some filmmakers experimented with "day and date" release (simultaneous distribution on multiple platforms). Diane Garrett, "Windows Rattled," *Variety*, 21 March 2006, http://www.variety.com/index.asp?layout=print_story&articleid=VR1117940116&categoryid=20.

5. Markets are generally the zones around major metropolitan areas reached by local stations, so although they tend to overlap with major cities, they exceed city boundaries.

6. A few shows were developed for first-run syndication, but these shows rarely aired during prime time. Mostly first-run syndication includes game (*Jeopardy*) and talk shows (*Oprah*), but also can include scripted series such as *Xena*. These shows operate under a much different financial model and typically do not rely on deficit financing. The shows are sold to stations in each market, usually for some combination of cash and advertising time.

7. Admittedly, proclaiming a singular first move is difficult. One could also argue that the practice of syndicating series with fewer and fewer episodes marked an earlier procedural shift, but to the degree that this was a slight alteration in an existing norm, this seems less significant than original-run repurposing.

8. Previously a series needed a bank of roughly a hundred episodes before it could be sold in syndication. This meant that it would take four or five years before a series could begin a syndicated run.

9. Notably, this occurred over five years before the NBC/Universal merger that brought Studios USA into the NBC conglomerate. See Joe Schlosser, "Kissinger Tops USA Network TV," *Broadcasting & Cable*, 26 April 1999, 28.

10. See Deborah McAdams and Joe Schlosser, "*Once* Again on Lifetime," *Broadcasting & Cable*, 27 September 1999, 8.

11. John Dempsey, "Shared Runs: Cachet over Cash," *Variety*, 12 June 2001, 11, 54.

12. The deal between NBC and DirecTV for the fourth and fifth seasons of *Friday Night Lights* is related to this phenomenon, but since it aired on both entities, likely has more in common with the first version of reallocation.

13. John Lippman, "New Shows Try Their Hand at Copying Fox Hit *24*," *Wall Street Journal*, 24 March 2006, W7.

14. Lippman, "New Shows"; David Koeppel, "For Those of You Who Wonder How That TV Show Began," *New York Times*, 21 March 2005, C2.

15. Koeppel, "For Those of You Who Wonder."

16. See Derek Kompare, "Acquisition Repetition: Home Video and the Television Heritage," in *Rerun Nation: How Repeats Invented American Television* (New York: Routledge, 2005), 197–220.

17. Susanne Ault, "Fallen *Wonderfalls* Is a DVD Wonder," *Video Business*, 11 February 2005, http://www.videobusiness.com/article/CA612067.html?text=wonderfalls.

18. Anthony Breznican, "*Firefly* Alights on Big Screen as *Serenity*," *USA Today*, 21 September 2005, http://www.usatoday.com/life/movies/news/2005-09-21-serenity_x.htm.

19. The show was also syndicated on cable's Cartoon Network as part of the Adult Swim programming block, where it also performed successfully.

20. Meg James, "Fox Reuniting Itself with *Family Guy*," *Los Angeles Times*, 13 April 2005, http://www.latimes.com/business/la-fi-family13Apr13,0,43681.story?coll=la-home-business; Jill Vejnoska, "The DVD Effect," *Atlanta Journal-Constitution*, 3 May 2005, 1E.

21. John M. Higgins, "Fast-Forward: With Scant Notice, TV-DVD Sales Top $1B and Begin to Affect Scheduling, Financing," *Broadcasting & Cable*, 22 December 2003, 1.

22. Ibid.

23. R. Thomas Umstead, "DVDs, Video Games Boost Net Brands," *Multichannel News*, 19 July 2004, 66.

24. Quoted in ibid.

25. The industry refers to entities such as Netflix as SVOD (subscription video on demand). I distinguish them as broadband-delivered in order to highlight that this content comes into the home over the "Internet" rather than conventional cable.

26. Chris Anderson, "The Long Tail," *Wired*, October 2004, 171–79.

27. Daisy Whitney, "VOD Came Alive in 2005," *Television Week*, 2 January 2006, 12.

28. In many cases, however, the studio is owned by a conglomerate that also owns the originally licensing network, so it could be argued that the revenue goes to a common source. In practice, conglomerates evaluate their divisions independently of how one might help another, which leads to little intra-conglomerate collaboration.

29. Andrew Wallenstein, "Netflix Flexes New Muscle with *Breaking Bad* Ratings Boom," *Variety*, 12 August 2013, http://variety.com/2013/digital/news/netflix-flexes-new-muscle-with-breaking-bad-ratings-boom-1200577029/.

30. Media Industries Project, "An Interview with Joe Flint," 4 September 2013, http://www. carseywolf.ucsb.edu/mip/article/interview-joe-flint.

31. Dawn C. Chmielewski, "Studios Not Sure Whether Web Video Innovator Is Friend or Foe," *Los Angeles Times*, 10 April 2006, http://www. latimes.com/business/la-fi-youtube10apr10,1,6694137,print.story?coll=laheadlines-business.

32. "A Video Discourse from Three Views," *Television Week*, 31 July 2006, 12, 51.

33. Moonves, quoted in Derek Thompson, "What Happens if TV Goes the Way of Music and Newspapers?" *Atlantic*, 19 October 2012, http://www.theatlantic.com/business/archive/2012/10/what-happens-if-tv-goes-the-way-of-music-and-newspapers/263895/.

34. Hastings, quoted in Alex Byers, "Netflix's Hastings: Online Steering TV's Future," *Politico*, 29 January 2013, http://www.politico.com/story/2013/01/netflixs-hastings-online-steering-tvs-future-86894.html.

35. Half-hour situation comedies typically retained an exclusive first syndication run on broadcast affiliates, while studios normally distributed hour-long dramas first to cable. The number of windows and potential revenue available to a given program depended significantly on its genre and narrative structure. Episodic dramas such as those in the *Law & Order* and *CSI* franchises earned considerable revenue in cable syndication because of their closed-ended stories that allowed audiences to view them intermittently and cable operators to play them out of order. As an illustration of this discrepancy, *Law & Order: Criminal Intent* set a new syndication fee record for a drama in 2004 when it was sold for $1.92 million per episode. The heavily serialized shows *Alias* and *24* earned only about $250,000 per episode. See Denise Martin and John Dempsey, "*Sopranos* Reruns Stir Mob Scene," *Variety*, 9 January 2005, http://www.variety.com/index.asp?layout=print_story&articleid =VR1117915991&categoryid=14. More serialized dramas such as *Alias* and *24* particularly benefited from the DVD sell-through markets, as did "cult" hits such as *Family Guy*. Previously, the international market would have been the only window through which producers of many shows would have earned substantial additional revenue.

36. Paul Toscano, "End of Cable Bundle Inevitable, with or without Aereo: CEO," CNBC, 13 June 2013, http://www.cnbc.com/id/100813523.

37. Susan Crawford, *Captive Audience: The Telecom Industry and Monopoly Power in the New Gilded Age* (New Haven: Yale University Press, 2013), 161.

38. Amy Schatz, "Verizon, Comcast Defend Spectrum-Purchase Plan," *Wall Street Journal*, 22 March 2012, http://online.wsj.com/article/SB10001424052702304636404577295703761555914.html.

39. Marguerite Reardon, "Competitive Wireless Carriers Take on AT&T and Verizon," CNET News, 10 September 2012, http://news.cnet.com/8301-1035_3-57505803-94/competitive-wireless-carriers-take-on-at-t-and-verizon/.

40. Mark Sullivan, "Google to Big ISPs: Fiber Is Good for You," *Tech Hive*, 1 November 2013, http://www.techhive.com/article/2059563/google-to-big-isps-fiber-is-good-for-you.html.

41. Moffett, cited in Matt Richtel and Brian Stelter, "In the Living Room, Hooked on Pay TV," *New York Times*, 23 August 2010, http://www.nytimes.com/2010/08/23/business/media/23couch.html?pagewanted=all&_r=0.

42. Crawford, *Captive Audience*, 64–65.

43. Ibid., 185.

44. Nielsen, "Free to Move between Screens: The Cross-Platform Report, Q4 2012," March 2013.

45. Brian Fuhrer, "Nielsen's Brian Fuhrer on Digital Video Trends," 2012 Interactive Advertising Bureau Digital Video Marketplace conference, 10 April 2012, http://www.youtube.com/watch?v=EHWZX4JGtMM.

46. Todd Spangler, "Pay-TV Prices Are at the Breaking Point—and They're Only Going to Get Worse," *Variety*, 29 November 2013, http://variety.com/2013/biz/news/pay-tv-prices-are-at-the-breaking-point-and-theyre-only-going-to-get-worse-1200886691/.

47. Evan Shapiro, personal communication, 2 April 2013.

48. Another regulatory issue that emerged in 2011, but was quickly resolved, involved the question of just what kind of use subscription to a cable service allowed. As Comcast, Cablevision, and Time Warner offered "TV Everywhere" apps to subscribers that would enable them to watch the content available on their living room screen on devices in and then out of the house, the content rights holder Viacom quickly filed suit, though parties settled just as quickly. A main concern among rights holders involved the MVPDs developing authentication systems that would be adequate to ensure that the user was a paying customer.

49. John McMurria, "Long-Format TV: Globalization and Network Branding in a Multi-Channel Era," in *Quality Popular Television*, ed. Mark Jancovich and James Lyons (London: British Film Institute, 2001), 65–87.

50. Quoted in Daisy Whitney, "Hopped Up on Hope," *Television Week*, 19 June 2006, 1.

51. Rupert Murdoch, "The Dawn of a New Age of Discovery: Media 2006," speech given for the Annual Livery Lecture at the Worshipful Company of Stationers and Newspaper Makers, 13 March 2006, London, http://www.newscorp.com/news/news_285.html.

52. Quoted in David S. Cohen, "Wolf Sounds Alarm," *Variety*, 20 March 2006, http://www.variety.com/index.asp?layout=print_story&articleid=VR111794006 5&categoryid=14.

53. Rebecca Greenfield, "The Economics of Netflix's $100 Million New Show," *Atlantic Wire*, 1 February 2013, http://www.theatlanticwire.com/technology/2013/02/economics-netflixs-100-million-new-show/61692/.

5. THE NEW ECONOMICS OF TELEVISION

1. Paul Keegan, "The Man Who Can Save Advertising," *Business 2.0*, November 2004, http://www.business2.com/b2/subscribers/articles/print/0,17925,704067,00.html.

2. Cited in Scott Donaton, *Madison & Vine: Why the Entertainment and Advertising Industries Must Converge to Survive* (New York: McGraw-Hill, 2004), 60.

3. For more, see William Boddy, "Redefining the Home Screen: The Case of the Digital Video Recorder," in *New Media and Popular Imagination: Launching Radio, Television, and Digital Media in the United States* (New York: Oxford University Press, 2004), 100–107.

4. The publication of Malcolm Gladwell's book *The Tipping Point: How Little Things Can Make a Big Difference* (Boston: Little, Brown, 2002) has led the television industry and many others to think about the process of change and the existence of a "tipping point" in a very particular way.

5. Claudia Deutsch, "Study Details Decline in Spending on Ads," *New York Times*, 5 September 2001, C5.

6. Stuart Elliott, "An Agency Giant Is Expected to Warn of Lower Profits, and Analysts Darken Their Outlook," *New York Times*, 2 October 2001, C2; Stuart Elliott, "Networks Watch Prices for Commercial Time Drop," *New York Times*, 25 September 2001.

7. Reported in Laura Martin and Dan Medina, "The Future of TV," *Needham Insights*, 11 July 2013, 16, http://capknowledge.com/research_reports/disclaimer_thm_future_of_tv.html.

8. Data presented at SNL Kagan Multichannel Summit, 11 November 2013, New York, NY.

9. For more, see Henry Jenkins, *Convergence Culture: Where Old and New Media Collide* (New York: New York University Press, 2006), 61–64.

10. Joe Flint, "CBS to Collect $1 Billion a Year in Distribution Fees by 2016," *Los Angeles Times*, 11 September 2012, http://articles.latimes.com/2012/sep/11/entertainment/la-et-ct-cbs-20120911.

11. Susan Crawford, *Captive Audience: The Telecom Industry and Monopoly Power in the New Gilded Age* (New Haven: Yale University Press, 2013), 143.

12. Ibid., 111.

13. Michael Learmonth, "YouTube Drops Price for Upfront Packages to Lure TV Dollars," *Advertising Age*, 26 April 2013, http://adage.com/article/digital/youtube-drops-price-upfront-packages-lure-tv-dollars/241137/.

14. Matthew Garrahan, "Google Invests in YouTube Studio in LA," *Financial Times*, 12 July 2013, http://www.ft.com/intl/cms/s/2/3f4c846a-e9c1-11e2-bf03-00144feabdc0.html#slide12; Martin and Medina, "The Future of TV," 11.

15. Martin and Medina, "The Future of TV," 11.

16. Amir Efrati, "YouTube Unveils Paid Subscription Channels," *Wall Street Journal*, 9 May 2013, http://online.wsj.com/article/SB10001424127887324744104578473230854933060.html.

17. Martin and Medina, "The Future of TV," 11.

18. Brad Adgate, "The Next Big Ad War: Netflix vs. Hulu vs. Amazon," *Advertising Age*, 30 August 2013, http://adage.com/article/viewpoint-editorial/big-ad-war-netflix-hulu-amazon/243927/.

19. Ibid., 13.

20. Sponsorship continued into the 1960s, although by that point it was increasingly a residual rather than dominant practice.

21. William Boddy, *Fifties Television: The Industry and Its Critics* (Urbana: University of Illinois Press, 1993), 95–97.

22. Data drawn from Boddy, *Fifties Television*, citing U.S. Federal Communications Commission, Office of Network Study, "Second Interim Report: Television Network Procurement," part 2, docket no. 12782 (Washington, DC: U.S. Government Printing Office, 1965), 736.

23. Boddy, *Fifties Television*, 159.

24. W. L. Bird, "Advertising, Company Voice," in *Encyclopedia of Television*, 2nd ed., ed. Horace Newcomb (New York: Routledge, 2004), 34–37.

25. Christopher Anderson, *Hollywood TV* (Austin: University of Texas Press, 1993).

26. For critical assessments of network and studio behavior in the era, see ibid.

27. "Cable's Bucks," *Broadcasting & Cable*, 21 April 2003, 28.

28. Scatter prices can be lower, although broadcasters avoid this in order to avoid angering advertisers that purchase upfront and that the networks will need to sell to again in the next year.

29. Nicole Laporte, "High Prices, Competish Pump a Record Upfront," *Variety*, 26 May 2003, http://www.variety.com/index.asp?layout=print_story&articleid=VR1117886871&categoryid=14.

30. Erwin Ephron, "How the TV Nets Got the Upfront," *Advertising Age*, 14 May 2001.

31. Erwin Ephron, "A Short History of the Upfront," *Jack Myers Report*, 2 May 2003, 1–2.

32. On some occasions, however, the networks have guaranteed scatter rates to help increase rates and purchase volume.

33. DVD and syndication sales contributed 20 percent of HBO's operating income by 2004. John M. Higgins and Allison Romano, "The Family Business," *Broadcasting & Cable*, 1 March 2004, 1, 6.

34. "US Ad Spending: Mid-2013 Forecast and Comparative Estimates," eMarketer, http://www.emarketer.com/corporate/reports.

35. Suzanne Vranica, "A 'Crisis' in Online Ads: One-Third of Traffic Is Bogus," *Wall Street Journal*, 23 March 2014, http://online.wsj.com/news/articles/SB10001424052702304026304579453253860786362.

36. Joseph Turow attends briefly to the consolidation of the industry in *The Daily You: How the New Advertising Industry Is Defining Your Identity and Your Worth* (New Haven: Yale University Press, 2011).

37. R. Craig Endicott and Kenneth Wylie, "*Ad Age* Agency Report," *Advertising Age*, 30 April 2006, http://adage.com/article?article_id=108906.

38. "Agency Family Trees 2012," *Advertising Age*, 30 April 2012, http://adage.coverleaf.com/advertisingage/20120430/?pg=13#pg13.

39. Jack Myers, "Communications Planning Will 'Sweep the Industry' in Next 36 Months," *Jack Myers Report*, 25 January 2005.

40. John Consoli, "Shops Form Units for Product Placement," *Adweek*, 14 February 2005, http://www.adweek.com/aw/national.article_display.jsp?vnu_content_id=1000799577.

41. For more on the basic operation of the advertising industry both historically and at this time, see Joe Cappo, *The Future of Advertising: New Media, New Clients, New Consumers in the Post-Television Age* (Chicago: McGraw-Hill, 2003).

42. Jack Myers, "Creativity Moves to Forefront at Media Buying Groups," *Jack Myers Report*, 4 January 2005.

43. Fifteen percent was an old—but eroding—industry standard fee. Jack Myers, "Restructuring the Agency/Client Compensation Model," *Jack Myers Report*, 26 January 2005.

44. The introduction of accounting requirements of the Sarbanes/Oxley legislation resulted in micromanagement of budgetary details and contributed to advertisers' demands for definitive information about the return they achieved on their investments in advertising.

45. James Twitchell, *AdCult USA: The Triumph of Advertising in American Culture* (New York: Columbia University Press, 1996).

46. Regulatory prohibitions focused on advertising aimed at children and required disclosure of paid promotion. See Mary-Lou Galician, ed., "Product Placement," special issue, *Journal of Promotion Management* 10, nos. 1–2 (2004).

47. The norms for cable and broadcast differed until recently. Even as late as 2000, a number of cable channels featured a fair amount of sponsor-produced programming. See Jim Forkan, "On Some Cable Shows, the Sponsors Take Charge," *Multichannel News*, 4 June 2001, 53.

48. Donaton, *Madison & Vine*.

49. The automaker Ford is a brand, while a specific automobile is a product. A client may choose to emphasize either, depending on the goal of his or her campaign. For example, Johnson and Johnson sponsors annual films on TNT that help reestablish the company's desired brand identification as something safe and reliable for the family. This is different from when it promotes a specific product, such as Acuvue contact lenses. Either a brand or a product might be placed or integrated into television programming.

50. Twitchell, *AdCult USA*, 18.

51. For more detailed history and other terms, see Donaton, *Madison & Vine*, 13–16.

52. Significantly, despite the ubiquitous placement of the iPod, Apple claims that it has never paid for iPod placement because "Apple is cool." The company is well known to have given away millions of dollars' worth of free computers to the Hollywood community, which has created substantial goodwill for the company. See Gail Schiller, "Brands Take Buzz to Bank through Free TV Placement," *Washington Post* online, 13 April 2006, http://www.washingtonpost.com/wp-dyn/content/article/2006/04/13/AR2006041300272_pf.html.

53. PQ Media, "Global Product Placement Spending Up 10% to $7.4 Billion in 2011," http://www.pqmedia.com/about-press-201212.html.

54. James A. Karrh reviews existing research on the effectiveness and value of placement in "Brand Placement: A Review," *Journal of Current Issues and Research in Advertising* 20, no. 2 (1998): 31–49.

55. Even here, though, differences between organic and inorganic may not be clear-cut. Take, for instance, a 2004 episode of the spy drama *Alias*, in which an early scene featured the heroine and her partner running from the bad guy du jour. From off-screen, the audience heard the protagonist, Sydney Bristow, shout to her partner, "Quick, to the F-150," and a car chase scene featuring the named vehicle with its logo prominently displayed on its grill ensued. This use was jarring and disruptive, although there was nothing that made this use particularly "inorganic"—by definition, car chases require vehicles. The step of including the brand name in the dialogue and the fact the character was off-screen when she used the line—which suggests it was edited in during post-production—caused the break in narrative that drew attention to the placement. Importantly, it was because this placement failed to be organic that I, for one, noticed it and could recall it over a year later—in fact, it might be the only truck I know by name. A more successful example of placement within *Alias* involved the spy's use of a Nokia phone: while much of the country was adopting personalized ring tones, Sydney continued to program her phone with the Nokia-brand ring, which functioned unobtrusively within the narrative while nonetheless remaining identifiable.

56. Wayne Friedman and Jack Neff, "Eagle-Eye Marketers Find Right Spot, Right Time," *Advertising Age*, 22 January 2001, 2–4.

57. In many cases, products' names must be stripped out of programs syndicated internationally.

58. See Forkan, "On Some Cable Shows," 53, for one of the few examinations of this phenomenon before the resurgence of product placement.

59. At a conference of the National Association of Television Program Executives in 2005, Debbie Myers, vice president of media services for Taco Bell, reported that producers of *The Apprentice* approached the company with an offer to focus an episode on the company for $5 million. Taco Bell declined; trade press reports locate that figure at $1 million for early seasons of *The Apprentice* and $2–3 million by its third edition in 2005. Still, by 2005, NBC scheduled four, thirteen-episode editions of the series, so that integration might offer the network $260 million a season. An industry source reported that the primary product association deals in *American Idol* cost as much as $40 million per season. Wayne Friedman, "Placement Bonanza Remains Elusive," *Television Week*, 11 October 2004, 22.

60. Quoted in Gary Levin, "The Newest Characters on TV Shows: Product Plugs," *USA Today*, 20 September 2006, A1.

61. Importantly, in industry discussions, "branded entertainment" is sometimes used as an umbrella term encompassing integration, placement, and the forms I discuss here.

Each of these strategies is significantly different, and more precision is required for this discussion.

62. The fashion show indicates multiple strategies at work. The year ABC aired the fashion show, it included models in that week's episode of *Spin City*, featured a *Who Wants to Be a Millionaire?* supermodel edition, and showed the models on *The View*, suggesting a sort of product placement throughout the week.

63. Alice Z. Cuneo and Wayne Friedman, "Spreading Secrets," *Advertising Age*, 22 October 2001, 3. Interview with Ed Razek, creative director, Victoria's Secret, in John Watkin and Eamon Harrington, dirs., *A Day in the Life of Television*, produced by the Museum of Radio and Television and Planet Grande, aired on CBS 2 September 2006.

64. Quoted in Cuneo and Friedman, "Spreading Secrets."

65. Donaton, *Madison & Vine*, 101.

66. Ibid.

67. Ibid.

68. Rae Ann Fera, "How Kmart Used Social Listening (and Some Nerve) to Create a Ship-My-Pants Funny Viral Hit," *Co.Create*, http://www.fastcocreate.com/1682826/how-kmart-used-social-listening-and-some-nerve-to-create-a-ship-my-pants-funny-viral-hit.

69. Andy Meisler, "Not Even Trying to Appeal to the Masses," *New York Times*, 4 October 1998, 45.

70. Frances Croke Page, personal communication with the author, 1 August 2005, New York, NY.

71. "Loyalty Cards Idea for TV Addicts," *BBC News*, 28 October 2004, http://news.bbc.co.uk/go/pr/fr/-/1/hi/technology/3958855.stm.

72. Michael Learmonth, "Hulu's New Guarantee: Someone Watched Your (Whole) Ad," *Advertising Age*, 17 April 2012, http://adage.com/article/special-report-digital-conference/hulu-s-guarantee-watched-ad/234164/.

73. Ibid.

74. Quoted in Amol Sharma and Suzanne Vranica, "On Demand: Quick Ad Switch—New Technology Allows Networks to Swiftly Replace Commercials on VOD Services," *Wall Street Journal*, 28 May 2013, B6.

75. Ibid.

76. Critical approaches to advertising have become multivocal in their outlook on the meaning of consumption. One (arguably more traditional) approach can be found in the work of Sut Jhally and Juliet Schor, who link the rampant consumerism of post-Fordist America with the attendant horrific consequences on the environment and quality of life. Noting that citizens work endless hours and take on credit to purchase unnecessary goods sold to them through lifestyle advertising, they explore the growing economic disparity in the United States as a disproportionately small segment of the population uses incredible amounts of resources. This critical perspective, which identifies consumption as having an all-encompassing role, focuses on the degree to which people view social power as being accessible primarily through the purchase of goods and services. Hence, this perspective is in accord with social theory that has dichotomized the identities of consumer and citizen and denigrated the activity of the former, arguing that it often comes at the expense of the latter. See Sut Jhally, dir., *Advertising and the End of the World* (Amherst: Media Education Foundation, 1997); Juliet B. Schor, *The Overspent American: Upscaling, Downshifting, and the New Consumer* (New York: Basic Books, 1998).

By contrast, other critical scholars embrace consumption as an expression of agency and theorize participation in commercial culture as an activity with resistant dimensions. In response to a legacy of criticism heavily influenced by Marxist thought that dismissed advertising and consumerism as insignificant or the terrain of cultural dupes, scholars including Meaghan Morris and Hilary Radner, among others, have considered shopping and brand selection as meaningful cultural activities. Their work, which was part of the effort to expand the field of cultural studies to the study of the everyday, treats shopping and consumption as important aspects of life in industrial cultures, as well as noting the gendered dimensions of these activities. See Meaghan Morris, "Things to Do with Shopping Centres," in *The Cultural Studies Reader*, ed. Simon During (London: Routledge, 1993), 295–319; W. S. Kowinski, *The Malling of America: An Inside Look at the Great Consumer Paradise* (New York: Pantheon, 1985); Mica Nava, "Consumerism and Its Contradictions," *Cultural Studies* 1, no. 2 (1987): 204–10; Judith Williamson, *Consuming Passions: The Dynamics of Popular Culture* (London: Marion Boyers, 1986); Hilary Radner, *Shopping Around: Feminine Culture and the Pursuit of Pleasure* (New York: Routledge, 1995).

An alternative critical approach has also developed that intervenes in these two trajectories of scholarship that tend to discount or embrace consumption. This more middle-ground perspective acknowledges consumption as part of citizenship in post-Fordist culture without arguing that it provides such an empowering form of agency. Thus, for example, Thomas Frank uses cultural texts and documents to trace the increasing acceptance of the logic of capital markets during the 1980s and 1990s to the point that the beneficence and propriety of markets become part of a dominant American ideology. For his part, Martin Davidson provides more evaluative analysis that acknowledges the complex relationships among consumption and critical theory. See Néstor García Canclini, *Consumers and Citizens: Globalization and Multicultural Conflicts* (Minneapolis: University of Minnesota Press, 2001); Thomas Frank, *One Market under God: Extreme Capitalism, Market Populism, and the End of Economic Democracy* (New York: Doubleday, 2000); Martin Davidson, *The Consumerist Manifesto: Advertising in Postmodern Times* (London: Routledge, 1992).

Authors such as Frank and Davidson are not diametrically opposed to Jhally and Schor, but their work intervenes in different critical histories. Davidson identifies the cultural significance of branding and the importance of associating a lifestyle with a brand to argue for the study of consumption as a part of life because of its gross absence in much preceding scholarship and a tendency for critical scholarship to take a fairly unsophisticated and naïve view toward advertising and consumption that does not reflect the lived experience of many in late industrial societies (Davidson, 175). Likewise, Frank traces the infiltration and acceptance of the logic of the market fairly dispassionately, while reserving his criticism for a particular version of cultural studies that embraces consumption as an expression of agency. Although written in 1992, Davidson's assessment of the interpenetration of commerce and culture and the attendant impossibility of macro-theoretical explanations of this complexity remains relevant and unimproved nearly fifteen years later.

Here, it is particularly relevant to my discussion to note that Frank spends a chapter exploring and critiquing cultural studies in a manner uncommon for a book targeted to a popular audience. Frank is exceedingly critical of a branch of cultural studies that argues that activities such as fanship and consumerism can provide agency. While aspects of Frank's critique are sound, he allows work that others might define as marginal to cultural studies to define the field in its totality in a manner that makes his point, but in the process misrepresents an intellectual field that is much broader and more varied than he allows.

77. Joseph Turow, *Breaking Up America: Advertisers and the New Media World* (Chicago: University of Chicago Press, 1997); Arlene Dávila, *Latinos, Inc.: The Marketing and Making of a People* (Berkeley: University of California Press, 2001).

78. Importantly, these studies combine industrial analyses and interviews with industry workers to explain the complex practices involved in this transition, as well as considering the advertising messaging produced as a consequence.

79. Turow, *Breaking Up America*, 2.

80. Dávila, *Latinos, Inc.*, 2.

81. García Canclini, *Consumers and Citizens*.

82. Joseph Turow, "Unconventional Programs on Commercial Television: An Organizational Perspective," in *Mass Communicators in Context*, ed. D. Charles Whitney and James Ettema (Beverly Hills: Sage, 1982), 107–29.

83. Joseph Turow, *Media Systems in Society: Understanding Industries, Strategies and Power* (White Plains, NY: Longman, 1997), 174–235.

6. RECOUNTING THE AUDIENCE

1. Jon Gertner, "Our Ratings, Ourselves," *New York Times Magazine*, 10 April 2005.

2. Susan Whiting, personal communication, 16 April 2005.

3. Joseph R. Dominick, Barry L. Sherman, and Fritz Messere, *Broadcasting, Cable, the Internet, and Beyond*, 4th ed. (Boston: McGraw-Hill, 2000), 259; Nielsen Media Research, *2000 Report on Television: The First Fifty Years* (New York: Nielsen Media Research, 2000).

4. Hugh Malcolm Beville, *Audience Ratings: Radio, Television, Cable*, rev. student ed. (Hillsdale, NJ: Lawrence Erlbaum Associates, 1988), 72.

5. Steve Behrens, "People Meters vs. the Gold Standard," *Channels* 7 (September 1987): 72; Steve Behrens, "A Finer Grind from the Ratings Mill," *Channels* 7 (December 1987): 10–16.

6. Peter J. Boyer, "Networks Fight to Delay New Ratings Method," *New York Times*, 17 April 1986, C29.

7. Brian Dumaine, "Who's Gypping Whom in TV Ads?," *Fortune*, 6 July 1987, 78–79.

8. Joe Mandese, "Prime-Time Rating Points Valued at Nearly $400 Million," *Media Daily News*, 11 October 2006, http://publications.mediapost.com/index.cfm?fuseaction=Articles.san&s=49458&Nid=24151&p=368626. A rating point in the eighteen- to forty-nine-year-old demographic was estimated at $763.9 million.

9. Steve McClellan, "Nielsen: We Just Got Better," *Broadcasting & Cable*, 1 December 2003, 26.

10. Dan Trigoboff, "Nielsen Grows Local People Meter," *Broadcasting & Cable*, 3 March 2003, 14.

11. Nielsen Media Research, "The Facts on Nielsen and Local People Meters," http://everyonecounts.tv, accessed 14 November 2004.

12. Ibid.

13. Nielsen Media Research, "Media Rating Council Grants Accreditation to Nielsen's 25 Local People Meter Markets," press release, 14 July 2010, http://www.nielsen.com/us/en/press-room/2010/Media_Rating_Council_Grants_Accreditation_to_Nielsens_25_Local_People_Meter_Markets.html.

14. Nielsen Media Research, "Nielsen Begins Largest Ever Expansion of Its National U.S. Television Ratings Panel," *PR Newswatch*, 26 September 2007, http://www.prnewswire.com/news-releases/nielsen-begins-largest-ever-expansion-of-its-national-us-television-ratings-panel-58296822.html.

15. Steve McClellan, "Nielsen Gives It the New College Try," *Broadcasting & Cable*, 27 January 2003, 8.

16. Michele Greppi, "On-Campus Viewing High in Nielsen Test," *Television Week*, 21 March 2005, 6.

17. Louise Story, "At Last, Television Ratings Go to College," *New York Times*, 29 January 2007, C1.

18. Brian Hughes, "Nielsen's DVR Impact Assessment Study," *Media Insights: A Publication of Magna Global*, September 2005. VCRs could be used in this same manner, but Nielsen studies indicated that this was not a regular viewer behavior, except for soap opera taping, which accounted for seven of the top ten most-taped shows by the mid-1980s. See James Traub, "The World according to Nielsen," *Channels* 4, no. 1 (January–February 1985): 26–32, 70–72.

19. Wayne Friedman, "DVR Measurement Hangs in the Balance," *Television Week*, 28 February 2005, 24.

20. Jack Myers, "Upfront Chronicles 2005: Part Two," *Jack Myers Report*, 25 April 2005.

21. World Federation of Advertisers and European Association of Communications Agencies, "The WFA/EACA Guide to the Organisation of Television Audience Research," January 2001, http://www.wfanet.org/pdf/WFA_guideOrgofTVaudresearch.pdf.

22. David Goetzl, "Rentrak Deal Offers Juice for Set-Top-Box Data Advocates," MediaPost. com, 8 August 2013, http://www.mediapost.com/publications/article/206436/rentrak-deal-offers-juice-for-set-top-box-data-adv.html#axzz2cQQ4axcL; Janet Stilson, "Rentrak's Influence Growing in Ratings Wars," TVNewsCheck.com, 13 December 2011, http://www.tvnewscheck.com/article/56019/rentraks-influence-growing-in-ratings-wars.

23. "NBC Strikes Response Measurement Deal with Toyota," *Ad Business Report* (e-mail newsletter), received 24 July 2006.

24. Quoted in Jack Myers, "Nielsen Losing Value as Custom Insights Gain Status, Says Uva," *Jack Myers Report*, 28 April 2005.

25. Future of TV Seminar, 23 September 2004, Los Angeles, CA.

26. Leo Bogart, "Buying Services and the Media Marketplace," *Journal of Advertising Research*, September–October 2000, 37–41.

27. David Poltrack and Kevin Bowen, "The Future Is Now: In Pursuit of a More Efficient and Effective Media Strategy," presented at the Advertising Research Foundation, 2011, http://www.slideshare.net/TheARF/the-future-is-now-in-pursuit-of-a-more-efficient-and-effective-media-strategy.

28. Nielsen Media Research, *2000 Report on Television*.

29. Joseph Turow, "Audience Construction and Culture Production: Marketing Surveillance in the Digital Age," *American Academy of Political and Social Science* 597 (January 2005): 103–21.

7. TELEVISION STORYTELLING POSSIBILITIES AT THE BEGINNING OF THE POST-NETWORK ERA

1. A. J. Jacobs, "Let's Talk about *Sex*," *Entertainment Weekly*, 5 June 1998, 32.

2. For more, see Amanda D. Lotz, "If It Is Not TV, What Is It? The Case of U.S. Subscription Television," in *Cable Visions: Television beyond Broadcasting*, ed. Sarah Banet-Weiser, Cynthia Chris, and Anthony Freitas (New York: New York University Press, 2007).

3. Except in season 5, in which the pregnancy of the actor Sarah Jessica Parker led to the production of just eight episodes.

4. Kim Potts, "Women Love *Sex*: HBO's Bawdy Comedy Not Just a Pretty Face," *Daily Variety*, 17 September 1999, A1.

5. Toni Fitzgerald, "Where the Real $ Is for Sexy HBO," *Media Life*, 26 February 2004, http://www.medialifemagazine.com/news2004/feb04/feb23/4_thurs/news2thursday.html.

6. Previous original HBO series such as *The Larry Sanders Show* and *Dream On* had sold in various other distribution windows, but none earned nearly the revenue of *Sex and the City*.

7. John Dempsey, "HBO Sells *Sex* Reruns to TBS Net," *Daily Variety*, 30 September 2003, 6; John Dempsey and Meredith Amdur, "Tribune Spices Up HBO's *Sex*," *Variety*, 10 September 2003.

8. Gail Schiller, "It's Not a Plug, It's HBO," *Hollywood Reporter*, 10 December 2004.

9. Claire Atkinson, "Absolut Nabs Sexy HBO Role," *Advertising Age*, 4 August 2003, 6.

10. Michael McCarthy, "HBO Shows Use Real Brands," *USA Today*, 3 December 2002, 3B.

11. Ibid.

12. Chad Raphael, "The Political Economic Origins of Reali-TV," in *Reality TV: Remaking Television Culture*, ed. Susan Murray and Laurie Ouellette (New York: New York University Press, 2004), 119–36.

13. Bill Carter, *Desperate Networks* (New York: Doubleday, 2006), 67–89.

14. Raphael, "The Political Economic Origins of Reali-TV."

15. Bill Carter, "Survival of the Pushiest," *New York Times*, 28 January 2001; also see Carter, *Desperate Networks*, 67–89, for a more detailed version of this story.

16. Anne Becker and Allison Romano, "*Fear Factor* Soars on FX," *Broadcasting & Cable*, 13 September 2004, 2.

17. Reported by Bhuvan Lall, managing director, Empire Entertainment Private Limited, at 2004 NAPTE Faculty Seminar, Las Vegas, 17 January 2004.

18. Herbert Schiller, *Communication and Cultural Domination* (White Plains, NY: International Arts and Sciences Press, 1976).

19. See Albert Moran, *Copycat TV: Globalisation, Program Formats and Cultural Identity* (Luton, UK: University of Luton Press, 1998), for a detailed examination of some of the early aspects of format exports.

20. Melissa Grego, "Burnett's New Studio Model," *Television Week*, 15 August 2005, 1.

21. Bill Carter, "Mark Burnett Makes Producing Reality TV Hits Look Easy," *New York Times*, 26 June 2011, http://www.nytimes.com/2011/06/27/business/media/27burnett.html?pagewanted=all&_r=1&.

22. SciFi's *Battlestar Galactica* and Comedy Central's *South Park* also received Peabody Awards in 2006. Although this distinction is particular to live-action shows, Comedy Central's *Dr. Katz: Professional Therapist*, various animated Nickelodeon shows, and *Mystery Science Theater 3000* had won previously.

23. Interview with Shawn Ryan, http://www.fxnetworks.com/shows/originals/the_shield/interviews/2.htlm, accessed 15 March 2006.

24. Michael Freeman, "An HBO Kind of Respect," *Electronic Media*, 8 July 2002, 22.

25. "TV Out of the Box," Trio Network, 2003.

26. Chiklis was known for a very different police role as the jovial title character of *The Commish*.

27. Quoted in Anthony Crupi, "Young Blood: FX Pulls in 18–49 Demo with Gritty Fare," *Mediaweek*, 16 January 2006, http://mediaweek.com/mw/news/cabletv/article_display.jsp?vnu_content_id=1001844104.

28. John M. Higgins, "Edgy Fare Drives FX," *Broadcasting & Cable*, 12 September 2004, 4.

29. Michael Curtin, "Feminine Desire in the Age of Satellite Television," *Journal of Communication* 49, no. 2 (1999): 55–70; Thomas Streeter, "Media: The Problem of Creativity,"

paper presented at the International Communication Association annual conference, Washington, DC, 26 May 2001; Michael Curtin and Thomas Streeter, "Media," in *Culture Works: The Political Economy of Culture*, ed. Richard Maxwell (Minneapolis: University of Minnesota Press, 2001), 225–50.

30. A. J. Frutkin, "One Tough Show," *Mediaweek*, 16 December 2002; Megan Larson, "Out of the Foxhole," *Mediaweek*, 19 May 2003.

31. Denise Martin, "*Shield* Cops a 7th Season," *Variety*, 5 June 2006, http://www.variety.com/index.asp?layout=print_story&articleid=VR1117944652&categoryid=14.

32. David Lieberman, "Could Tony on A&E Bring Restrictions to Cable?," *USA Today*, 1 December 2006, http://www.usatoday.com/printedition/money/20061201/sleazecov.art.htm.

33. Derek Baine, quoted in Larson, "Out of the Foxhole."

34. Larson, "Out of the Foxhole."

35. SNL Kagan figures, cited in Rick Kissell, "Hit Cable Series on Rise as Broadcasters Suffer Record-Low Ratings," *Variety*, 26 March 2013, http://variety.com/2013/tv/news/hit-cable-series-on-the-rise-as-broadcasters-suffer-record-low-ratings-1200329213/.

36. News Corp, the conglomerate owning FX, created FXX from a rebranding of the fledgling Fox Soccer.

37. Freeman, "An HBO Kind of Respect," 22.

38. Shawn Ryan, phone interview, 4 May 2006.

39. A&E, however, had purchased *The Sopranos*, illustrating basic cable purchase of a subscription cable series.

40. John Dempsey and Denise Martin, "Spike Wields *Shield* in Big Deal," *Daily Variety*, 28 July 2005, 5; Christopher Lisotta, "Sony Sticking to the Flight Plan," *Television Week*, 10 April 2006, 1, 48, 48.

41. Lisotta, "Sony Sticking to the Flight Plan," 48.

42. Larson, "Out of the Foxhole."

43. Ari Posner, "Can This Man Save the Sitcom?," *New York Times*, 1 August 2004, sec. 2, p. 1.

44. Nellie Andreeva, "Fox Pays Big to Get *Arrested*," *Hollywood Reporter*, 26 September 2002.

45. Posner, "Can This Man Save the Sitcom?"

46. Comedy Central had success with *South Park*, *The Daily Show*, and *The Colbert Report*, but their animation and non-narrative form distinguished these shows from traditional broadcast comedy forms. This is likewise the case for various MTV programs such as *The Tom Green Show* and *Jackass*. A live-action narrative show such as *Arrested Development* would require a substantially higher budget than any of these shows.

47. Michael Schneider, "New Development," *Daily Variety*, 26 September 2002, 5.

48. Josef Adalian, "Hurwitz Takes a Hike," *Variety*, 27 March 2006, http://www.variety.com/index.asp?layout=print_story&articleid= VR1117940467&categoryid=1417.

49. News Corp, "Emmy Award Winning Fox Comedy *Arrested Development* Finds Post Broadcast Home Online, on Hi-Def TV and on Basic Cable," *Business Wire*, 26 July 2006, http://home.businesswire.com/portal/site/google/index.jsp?ndmViewId=news_view&newsId=20060726005574&newsLand= en.

50. Brent Renaud, phone interview, 26 April 2006.

51. Daisy Whitney, "Channel Schedules Limited Series," *Television Week*, 6 June 2005, 22.

52. Ibid.

53. Brent Renaud, phone interview, 26 April 2006.

54. Ibid.

CONCLUSION

1. David Carr, "Taken to a New Place, by a TV in the Palm," *New York Times*, 18 December 2005, sec. 4, p. 3.
2. For another account of use at this time, see Mitch Oscar, "Does Anybody Really Know How People Watch TV?," *Media Post*, 16 January 2007, http://publications.mediapost.com/index/cfm?fuseaction=Articles.showArticleHomePage&art_aid=5396.
3. Diane Mermigas, "Searching for Success in the Interactive Age," *Hollywood Reporter*, 17 January 2006, http://www.insidebranded entertainment.com/bep/article_display.jsp?JSESSIONID=DT%GZpzFGJChBJLTPTh1ZKvr8I9TsTWdBy42d3vJmW14r7T83nRn!135563654 4&vnu_content_id=1001844514.
4. Nicholas Negroponte, *Being Digital* (New York: Vintage, 1995), 48.
5. Tim Wu, "Netflix's War on Mass Culture," *New Republic*, 4 December 2013, http://www.newrepublic.com/article/115687/netflixs-war-mass-culture.
6. Chris Anderson, "A Problem with the Long Tail," talk given at iConference, 16 October 2006, Ann Arbor, MI.
7. Scott W. Campbell and Tracy C. Russo, "The Social Construction of Mobile Telephony: An Application of the Social Influence Model to Perceptions and Uses of Mobile Phones within Personal Communication Networks," *Communication Monographs* 70, no. 4 (2003): 317–34.
8. The first use of this comparison that I saw was made by a group of faculty who were tasked to represent the interest of local affiliates in a 2004 International Radio and Television Society case study competition, and offered a well-conceived future path for the affiliates.
9. Joseph Turow, *Breaking Up America: Advertisers and the New Media World* (Chicago: University of Chicago Press, 1997), provides a valuable exception. W. Russell Neuman, *The Future of the Mass Audience* (Cambridge: Cambridge University Press, 1991), also perceived these potential developments before they became clearly manifest.
10. Manuel Castells, *The Rise of the Network Society* (New York: Blackwell, 1996).
11. Todd Spangler, "Pay-TV Prices Are at the Breaking Point—and They're Only Going to Get Worse," *Variety*, 29 November 2013, http://variety.com/2013/biz/news/pay-tv-prices-are-at-the-breaking-point-and-theyre-only-going-to-get-worse-1200886691/.
12. Moffett, cited in Tom Lowry, "Poverty a Problem for Pay TV," *Variety*, 31 May 2011, http://variety.com/2011/tv/news/poverty-a-problem-for-pay-tv-1118037755/.
13. Ibid.
14. Horace Newcomb, *TV: The Most Popular Art* (Garden City: Anchor, 1974).

Adalian, Josef. "Hurwitz Takes a Hike: *Arrested* Creator Bails as Showrunner." *Variety*, March 27, 2006. http://variety.com/2006/scene/news/hurwitz-takes-a-hike-1117940467/.

Albiniak, Paige. "Deficit Disorder: Why Some Good Pilots Never Get Produced." *Broadcasting & Cable*, May 5, 2003, 27.

———. "Sunburned: Cable, Broadcast Each Had Summers That Hurt." *Broadcasting & Cable*, August 31, 2003. http://www.broadcastingcable.com/article/150614-Sunburned.php.

Althusser, Louis. "Ideology and Ideological State Apparatuses." In *Lenin and Philosophy and Other Essays*, trans. Ben Brewster, 127–88. London: Monthly Review Books, 1971.

Alvey, Mark. "The Independents: Rethinking the Television Studio System." In *Television: The Critical View*, 6th ed., ed. Horace Newcomb, 34–51. New York: Oxford University Press, 2000.

Anderson, Chris. *The Long Tail: Why the Future of Business Is Selling Less of More*. New York: Hyperion, 2006.

Anderson, Christopher. *Hollywood TV*. Austin: University of Texas Press, 1993.

Anderson, Sam. "The Joys of Rising from the Cultural Dead." *Slate*, April 6, 2006. http://www.slate.com/articles/arts/dvdextras/2006/04/the_joys_of_rising_from_the_cultural_dead.html.

Andreeva, Nellie. "Fox Pays Big to Get *Arrested*." *Hollywood Reporter*, September 26, 2002.

Atkinson, Claire. "Absolut Nabs Sexy HBO Role." *Advertising Age*, August 4, 2003. http://adage.com/article/news/absolut-nabs-sexy-hbo-role/95660/.

———. "Marketers Hunger for *Idol* Reprise." *Advertising Age*, May 28, 2006. http://adage.com/article/media/marketers-hunger-idol-reprise/109484/.

Aufderheide, Patricia. *Communications Policy and the Public Interest*. New York: Guilford, 1999.

Auletta, Ken. *Three Blind Mice: How the TV Networks Lost Their Way*. New York: Vintage, 1992.

Ault, Susanne. "Fallen *Wonderfalls* Is a DVD Wonder." *Video Business*, February 11, 2005, 6.

Bagdikian, Ben H. *The New Media Monopoly*. Boston: Beacon, 2004.

Ballaster, Ros, Margaret Beetham, Elizabeth Frazier, and Sandra Hebron. *Women's Worlds: Ideology, Femininity, and Women's Magazines*. London: Macmillan, 1991.

Barclay, Paris. "Paris Barclay: Director, President of the Directors Guild of America." Interview. Media Industries Project, June 27, 2013. http://www.carseywolf.ucsb.edu/mip/article/paris-barclay.

Barnes, Brooks. "How Old Media Can Survive In a New World." *Wall Street Journal*, May 23, 2005. http://online.wsj.com/public/article/0,,SB111643067458336994,00.html?mod=todays_free_feature.

Barnouw, Erik. *Tube of Plenty: The Evolution of American Television*. 2nd rev. ed. New York: Oxford University Press, 1990.

Bauder, David. "Study: Clutter of Advertising Soaring on Prime-Time Television." *Associated Press*, April 12, 1999.

Becker, Anne, and Allison Romano. "*Fear Factor* Soars on FX." *Broadcasting & Cable*, September 12, 2004. http://www.broadcastingcable.com/article/154601-Fear_Factor_Soars_on_FX.php.

Bedell, Sally. *Up the Tube: Prime-Time TV and the Silverman Years*. New York: Viking, 1981.

Benjamin, Louise. "At the Touch of a Button: A Brief History of Remote Control Devices." In *The Remote Control in the New Age of Television*, ed. James R. Walker and Robert V. Bellamy Jr., 15–22. Westport, CT: Praeger, 1993.

Benner, Katie. "Meet Content Partners: The Investors Who Bought *CSI*." *CNN Money*, March 11, 2013. http://finance.fortune.cnn.com/2013/03/11/content-partners-csi/.

Bennett, James. "Introduction: Television as Digital Media." In *Television as Digital Media*, ed. James Bennett and Niki Strange, 1–30. Durham: Duke University Press, 2011.

Bennett, Jim. "The Cinematch System: Operation Scale, Coverage, Accuracy, Impact." Presentation made at "The Present and Future of Recommender Systems" conference, Bilboa, Spain, September 13, 2006. http://blog. recommenders06.com/wp-content/uploads/2006/09/bennett.pdf.

Benson, Jim. "Debmar-Mercury Secures $200 Million Distribution Deal for *Tyler Perry's House of Payne*." *Broadcasting & Cable*, August 23, 2006. http://www.broadcastingcable.com/article/105512-Debmar_Mercury_Secures_200_Million_Distribution_Deal_for_Tyler_Perry_s_House_of_Payne.php.

Beville, Hugh Malcolm. *Audience Ratings: Radio, Television, Cable*. Rev. student ed. Hillsdale, NJ: Lawrence Erlbaum Associates, 1988.

Blackledge, Chris. "They Deserve a Fat Promotion: Nets Angle for Fall Viewers." *NATPE News*, September 2005. http://www.natpe.org/memberresources/natpenews/articles/story.jsp?id_string=200023:X58IMn7ClV8Ry8yPMfzucQ**.

Blumenthal, Howard J., and Oliver R. Goodenough. *This Business of Television*. 2nd ed. New York: Billboard Books, 1998.

Boddy, William. *Fifties Television: The Industry and Its Critics*. Urbana: University of Illinois Press, 1993.

———. "Redefining the Home Screen: The Case of the Digital Video Recorder." In *New Media and Popular Imagination: Launching Radio, Television, and Digital Media in the United States*, 100–107. New York: Oxford University Press, 2004.

Borland, John, and Evan Hansen. "The TV Is Dead. Long Live the TV." *Wired*, April 6, 2007. http://www.wired.com/entertainment/hollywood/news/2007/04/tvhistory_0406.

Breznican, Anthony. "*Firefly* Alights on Big Screen as *Serenity*." *USA Today*, September 21, 2005. http://usatoday30.usatoday.com/life/movies/news/2005-09-21-serenity_x.htm.

Brinkley, Joel. *Defining Vision: The Battle for the Future of Television*. San Diego: Harcourt Brace, 1997.

Brown, Les. *Television: The Business behind the Box*. New York: Harcourt Brace Jovanovich, 1971.

Browne, Nick. "The Political Economy of the Television (Super) Text." *Quarterly Review of Film Studies* 9, no. 3 (Summer 1984): 174–82.

Brunsdon, Charlotte. "Lifestyling Britain: The 8–9 Slot on British Television." In *Television after TV: Essays on a Medium in Transition*, ed. Lynn Spigel and Jan Olsson, 75–92. Durham: Duke University Press, 2004.

———. "What Is the 'Television' of Television Studies?" In *The Television Studies Book*, ed. Christine Geraghty and David Lusted, 95–113. London: Arnold, 1998.

Byars, Jackie, and Eileen R. Meehan. "Once in a Lifetime: Constructing the 'Working Woman' through Cable Narrowcasting." In "Lifetime: A Cable Network for Women," ed. Julie d'Acci, special issue, *Camera Obscura* 33–34 (1994): 12–41.

Byers, Alex. "Netflix's Hastings: Online Steering TV's Future." *Politico*, January 29, 2013. http://www.politico.com/story/2013/01/netflixs-hastings-online-steering-tvs-future-86894.html.

Calabrese, Andrew, and Colin Sparks, eds. *Toward a Political Economy of Culture: Capitalism and Communication in the Twenty-First Century*. Lanham, MD: Rowman and Littlefield, 2004.

Caldwell, John Thornton. *Televisuality: Style, Crisis, and Authority in American Television*. New Brunswick: Rutgers University Press, 1995.

Campbell, Scott W., and Tracy C. Russo. "The Social Construction of Mobile Telephony: An Application of the Social Influence Model to Perceptions and Uses of Mobile Phones within Personal Communication Networks." *Communication Monographs* 70, no. 4 (2003): 317–34.

Cappo, Joe. *The Future of Advertising: New Media, New Clients, New Consumers in the Post-Television Age*. Chicago: McGraw-Hill, 2003.

Carey, James. "Abolishing the Old World Spirit." *Critical Studies in Mass Communication* 12, no. 1 (1995): 82–89.

Carr, David. "Taken to a New Place, by a TV in the Palm." *New York Times*, December 18, 2005. http://www.nytimes.com/2005/12/18/weekinreview/18carr.html?_r=1.

Carter, Bill. *Desperate Networks*. New York: Doubleday, 2006.

———. "Here Comes the Judge." *New York Times*, March 12, 2006, sec. 2, p. 1.

———. "NBC to Pay Outsiders for Blocks of Programs." *New York Times*, December 3, 2007. http://www.nytimes.com/2007/12/03/business/media/03nbc.html?_r=1&.

———. "Reality TV Alters the Way TV Does Business." *New York Times*, January 25, 2003. http://www.nytimes.com/2003/01/25/business/reality-shows-alter-the-way-tv-does-business.html?pagewanted=all.

———. "Survival of the Pushiest." *New York Times*, January 28, 2001. http://www.nytimes.com/2001/01/28/magazine/survival-of-the-pushiest.html?pagewanted=all.

Castells, Manuel. *The Rise of the Network Society*. New York: Blackwell, 1996.

Chmielewski, Dawn C. "Studios Not Sure Whether Web Video Innovator Is Friend or Foe." *Los Angeles Times*, April 10, 2006. http://articles.latimes.com/2006/apr/10/business/fi-youtube10.

Cohen, David S. "Wolf Sounds Alarm." *Variety*, March 20, 2006. http://variety.com/2006/more/news/wolf-sounds-alarm-1117940065/.

ComScore. "ComScore Reports February 2013 U.S. Smartphone Subscriber Market Share." News release, April 4, 2013. ComScore Reports. http://www.comscore.com/Insights/Press_Releases/2013/4/comScore_Reports_February_2013_U.S._Smartphone_Subscriber_Market_Share.

Consumer Electronics Association. "One in Five Owners Bought Their HDTV to Watch the Super Bowl." News release, January 31, 2013. http://www.ce.org/News/News-Releases/Press-Releases/2013-Press-Releases/One-in-Five-Owners-Bought-Their-HDTV-to-Watch-the.aspx.

Crawford, Susan P. *Captive Audience: The Telecom Industry and Monopoly Power in the New Gilded Age*. New Haven: Yale University Press, 2013.

Crupi, Anthony. "Young Blood: FX Pulls in 18-49 Demographic with Gritty, Graphic Fare." *Mediaweek*, January 16, 2006, 6. General OneFile.

Curtin, Michael. "Feminine Desire in the Age of Satellite Television." *Journal of Communication* 49, no. 2 (1999): 55–70.

———. "On Edge: Culture Industries in the Neo-Network Era." In *Making and Selling Culture*, ed. Richard Ohmann, Gage Averill, Michael Curtin, David Shumway, and Elizabeth Traube, 181–202. Hanover, NH: Wesleyan University Press, 1996.

Curtin, Michael, and Thomas Streeter. "Media." In *Culture Works: The Political Economy of Culture*, ed. Richard Maxwell, 225–50. Minneapolis: University of Minnesota Press, 2001.

D'Acci, Julie. "Cultural Studies, Television Studies, and the Crisis in the Humanities." In *Television after TV: Essays on a Medium in Transition*, ed. Lynn Spigel and Jan Olsson, 418–45. Durham: Duke University Press, 2004.

Davidson, Martin. *The Consumerist Manifesto: Advertising in Postmodern Times*. London: Routledge, 1992.

Dávila, Arlene. *Latinos, Inc.: The Marketing and Making of a People*. Berkeley: University of California Press, 2001.

Dayan, Daniel, and Elihu Katz. *Media Events: The Live Broadcasting of History*. Cambridge: Harvard University Press, 1992.

Dempsey, John. "Cable Aud's Now Bigger Than B'Cast." *Variety*, May 20, 2004. http://variety.com/2004/scene/news/cable-aud-s-now-bigger-than-b-cast-1117905396/.

———. "HBO Sells *Sex* Reruns to TBS." *Daily Variety*, September 30, 2003. http://variety.com/2003/tv/news/hbo-sells-sex-reruns-to-tbs-1117893187/.

———. "Shared Runs: Cachet over Cash." *Variety*, June 12, 2001. http://variety.com/2001/tv/news/shared-runs-cachet-over-cash-1117801281/.

Dempsey, John, and Denise Martin. "Spike Wields *Shield* in Big Deal." *Daily Variety*, July 28, 2005.

Deutsch, Claudia. "Study Details Decline in Spending on Ads." *New York Times*, September 5, 2001. http://www.nytimes.com/2001/09/05/business/the-media-business-advertising-addenda-study-details-decline-in-spending-on-ads.html.

Dominick, Joseph R., Barry L. Sherman, and Fritz Messere. *Broadcasting, Cable, the Internet, and Beyond*. 4th ed. Boston: McGraw-Hill, 2000.

Donaton, Scott. *Madison & Vine: Why the Entertainment and Advertising Industries Must Converge to Survive*. New York: McGraw-Hill, 2004.

Dover, Sara. "Study: Number of Smartphone Users Tops 1 Billion." *CBS News*, October 17, 2012. http://www.cbsnews.com/8301-205_162-57534583/study-number-of-smartphone-users-tops-1-billion/.

Doyle, Gillian. *Understanding Media Economics*. Thousand Oaks, CA: Sage, 2002.

Du Gay, Paul, Stuart Hall, Linda Janes, Hugh Mackay, and Keith Negus. *Doing Cultural Studies: The Story of the Sony Walkman*. London: Sage, 1997.

Eastman, Susan T. "Orientation to Promotion and Research." In *Research in Media Promotion*, ed. Susan T. Eastman, 3–18. Mahwah, NJ: Lawrence Erlbaum Associates, 2000.

Eastman, Susan T., and Gregory D. Newton. "The Impact of Structural Salience within On-Air Promotion." *Journal of Broadcasting & Electronic Media* 42 (1998): 50–79.

Edwards, Cliff. "Netflix Declines Most since 2004 after Losing 800,000 U.S. Subscribers." *Bloomberg News*, October 25, 2011. http://www.bloomberg.com/news/2011-10-24/netflix-3q-subscriber-losses-worse-than-forecast.html.

Elliott, Stuart. "An Agency Giant Is Expected to Warn of Lower Profits, and Analysts Darken Their Outlook." *New York Times*, October 2, 2001. http://www.nytimes.com/2001/10/02/business/media-business-advertising-agency-giant-expected-warn-lower-profits-analysts.html.

———. "Networks Watch Prices for Commercial Time Drop." *New York Times*, September 25, 2001. http://www.nytimes.com/2001/09/25/business/media/25ADCO.html.

———. "Watching a Show Live, with 72 Hours to Do It." *New York Times*, May 17, 2007. http://www.nytimes.com/2007/05/17/business/media/17adco.html?_r=1.

Fera, Rae Ann. "How Kmart Used Social Listening (and Some Nerve) to Create a Ship-My-Pants Funny Viral Hit." *Co.Create*. http://www.fastcocreate.com/1682826/how-kmart-used-social-listening-and-some-nerve-to-create-a-ship-my-pants-funny-viral-hit.

Fernandez, Maria Elena. "*CSI: Miami* Goes Global." *Los Angeles Times*, September 18, 2006.

Fidgeon, Robert. "FOX on the Run." *Herald Sun* (Sydney), July 26, 2000, 1st ed., Home Entertainment sec.

File, Thom. *Computer and Internet Use in the United States: Population Characteristics.* U.S. Census Bureau Report, May 2013. http://www.census.gov/prod/2013pubs/p20-569.pdf.

Filmscape News Center. "1998 Hollywood Writers Strike." April 30, 2001. http://www.filmscape.co.uk/news/fullnews.cgi?newsid988621880,79143.

Fine, Jon, and Tobi Elkin. "On the House: AOL Time Warner's Bob Pittman Is Doing a Great Job Selling Ads—to Himself." *Advertising Age*, March 18, 2002.

Fischetti, Mark. "The Future of TV." *Technology Review*, November 2001, 35–40.

Fisher, Max. "Cable TV Is Doomed." *Atlantic*, March 18, 2010. http://www.theatlantic.com/business/archive/2010/03/cable-tv-is-doomed/37675/.

Fiske, John, and John Hartley. *Reading Television*. London: Methuen, 1978.

Fitzgerald, Toni. "Where the Real $ Is for Sexy HBO." *Media Life Magazine*, February 26, 2004. http://www.medialifemagazine.com/news2004/feb04/feb23/4_thurs/news2thursday.html.

Flint, Joe. "CBS to Collect $1 Billion a Year in Distribution Fees by 2016." *Los Angeles Times*, September 11, 2012. http://articles.latimes.com/2012/sep/11/entertainment/la-et-ct-cbs-20120911.

———. "*I Love Lucy* Still a Cash Cow for CBS." *Los Angeles Times*, September 20, 2012. http://articles.latimes.com/2012/sep/20/entertainment/la-et-ct-cbslucy-20120920.

Frank, Thomas. *One Market under God: Extreme Capitalism, Market Populism, and the End of Economic Democracy.* New York: Doubleday, 2000.

Frankel, Daniel. "Ancillary Waters Run Deep." *Variety*, July 9, 2006. http://variety.com/2006/scene/markets-festivals/ancillary-waters-run-deep-1200340374/.

Freeman, Michael. "An HBO Kind of Respect." *Electronic Media*, July 8, 2002.

Friedman, Wayne, and Jack Neff. "Eagle-Eye Marketers Find Right Spot, Right Time." *Advertising Age*, January 22, 2001. http://adage.com/article/required-reading/eagle-eye-marketers-find-spot-time/55659/.

Frutkin, A. J. "One Tough Show: While FX's *The Shield* Remains a Hard Sell, CPMs Are Up for Second Season." *Mediaweek*, December 16, 2002, 8.

Fry, Jason. "O Pioneers! Watching a Child Who's Grown Up with TiVo Leads to Questions about the Future of TV." *Wall Street Journal*, December 12, 2005. http://online.wsj.com/article/SB113398924199016540.html?mod=technology_main_promo_left.

Fuhrer, Brian. "Nielsen's Brian Fuhrer on Digital Video Trends." IAB Digital Video Marketplace conference, April 10, 2012. http://youtube/EHWZX4JGtMM.

Galician, Mary-Lou, ed. "Product Placement." Special issue, *Journal of Promotion Management* 10, nos. 1–2 (2004).

García Canclini, Néstor. *Consumers and Citizens: Globalization and Multicultural Conflicts.* Minneapolis: University of Minnesota Press, 2001.

Garnham, Nicholas. "Political Economy and Cultural Studies: Reconciliation or Divorce?" *Critical Studies in Mass Communication* 12, no. 1 (1995): 62–71.

Garrett, Diane. "Windows Rattled." *Variety*, March 21, 2006. http://www.variety.com/index.asp?layout=print_story&articleid=VR1117940116&categoryid=20.

Gitelman, Lisa. *Always Already New: Media, History and the Data of Culture.* Cambridge: MIT Press, 2006.

Gitlin, Todd. *Inside Prime Time*. New York: Pantheon, 1983.

———. "Prime Time Ideology: The Hegemonic Process in Television Entertainment." In *Television: A Critical View*, 5th ed., ed. Horace Newcomb, 516–36. New York: Oxford University Press, 1994.

Gladwell, Malcolm. *The Tipping Point: How Little Things Can Make a Big Difference*. Boston: Little, Brown, 2002.

Goolsbee, Austan. "The BOOB Tube Won't Make Your Kid a Boob." *Chicago Sun-Times*, March 5, 2006.

Gough-Yates, Anna. *Understanding Women's Magazines: Publishing, Markets and Readerships*. New York: Routledge, 2003.

Greenfield, Rebecca. "The Economics of Netflix's $100 Million New Show." *Atlantic Wire*, February 1, 2013. http://www.theatlanticwire.com/technology/2013/02/economics-netflixs-100-million-new-show/61692/.

Grego, Melissa. "Burnett's New Studio Model." *Television Week*, August 15, 2005.

Gripsrud, Jostein. "Broadcast Television: The Chances of Its Survival in a Digital Age." In *Television after TV: Essays on a Medium in Transition*, ed. Lynn Spigel and Jan Olsson, 210–24. Durham: Duke University Press, 2004.

Grossberg, Lawrence. "Cultural Studies vs. Political Economy: Is Anybody Else Bored with This Debate?" *Critical Studies in Mass Communication* 12, no. 1 (1995): 72–81.

Haralovich, Mary Beth. "Sitcoms and Suburbs: Positioning the 1950s Homemaker." In *Private Screenings: Television and the Female Consumer*, ed. Lynn Spigel and Denise Mann, 111–42. Minneapolis: University of Minnesota Press, 1992.

Hardawar, Devindra. "The Magic Moment: Smartphones Now Half of All U.S. Mobiles." *Venture Beat*, March 29, 2012. http://venturebeat.com/2012/03/29/the-magic-moment-smartphones-now-half-of-all-u-s-mobiles/.

———. "Netflix Now Accounts for 25% of North American Internet Traffic." *Venture Beat*, May 17, 2011. http://venturebeat.com/2011/05/17/netflix-north-america-traffi/.

Harries, Dan. "Watching the Internet." In *The New Media Book*, ed. Dan Harries, 171–82. London: British Film Institute, 2002.

Hartley, John. *Uses of Television*. London: Routledge, 1999.

Havens, Timothy. "'It's Still a White World Out There': The Interplay of Culture and Economics in International Television Trade." *Critical Studies in Media Communication* 19, no. 4 (2002): 377–97.

Helm, Burt. "Why TV Will Never Be the Same." *Business Week*, November 22, 2004. http://www.businessweek.com/stories/2004-11-22/why-tv-will-never-be-the-same.

Hesmondhalgh, David. *The Cultural Industries*. London: Sage, 2002.

Higgins, John M. "Edgy Fare Drives FX: *Nip/Tuck, Rescue Me* Boost Ratings but Alienate Some Advertisers." *Broadcasting & Cable*, September 12, 2004. http://www.broadcastingcable.com/article/154585-Edgy_Fare_Drives_FX.php.

———. "Empty Screens: If Cable's Video-on-Demand Is So Hot, Where Are All the Good Shows?" *Broadcasting & Cable*, September 19, 2005. http://www.broadcastingcable.com/article/158178-Empty_Screens.php.

———. "Fast Forward: With Scant Notice, TV-DVD Sales Top $1B and Begin to Affect Scheduling, Financing." *Broadcasting & Cable*, December 22, 2003, 1.

———. "It's Not All in the Family: In-House Still Rules, but Networks Look outside Their Companies, Too." *Broadcasting & Cable*, May 22, 2013. http://www.broadcastingcable.com/article/157221-It_s_Not_All_In_the_Family.php.

Hilmes, Michele. "Cable, Satellite and Digital Technologies." In *The New Media Book*, ed. Dan Harries, 3–16. London: British Film Institute, 2002.

Howe, Neil, and William Strauss. *Millennials Rising: The Next Great Generation*. New York: Vintage, 2000.

Hu, Jim. "Baseball Officials Plan Live Video Streaming." CNET News, October 30, 2001. http://news.cnet.com/2100-1023-275123.html.

IBM Business Consulting Services. "The End of Television as We Know It." Executive Brief, March 27, 2006. http://www-935.ibm.com/services/us/imc/pdf/ge510-6248-end-of-tv-full.pdf.

Jacobs, A. J. "Let's Talk about *Sex*." *Entertainment Weekly*, June 5, 1998.

Jacobs, Jason. "Television, Interrupted: Pollution or Aesthetic?" In *Television as Digital Media*, ed. James Bennett and Niki Strange, 255–82. Durham: Duke University Press, 2011.

James, Meg. "TV in Your Pocket Is the Next Small Thing." *Los Angeles Times*, November 1, 2005. http://articles.latimes.com/2005/nov/01/business/fi-mobile1.

———. "TV Networks Pursue the 'Super Fan.'" *Los Angeles Times*, September 19, 2005. http://articles.latimes.com/2005/sep/19/business/fi-tvbuzz19.

Jayson, Sharon. "Totally Wireless on Campus." *USA Today*, October 2, 2006. http://usatoday30.usatoday.com/life/2006-10-02-gennext-tech_x.htm.

Jenkins, Henry. *Convergence Culture: Where Old and New Media Collide*. New York: New York University Press, 2006.

———. "Interactive Audiences?" In *The New Media Book*, ed. Dan Harries, 157–70. London: British Film Institute, 2002.

Kantor, Jodi. "The Extra-Large, Ultra-Small Medium." *New York Times*, October 30, 2005. http://www.nytimes.com/2005/10/30/arts/television/30kant.html?pagewanted=all.

Karrh, James A. "Brand Placement: A Review." *Journal of Current Issues and Research in Advertising* 20, no. 2 (1998): 31–49.

Katzmaier, David. "Quick Guide: HDTV Resolution Explained." CNET, September 12, 2006, http://reviews.cnet.com/hdtv-resolution/?tag=tnav.

Keegan, Paul. "The Man Who Can Save Advertising." *Business 2.0*, November 2004, http://www.business2.com/b2/subscribers/articles/print/0,17925,704067,00.html.

Kellner, Douglas. "Overcoming the Divide: Cultural Studies and Political Economy." In *Cultural Studies in Question*, ed. Marjorie Ferguson and Peter Golding, 102–20. London: Sage, 1997.

Klein, Paul. "Why You Watch When You Watch." In *TV Guide, the First 25 Years*, ed. Jay S. Harris, 186–88. New York: Simon and Schuster, 1978.

Klinger, Barbara. *Beyond the Multiplex: Cinema, New Technologies, and the Home*. Berkeley: University of California Press, 2006.

Klopfenstein, Bruce C. "From Gadget to Necessity: The Diffusion of Remote Control Technology." In *The Remote Control in the New Age of Television*, ed. James R. Walker and Robert V. Bellamy Jr., 23–39. Westport, CT: Praeger, 1993.

Knowledge Networks Statistical Research. "The Home Technology Monitor: Spring 2005 Ownership and Trend Report." Crawford, NJ: Knowledge Networks SRI, 2005.

Koeppel, David. "For Those of You Who Wonder How That TV Show Began." *New York Times*, March 21, 2005. http://www.nytimes.com/2005/03/21/business/media/21dvd.html?_r=0.

Kompare, Derek. "Acquisition Repetition: Home Video and the Television Heritage." In *Rerun Nation: How Repeats Invented American Television*, 197–220. New York: Routledge, 2005.

Kowinski, W. S. *The Malling of America: An Inside Look at the Great Consumer Paradise*. New York: Pantheon, 1985.

Lall, Bhuvan. Address at the 2004 NATPE Faculty Seminar, Las Vegas, January 17, 2004.

Larson, Megan. "Out of the Foxhole: Peter Liguori and Kevin Reilly Are Ramping Up FX's Original Programming to Wean the News Corp. Cable Network from an Overreliance on Sibling Product." *Mediaweek*, May 19, 2003, 18.

Learmonth, Michael. "YouTube Drops Price for Upfront Packages to Lure TV Dollars." *Advertising Age*, April 26, 2013. http://adage.com/article/digital/youtube-drops-price-upfront-packages-lure-tv-dollars/241137/.

Levine, Stuart. "*Sons of Anarchy* Stretches Episode Length." *Variety*, November 26, 2012. http://variety.com/2012/tv/news/sons-of-anarchy-stretches-episode-length-1118062663/.

Levy, Jonathan, Marcellino Ford-Livene, and Anne Levine. "Broadcast Television: Survivor in a Sea of Competition." Report. Federal Communications Commission Office of Plans and Policy, September 2002.

Lieberman, David. "Could Tony on A&E Bring Restrictions to Cable?" *USA Today*, December 1, 2006. http://usatoday30.usatoday.com/money/media/2006-12-01-cable-broadcast-usat_x.htm.

Lippman, John. "New Shows Try Their Hand at Copying Fox Hit *24*." *Wall Street Journal*, March 24, 2006, W7.

Lipsitz, George. "The Meaning of Memory: Family, Class, and Ethnicity in Early Network Television Programs." In *Private Screenings: Television and the Female Consumer*, ed. Lynn Spigel and Denise Mann, 71–110. Minneapolis: University of Minnesota Press, 1992.

Lisotta, Christopher. "Sony Sticking to the Flight Plan." *Television Week*, April 10, 2006.

Lisotta, Christopher, and Jon Lafayette. "WB, Yahoo! Offer 'Super' Preview." *Television Week*, September 5, 2005, 1.

Littlefield, Warren, and T. R. Pearson. *Top of the Rock: Inside the Rise and Fall of Must See TV*. New York: Anchor, 2012.

Littleton, Cynthia. "Fast-Tracked Sitcom May Be Way of Future." *Variety*, June 26, 2012. http://variety.com/2012/tv/news/fast-tracked-sitcom-may-be-way-of-future-1118055951/.

Lotz, Amanda D., ed. *Beyond Prime Time: Television Programming in the Post-Network Era*. New York: Routledge, 2009.

———. "If It Is Not TV, What Is It? The Case of U.S. Subscription Television." In *Cable Visions: Television beyond Broadcasting*, ed. Sarah Banet-Weiser, Cynthia Chris, and Anthony Freitas. New York: New York University Press, 2007.

———. "Textual (Im)Possibilities in the U.S. Post-Network Era: Negotiating Production and Promotion Processes on Lifetime's *Any Day Now*." *Critical Studies in Media Communication* 21, no. 1 (2004): 22–43.

———. "Using 'Network' Theory in the Post-Network Era: Fictional 9/11 U.S. Television Discourse as a 'Cultural Forum.'" *Screen* 45, no. 4 (2004): 423–39.

"Loyalty Cards Idea for TV Addicts." *BBC News*, October 28, 2004. http://news.bbc.co.uk/2/hi/technology/3958855.stm.

MacDonald, J. Fred. *One Nation under Television: The Rise and Decline of Network TV*. New York: Pantheon, 1990.

Marich, Robert. "TV Distribs Caught in Sports Vise." *Variety*, March 1, 2013. http://variety.com/2013/tv/news/tv-distribs-caught-in-sports-vise-1200001893.

Martin, Denise. "*Shield* Cops a 7th Season." *Variety*, June 5, 2006. http://variety.com/2006/scene/news/shield-cops-a-7th-season-1200338389/.

Maynard, John. "With DVD, TV Viewers Can Channel Their Choices." *Washington Post*, January 30, 2004, C1.

McAdams, Deborah D., and Joe Schlosser. "*Once* Again on Lifetime." *Broadcasting & Cable*, September 27, 1999.

McCarthy, Anna. *Ambient Television: Visual Culture and Public Space*. Durham: Duke University Press, 2001.

McCarthy, Michael. "HBO Shows Use Real Brands." *USA Today*, December 3, 2002. http://usatoday30.usatoday.com/money/advertising/2002-12-02-sopranos_x.htm.

McChesney, Robert W. *The Problem of the Media: U.S. Communication Politics in the 21st Century*. New York: Monthly Review Press, 2004.

McDonald, Sue. "Jon Stewart's *Crossfire* Transcript Most Blogged News Item of 2004, Intelliseek Finds." *Business Wire*, December 15, 2004. http://www.businesswire.com/news/home/20041215005558/en/Jon-Stewarts-Crossfire-Transcript-Blogged-News-Item.

McMurria, John. "Long-Format TV: Globalization and Network Branding in a Multi-Channel Era." In *Quality Popular Television*, ed. Mark Jancovich and James Lyons, 65–87. London: British Film Institute, 2001.

McNary, Dave, and Ben Fritz. "Guilds out of the Online Loop." *Variety*, August 15, 2006. http://variety.com/2006/digital/news/guilds-out-of-the-online-loop-1200341978/.

Meisler, Andy. "Not Even Trying to Appeal to the Masses." *New York Times*, October 4, 1998. http://www.nytimes.com/1998/10/04/arts/television-radio-not-even-trying-to-appeal-to-the-masses.html?pagewanted=all.

Mermigas, Diane. "Searching for Success in the Interactive Age." *Hollywood Reporter*, January 17, 2006, 21.

Miège, Bernard. *The Capitalization of Cultural Production*. New York: International General, 1989.

Miller, Sean J. "'Tragic' Drop in L.A.'s TV Production in 2012." *Backstage*, January 8, 2013. http://www.backstage.com/news/tragic-drop-ls-tv-production-2012/.

Miniwatts Marketing Group. "Internet World Stats: Usage and Population Statistics." World Internet Users Statistics Usage and World Population Stats. February 17, 2013. http://www.internetworldstats.com/stats.htm.

Mittell, Jason. "TiVoing Childhood." *Flow*, February 24, 2006. http://flowtv.org/2006/02/tivoing-childhood/.

Montgomery, Kathryn C. *Target: Prime Time; Advocacy Groups and the Struggle over Television Entertainment*. New York: Oxford University Press, 1989.

Moran, Albert. *Copycat TV: Globalisation, Program Formats and Cultural Identity*. Luton, UK: University of Luton Press, 1998.

Morris, Meaghan. "Things to Do with Shopping Centres." In *The Cultural Studies Reader*, ed. Simon During, 295–319. London: Routledge, 1993.

Murdoch, Rupert. "The Dawn of a New Age of Discovery: Media 2006." *Guardian*, March 13, 2006. http://www.guardian.co.uk/technology/2006/mar/13/news.rupertmurdoch.

Nava, Mica. "Consumerism and Its Contradictions." *Cultural Studies* 1, no. 2 (1987): 204–10.

Negroponte, Nicholas. *Being Digital*. New York: Vintage, 1995.

News Corp. "Emmy Award Winning Fox Comedy *Arrested Development* Finds Post Broadcast Home Online, on Hi-Def TV and on Basic Cable." *Business Wire*, July 26, 2006. http://www.businesswire.com/portal/site/google/index.jsp?ndmViewId=news_view.

Nielsen Media Research. *2000 Report on Television: The First 50 Years*. New York: Nielsen Media Research, 2000.

———. "Average U.S. Home Now Receives a Record 118.6 TV Channels." News release, June 6, 2008. http://www.nielsen.com/us/en/press-room/2008/average_u_s_home.html.

———. "The Digital Consumer." February 2014. http://www.slideshare.net/tinhanhvy/the-digital-consumer-report-2014-nielsen.

————. "An Era of Growth: The Cross-Platform Report." March 2014. http://www.nielsen.com/us/en/reports/2014/an-era-of-growth-the-cross-platform-report.html

————. "High Definition Is the New Normal." News release, October 17, 2012. http://www.nielsen.com/us/en/newswire/2012/high-definition-is-the-new-normal.html.

————. "The Media Universe: State of the Media US Consumer Usage Report 2012." January 7, 2013. http://www.nielsen.com/content/dam/corporate/us/en/reports-downloads/2013%20Reports/Nielsen-US-Consumer-Usage-Report-2012-FINAL.pdf.

————. "State of the Media: The Cross-Platform Report: Quarter 2, 2012—US." January 11, 2013. http://www.nielsen.com/content/dam/corporate/us/en/reports-downloads/2012-Reports/Nielsen-Cross-Platform-Report-Q2-2012-final.pdf.

————. *TV Audience*. New York: Nielsen Media Research, 2003.

————. "Viewing on Demand: The Cross-Platform Report." September 2013. http://www.slideshare.net/mapleaikon/q2-2013crossplatformreport-27230278

Neuman, W. Russell. *The Future of the Mass Audience*. Cambridge: Cambridge University Press, 1991.

Newcomb, Horace. "Studying Television: Same Questions, Different Contexts." *Cinema Journal* 45, no. 1 (2005): 107–11.

————. "This Is Not Al Dente: *The Sopranos* and the New Meaning of Television." In *Television: The Critical View*, 7th ed., ed. Horace Newcomb, 561–78. New York: Oxford University Press, 2007.

————. *TV: The Most Popular Art*. Garden City, NJ: Anchor, 1974.

Newcomb, Horace, and Paul Hirsch. "Television as a Cultural Forum." In *Television: A Critical View*, 5th ed., ed. Horace Newcomb, 503–15. New York: Oxford University Press, 2004.

Nussbaum, Emily. "When TV Became Art: Good-Bye Boob Tube, Hello Brain Food." *New York* magazine, December 4, 2009. http://nymag.com/arts/all/aughts/62513/.

Oldenberg, Ann. "TV Goes to Blogs: Shows Add Extra Information as Treat for Fans." *USA Today*, April 5, 2006. http://usatoday30.usatoday.com/tech/news/2006-04-04-tv-show-blogs_x.htm.

Oscar, Mitch. "Does Anybody Really Know How People Watch TV?" *Media Post*, January 16, 2007. http://www.mediapost.com/publications/article/53961/#axzz2Z9NNDvoX.

Penenberg, Adam. "The Death of Television: Will the Internet Replace the Boob Tube?" *Slate*, October 17, 2005. http://www.slate.com/articles/technology/technology/2005/10/the_death_of_television.html.

Pew Internet and American Life Project. "Internet Use and Home Broadband Connections." July 24, 2012. http://pewinternet.org/Infographics/2012/Internet-Use-and-Home-Broadband-Connections.aspx.

Popcorn, Faith. *The Popcorn Report: Faith Popcorn on the Future of Your Company, Your World, Your Life*. New York: Doubleday, 1991.

Popcorn, Faith, and Lys Marigold. *Clicking: 16 Trends to Future Fit Your Life, Your Work, and Your Business*. New York: HarperCollins, 1996.

Posner, Ari. "Can This Man Save the Sitcom?" *New York Times*, August 1, 2004. http://www.nytimes.com/2004/08/01/arts/television-can-this-man-save-the-sitcom.html?pagewanted=all.

Potts, Kim. "Women Love *Sex*: HBO's Bawdy Comedy Not Just a Pretty Face." *Daily Variety*, September 17, 1999, A1.

PQ Media. "Global Product Placement Spending Up 10% to $7.4 Billion in 2011." Press release, http://www.pqmedia.com/about-press-201212.html.

Pursell, Chris, and Michael Freeman. "Studios USA Will Stay the Course." *Electronic Media*, October 29, 2001, 1.

Quigley, Eileen S., Aaron Dior Pinkham, and Dee Quigley, eds. *International Television & Video Almanac 2005*. 50th ed. New York: Quigley, 2005.

Radner, Hilary. *Shopping Around: Feminine Culture and the Pursuit of Pleasure*. New York: Routledge, 1995.

Raphael, Chad. "The Political Economic Origins of Reali-TV." In *Reality TV: Remaking Television Culture*, ed. Susan Murray and Laurie Ouellette, 119–36. New York: New York University Press, 2004.

Reardon, Marguerite. "Competitive Wireless Carriers Take on AT&T and Verizon." CNET News, September 10, 2012. http://news.cnet.com/8301-1035_3-57505803-94/competitive-wireless-carriers-take-on-at-t-and-verizon/.

Renaud, Brent. Telephone interview by author, April 26, 2006.

Reynolds, Mike. "TV Guide Net Fills Out Its Full-Screen Roster." *Multichannel News*, December 17, 2012, 9.

Richtel, Matt, and Brian Stelter. "In the Living Room, Hooked on Pay TV." *New York Times*, August 23, 2010. http://www.nytimes.com/2010/08/23/business/media/23couch.html?pagewanted=all&_r=0.

Rosenbloom, Stephanie. "*Lost* Weekend: A Season in One Sitting." *New York Times*, October 27, 2005. http://www.nytimes.com/2005/10/27/fashion/thursdaystyles/27dvd.html?n=Top%2FReference%2FTimes%20Topics%2FSubjects%2FT%2FTelevision.

Rosenthal, Phil. "On-Demand Deals a New Dawn for TV." *Chicago Tribune*, November 9, 2005.

Rutenberg, Jim. "Much in a Name." *New York Times*, August 15, 2001.

Ryan, Shawn. Telephone interview by author, May 4, 2006.

Sarandos, Ted. "Ted Sarandos, Chief Content Officer, Netflix." Interview. Carsey Wolf Center, Media Industries Project. Accessed April 22, 2013. http://www.carseywolf.ucsb.edu/mip/article/ted-sarandos.

Schatz, Amy. "Verizon, Comcast Defend Spectrum-Purchase Plan." *Wall Street Journal*, March 22, 2012. http://online.wsj.com/article/SB10001424052702304636404577295703761555914.html.

Schatz, Amy, and Brooks Barnes. "To Blunt Web's Impact, TV Tries Building Online Fences." *Wall Street Journal*, March 16, 2006, A1.

Schiller, Gail. "It's Not a Plug, It's HBO." *Hollywood Reporter*, December 10, 2004.

Schiller, Herbert I. *Communication and Cultural Domination*. White Plains, NY: International Arts and Sciences Press, 1976.

Schlosser, Joe. "Kissinger Tops USA Network TV." *Broadcasting & Cable*, April 26, 1999.

Schneider, Michael. "New Development." *Daily Variety*, September 26, 2002.

Schor, Juliet B. *The Overspent American: Upscaling, Downshifting, and the New Consumer*. New York: Basic Books, 1998.

Scocca, Tom. "The YouTube Devolution." *Observer*, July 31, 2006. http://observer.com/2006/07/the-youtube-devolution/.

Spigel, Lynn. *Make Room for TV: Television and the Family Ideal in Postwar America*. Chicago: University of Chicago Press, 1992.

———. "Portable TV: Studies in Domestic Space Travel." In *Welcome to the Dreamhouse: Popular Media and Postwar Suburbs*, 60–103. Durham: Duke University Press, 2001.

Spigel, Lynn, and Jan Olsson, eds. *Television after TV: Essays on a Medium in Transition*. Durham: Duke University Press, 2004.

Steinberg, Brian. "*Law & Order* Boss Dick Wolf Ponders the Future of TV Ads (Doink, Doink)." *Wall Street Journal*, October 18, 2006. http://online.wsj.com/article/SB116113866082296113. html.

Stelter, Brian. "A Long Wait Stirs Enthusiasm for Fox Show *Glee*." *New York Times*, September 1, 2009. http://www.nytimes.com/2009/09/02/business/media/02adco.html?_r=1&.

Sternberg, Steve. "Most Homes Have Only One Set On during Primetime." *Television Insights: A Publication of Magna Global*, September 12, 2006.

Sweeney, Anne. Address at the National Cable Show, Georgia World Congress Center, Atlanta, April 10, 2006.

Tartikoff, Brandon, and Charles Leerhsen. *The Last Great Ride*. New York: Turtle Bay, 1992.

Thompson, Derek. "Sports Could Save the TV Business—or Destroy It." *Atlantic*, July 17, 2013. http://www.theatlantic.com/business/archive/2013/07/sports-could-save-the-tv-business-or-destroy-it/277808/.

Tillson, Tamsen, and Elizabeth Guider. "CBS Eyes a *CSI* Buyout." *Variety*, December 20, 2006. http://variety.com/2006/scene/news/cbs-eyes-a-csi-buyout-1117956158/.

Tsukayama, Hayley. "YouTube Channel YOMYOMF Launches, Focus on Asian-American Pop Culture." *Washington Post*, June 15, 2012. http://articles.washingtonpost.com/2012-06-15/business/35462753_1_ryan-higa-youtube-channel-yomyomf.

Turner, Graeme, and Jinna Tay. *Television Studies after TV: Understanding Television in the Post-Broadcast Era*. London: Routledge, 2009.

Turow, Joseph. "Audience Construction and Culture Production: Marketing Surveillance in the Digital Age." *American Academy of Political and Social Science* 597 (January 2005): 103–21.

———. *Breaking Up America: Advertisers and the New Media World*. Chicago: University of Chicago Press, 1997.

———. *The Daily You: How the New Advertising Industry Is Defining Your Identity and Your Worth*. New Haven: Yale University Press, 2011.

———. *Media Systems in Society: Understanding Industries, Strategies and Power*. White Plains, NY: Longman, 1997.

———. "Unconventional Programs on Commercial Television: An Organizational Perspective." In *Mass Communicators in Context*, ed. D. Charles Whitney and James Ettema, 107–29. Beverly Hills: Sage, 1982.

Twitchell, James. *AdCult USA: The Triumph of Advertising in American Culture*. New York: Columbia University Press, 1996.

Umstead, R. Thomas. "DVDs, Video Games Boost Net Brands." *Multichannel News*, July 19, 2004. http://www.multichannel.com/news-article/dvds-video-games-boost-net-brands/89593.

U.S. Census Bureau. "Presence and Type of Computer for Households, by Selected Householder Characteristics: 2010." Table 1C. http://www.census.gov/hhes/computer/publications/2010. html.

Verrier, Richard. "Movies, Schmovies—TV's Taking Over L.A." *Los Angeles Times*, August 19, 2005, A1.

Walker, James R., and Robert V. Bellamy Jr. "The Remote Control Device: An Overlooked Technology." In *The Remote Control in the New Age of Television*, ed. James R. Walker and Robert V. Bellamy Jr., 3–14. Westport, CT: Praeger, 1993.

Walker, James, and Douglas Ferguson. *The Broadcast Television Industry*. Boston: Allyn and Bacon, 1998.

Wallenstein, Andrew. "Netflix Series Spending Revealed." *Variety*, March 8, 2013. http://variety.com/2013/digital/news/caa-agent-discloses-netflix-series-spending-1200006100/.

Warzel, Charlie. "YouTube Phenoms Raise Record Cash: FreddieW Shows Advertisers Why They Should Pay Attention to Original Web Series." *Adweek*, February 14, 2013. http://www.adweek.com/news/advertising-branding/youtube-phenoms-raise-record-cash-147287.

Waterston, Adriana. *VOD and DVR in the Home—A Glimpse of the Future.* July 7, 2005. http://www.horowitzassociates.com/white-papers/vod-and-dvr-in-the-home-a-glimpse-of-the-future.

Webster, James G. "Television Audience Behavior: Patterns of Exposure in the New Media Environment." In *Media Use in the Information Age: Emerging Patterns of Adoption and Consumer Use*, ed. Jerry L. Salvaggio and Jennings Bryant, 197–216. Hillsdale, NJ: LEA, 1989.

Webster, James G., Patricia F. Phalen, and Lawrence W. Lichty. *Ratings Analysis: The Theory and Practice of Audience Research.* 2nd ed. Mahwah, NJ: Lawrence Erlbaum Associates, 2000.

Werts, Diane. "A New Way to Watch TV." *Columbus Dispatch*, January 21, 2004, F1, 6.

Whitney, Daisy. "Channel Schedules Limited Series; *Off to War, America* on Discovery Lineup." *Television Week*, June 6, 2005, 22.

———. "Hopped Up on Hope." *Television Week*, June 19, 2006, 1.

———. "MySpace Video a Boon to *Mother*." *Television Week*, December 4, 2006, 1.

"William H. Macy Is Shameless on Showtime." Transcript. *Fresh Air*, WHYY-FM Philadelphia, January 30, 2013. http://www.npr.org/templates/transcript/transcript.php?storyId=170463939.

Williams, Raymond. *Television: Technology and Cultural Form.* New York: Schocken, 1974.

Williamson, Judith. *Consuming Passions: The Dynamics of Popular Culture.* London: Marion Boyers, 1986.

Winship, Janice. *Inside Women's Magazines.* London: Pandora, 1987.

Winston, Brian. *Media Technology and Society: A History; From the Telegraph to the Internet.* London: Routledge, 1998.

Wittel, Andreas. "Culture, Labor, and Subjectivity: For a Political Economy from Below." *Capital & Class*, no. 84 (Winter 1984): 11–30.

YouTube. *Statistics.* Accessed May 30, 2013. http://www.youtube.com/yt/press/statistics.html.

Zickuhr, Kathryn, and Aaron Smith. "Digital Differences." Pew Internet and American Life Project, April 13, 2012. http://pewinternet.org/Reports/2012/Digital-differences.aspx.

ABOUT THE AUTHOR

Amanda D. Lotz is Associate Professor of Communication Studies at the University of Michigan. She is the author of *Cable Guys: Television and Masculinities in the 21st Century*, *The Television Will Be Revolutionized*, and *Redesigning Women: Television after the Network Era*, co-author of *Understanding Media Industries* and *Television Studies*, and editor of *Beyond Prime Time: Television Programming in the Post-Network Era*.